Autonomous Vehicles and Systems

A Technological and Societal Perspective

RIVER PUBLISHERS SERIES IN AUTOMATION, CONTROL, AND ROBOTICS

Series Editors:

Ishwar K. Sethi
Oakland University, USA

Tarek Sobh
University of Bridgeport, USA

Feng Qiao
Shenyang JianZhu University, China

The "River Publishers Series in Automation, Control and Robotics" is a series of comprehensive academic and professional books which focus on the theory and applications of automation, control and robotics. The series focuses on topics ranging from the theory and use of control systems, automation engineering, robotics and intelligent machines.

Books published in the series include research monographs, edited volumes, handbooks and textbooks. The books provide professionals, researchers, educators, and advanced students in the field with an invaluable insight into the latest research and developments.

Topics covered in the series include, but are by no means restricted to the following:

- Robots and Intelligent Machines
- Robotics
- Control Systems
- Control Theory
- Automation Engineering

For a list of other books in this series, visit www.riverpublishers.com

Autonomous Vehicles and Systems

A Technological and Societal Perspective

Editor

Ishwar K. Sethi

Oakland University, USA

River Publishers

Routledge
Taylor & Francis Group
NEW YORK AND LONDON

Published 2023 by River Publishers
River Publishers
Alsbjergvej 10, 9260 Gistrup, Denmark
www.riverpublishers.com

Distributed exclusively by Routledge
605 Third Avenue, New York, NY 10017, USA
4 Park Square, Milton Park, Abingdon, Oxon OX14 4RN

Autonomous Vehicles and Systems / by Ishwar K. Sethi.

Routledge is an imprint of the Taylor & Francis Group, an informa business

ISBN 978-87-7022-885-5 (hardback)
ISBN 978-87-7004-052-5 (paperback)
ISBN 978-10-0381-067-4 (online)
ISBN 978-10-3262-953-7 (master ebook)

Contents

Preface

Automobile accidents represent a major public health concern not only in the United States but worldwide. Before the pandemic, there were more than 6.75 million reported crashes resulting in 33,487 fatalities and 1.9 million injuries in the United States in 2019. According to the World Health Organization, about 1.06 million road traffic fatalities are expected for the year 2022. Road traffic crashes have a significant economic impact estimated to be about 3% of the gross domestic product for most countries.

A vast majority of crashes are due to human error. While educating drivers and road users about safe driving practices and strict laws against drunk driving are important to reduce crashes, the Insurance Institute for Highway Safety (IIHS) suggests that nearly a third of crashes can be prevented by equipping all vehicles with driver assistance systems such as forward collision and lane departure warning systems, adaptive headlights, and systems for monitoring blind spots. These technologies are part of what is termed as autonomous vehicles that partially or fully replace the human driver while safely navigating to the desired destination. The US National Highway Traffic Safety Administration (NHTSA) defines five levels of vehicle autonomy, with level 1 designating autonomous vehicles equipped with one or more primary but independent automated features, for example cruise control. Level 5 designation implies fully autonomous vehicles capable of safely navigating on their own while monitoring the road conditions. Levels 2–4 represent vehicles with degrees of autonomy between levels 1 and 5.

Besides the potential of autonomous vehicles to reduce crashes, there are many other social benefits envisioned by the researchers studying the impacts of AV technology. These include access to mobility for disabled and older citizens, less congestion, the ability to share vehicles, and cost savings resulting from lower auto ownerships. With all these potential benefits, research and development in autonomous vehicle technologies have seen a quantum jump in the last decade. All established auto companies and many new entrants in the field have been making remarkable progress in this area.

The current book on autonomous vehicle systems is a tiny showcase of the AVs research activities taking place all over the world. The book contains contributions from researchers across the globe. It covers a range of topics from ethics, deep learning, controls, teleoperations, and software development for AVs. I sincerely hope the readers will find the book valuable.

List of Figures

List of Tables

List of Contributors

Loris, Belcastro, *Department of Informatics, Modeling, Electronics and Systems (DIMES), University of Calabria, Italy*

Brian Park, B., *University of Virginia, USA*

Cai, Lihua, *South China Normal University, China*

Chung, Chan-Jin, *Department of Math and Computer Science, Lawrence Technological University, USA*

DeRose, Guiseppe, *Department of Math and Computer Science, Lawrence Technological University, USA*

Dombecki, Justin, *Department of Math and Computer Science, Lawrence Technological University, USA*

Emirler, Mümin Tolga, *Yildiz Technical University, Turkey*

Fan, Jianping, *Lenovo Research, China*

Gasmi, Ines, *Artificial Intelligence Research Unit, National School of Computer Sciences, University of Manouba, Tunisia*

Guimin Dong, *Amazon.com*

Hürten, C., *University of Duisburg-Essen, Germany*

Kessentini, Marouane, *School of Engineering and Computer Science, Oakland University, USA*

Kocherovsky, Mark, *Department of Math and Computer Science, Lawrence Technological University, USA*

Kusari, Arpan, *University of Michigan Transportation Research Institute, USA*

Liu, Li, *Lenovo Research, China*

Mansouri, Nesrine, *Artificial Intelligence Research Unit, National School of Computer Sciences, University of Manouba, Tunisia*

Marozzo, Fabrizio, *Department of Informatics, Modeling, Electronics and Systems (DIMES), University of Calabria, Italy*

Mu, Zeyu, *University of Virginia, USA*

Orsino, Alessio, *Department of Informatics, Modeling, Electronics and Systems (DIMES), University of Calabria, Italy*

Paul, Nicholas, *Department of Math and Computer Science, Lawrence Technological University, USA*

Pleune, Mitchell, *Department of Math and Computer Science, Lawrence Technological University, USA*

Raj, Sunny, *Department of Computer Science and Engineering, Oakland University, USA*

Ren Zhong, *Department of Computer Science, Wayne State University, USA*

Rjeb, Imen Ben, *Artificial Intelligence Research Unit, National School of Computer Sciences, University of Manouba, Tunisia*

Rui, Yong, *Lenovo Research, China*

Schramm, Dieter, *University of Duisburg-Essen, Germany*

Schulte, Joseph, *Department of Math and Computer Science, Lawrence Technological University, USA*

Sethi, Ishwar K., *Computer Science and Engineering, Oakland University, USA*

Shan, Jinjun, *Department of Earth and Space Science, Lassonde School of Engineering, York University, Canada*

Shi, Zhongchao, *Lenovo Research, China*

Sidi Lu, *Department of Computer Science, William and Mary, USA*

Sieberg, P. M., *Schotte Automotive GmbH & Co. KG, Germany*

Soui, Makram, *School of Engineering and Computer Science Oakland University, USA*

Srinivasin, Karthic, *College of Computing and Informatics, Saudi Electronic University, Saudi Arabia*

Stockem Novo, A., *Institute of Computer Science, University of Applied Sciences. Germany*

Tang, Mingyue, *University of Virginia, USA*

Tian, Jiang, *Lenovo Research, China*

Timmis, Ian, *Department of Math and Computer Science, Lawrence Technological University, USA*

Yan, Runze, *University of Virginia, USA*

Weisong Shi, *Department of Computer & Information Sciences, University of Delaware, USA*

Yuan, Mingfeng, *Department of Earth and Space Science, Lassonde School of Engineering, York University, Canada*

List of Abbreviations

AAA	American automobile association
ACC	Adaptive cruise control
ADAS	Advanced driver assistance systems
ADS	Automated driving system
ADV	Autonomous driving vehicles
AEB	Automatic emergency braking
AI	Artificial intelligence
ANN	Artificial neural network
ANSYS	Analysis system
APC	Automatic passenger counting
AR	Augmented reality
ARIMA	Autoregressive integrated moving average
AV	Autonomous vehicles
AVS	Autonomous vehicles and systems
CA	Chain average
CAC	Chinese automation congress
CAD	Computer Aided Design
CARLA	Car Learning to Act
CCD	Charge-coupled device
CCPA	California consumer privacy act
CDR	CNN + DNN + RNN
CIG	Computational intelligence and games
CMOS	Complementary metal oxide semiconductor
CNN	Convolution neural network
DARPA	Defense Advanced Research Project Agency
DAVE	DARPA autonomous vehicle
DL	Deep learning
DNN	Densly Connected Neural Networks
DOF	Degrees of freedom
DPG	Deterministic policy gradient
DQN	Deep Q network

DRL	Deep reinforcement learning
DSO	Direct sparse odometry
ECCV	European conference on computer vision
EM	Electromagnetic
EMF	Electro Magnetic Force
EO	Edge orchestrator
ESP	Electronic stability program
FCNN	Fourier convolutional neural networks
FOV	Field-of-view
FP	False positive
GCNs	Graph convolutional neural networks
GGNN	Gated GNN
GHG	Greenhouse Gas
GNN	Graph neural network
GNSS	Global Navigation Satellite System
GO	Geometrical optics
GPS	Global positioning systems
GRN	Graph recurrent neural networks
GRU	Gated recurrent units
GT	Game theory
HAV	Highly Automated Vehicle
HD	High definition
HDV	Human-driven vehicles
HTTP	Historical trajectory-based time prediction
ICASSP	International conference on acoustics, speech and signal processing
ICE	Internal combustion engine
ICRA	International conference on robotics and automation
IGA	Immune genetic algorithm
IGVC	Intelligent Ground Vehicles Competition
IIHS	Insurance institute for highway safety
ILSVRC	ImageNet large scale visual recognition challenge
IoT	Internet of Things
IROS	Intelligent robots and systems
ISW	Importance sampling weights
ITE	Institution of transportation engineers

ITS	Intelligent transportation system
KNN	K-nearest neighbor
LARIA	Laboratory for Research in AI
LiDAR	Light detection and ranging
LMI	Linear Matrix Inequalities
LO	Lidar odometry
LQI	Linear quadratic integral
LQR	Linear quadratic regulator
LSTM	Long short-term memory
LTI	Linear time invariant
LTU	Lawrence Technological University
MAE	Mean Absolute Error
MDP	Markov decision process
MEMS	Microelectromechanical Systems
METI	Ministry of economy, trade and industry
MIPR	Multimedia information processing and retrieval
MITRE	Mitre Corporation
MKC	Modified K-means clustering
ML	Machine learning
MOOVE	Monitoring Outille Pourle Vehicule Dans Son environnement
MPC	Model predictive control
MSE	Mean Square Error
NBC	National Broadcasting Corporation
NES	NO ethical settings
NGTO	A Term for ε-Nash Equilibrium Strategies
NHTSA	National highway traffic safety administration
NIST	National institute of standards and technology
NLP	Natural language processing
NN	Neural network
OD	Original destination
ODD	Operational design domain
PEGASUS	Project for the establishment of generally accepted quality criteria, tools and methods as well as scenarios and situations
PER	Prioritized experience replay
PID	Proportional integral derivative
PP	Path prediction

RBF	Radial Basis Function
RCNN	Recurrent convolutional neural networks
ReLU	Rectifier Linear Unit
ResNet	Residual neural network
RFNN	Random forest based on near neighbor
RL	Reinforcement learning
RMSE	Root mean square error
RNN	Recurrent neural networks
RoI	Region of interest
ROS	Robotic operating system
RRT	Rapidly exploring Random Trees
RTK	Real-time Kinematic
SAE	Society of automotive engineers
SAKURA	Safety assurance KUdos for reliable autonomous
SBS	Sequential backward selection
SDK	Software Development Kit
SFFS	Sequential forward floating selection
SFS	Sequential forward selection
SLAM	Simultaneous localization and mapping
SMA	Successive moving average
SOTA	State-of-the-art
SSD	Single Shot Detector
StVO	The Road Traffic Act of Germany
SUMO	Simulation of Urban MObility
SUT	System under test
SV	Subject vehicle
SVM	Support vector machine
SVR	Support vector regression
SWA	Steering wheel angle
TC	Teleoperation consortium
ToF	Time of flight
TP	True positive
TraCI	Traffic control interface
UI	User Interface
V&V	Verification and validation
VCG	Visual geometry group
VEDECOM	An Institute for Energy Transition in France
ViT	Vision transformer

VM	Virtual machines
VOC	Visual Object Classes Challenge
WCICA	World congress on intelligent control and automation
WTP	Willingness-to-pay
YOLO	You only look once

1

Introduction

Ishwar K. Sethi

Computer Science and Engineering, Oakland University, USA

Abstract

The technological growth in autonomous vehicles and systems has been rapid in the last three decades. This introductory chapter provides a brief history of autonomous vehicles after summarizing the different levels of autonomy. Following that, it provides a brief overview of the chapters covering the different aspects of research and development toward realizing autonomous vehicles.

Keywords: Autonomous vehicles and systems, ethics, deep learning, sensors, simulation.

1.1 Introduction

The autonomous vehicles in recent years have been a major topic of research interest, given their potential to impact the society in multiple ways. An autonomous vehicle implies a moving platform capable of navigating from a starting location to the desired destination with little or no human control at all. The technological growth in this field has been rapid; many large and startup companies have banded together to compete and win the autonomous vehicles race. The early optimistic predictions of having such vehicles on roads in large numbers have sobered up, given the challenges of crafting such vehicles to operate in all kinds of weather conditions. The automobile manufacturers seem to be backing away from their earlier predictions to when the autonomous vehicles will be on the roads.

The autonomous vehicles are popularly termed as self-driving cars. Since different persons imply different capabilities when referring to autonomous vehicles, the Society of Automotive Engineers (SAE International) has created a six levels classification of driving automation to minimize the confusion. The characterization of autonomous vehicles at different levels are

- Level 0: No automation. The driver is in complete control of the primary vehicle functions at all times and is responsible for safe operation of the vehicle. Level 0 autonomous vehicles use sensors to help and warn the drivers while they are driving as well as use systems that provide momentary assistance, for example, automatic emergency braking.
- Level 1: Assisted ariving automation. Level 1 vehicles are equipped with primary automated features such as cruise control, parking sensors, and other assistive driving technologies. These safety features can provide a small degree of control over the vehicle's acceleration, braking, or steering, but the driver is still in control all the time. The different automated features work independently.
- Level 2: Partial automation. The level 2 category vehicles permit two or more assistive driving technologies to work together. The level 2 vehicles can assume control of certain functions in some limited conditions. The driver is still responsible for monitoring the roadway and safe operation and is expected to be ready to take control on short notice at all times. Most autonomous cars at present fall in this category.
- Level 3: Conditional automation. Level 3 vehicles allow the driver to relinquish control of vehicle's safety-critical functions under certain conditions. Once the environmental conditions ease, the control is transferred back to the driver. The most common instance of level three automation today is traffic jam assist.
- Level 4: High automation. Level 4 vehicles have a comprehensive set of features for autonomous driving technology in most conditions. The vehicle can perform all aspects of driving even if the driver does not respond to intervene.
- Level 5: Full automation. Level 5 vehicles are completely self-driving and do not need any assistance from the driver at any time.

At present, only a tiny fraction of the vehicles on the road can be classified at level 3 and above. Some of these autonomous vehicles are currently being deployed as shuttles on fixed routes at places such as college campuses. It is also now possible to ride a driverless car, that is robotaxi, in many cities, for example in San Francisco, USA. Other applications such as goods delivery are quickly emerging.

1.2 Brief History of Autonomous Vehicles

Some interesting observations about the history of autonomous vehicles can be had by using Google Ngram Viewer to check for the usage of the terms autonomous vehicles and self-driving cars. Figure 1.1 shows the Google Ngram output for these two terms from 1880 to 1980. The first observation is that the term self-driving cars appears to be a recent addition to the lexicon; second, the term autonomous vehicles first appeared in the literature over a small window of about 5 years, centered in the year 1900 before emerging again in 1950. The bump around the year 1900 is possibly related to the efforts of Gottlieb Daimler and Wilhelm Maybach [2] around that time toward building 1901 Mercedes considered to be the first modern automobile. In 1920s, the ideas for autonomous vehicles were taking shape, and the first such vehicle was introduced to the public in Dayton, Ohio [3]. It was a self driving car without a driver that was remotely controlled via a radio link. It was followed by another remote-controlled car, *American Wonder* driven on Broadway, New York in 1925 [3]. Later in the 1930s, such remote-controlled cars were used for advertising and transport safety campaigns.

During 1950s, the idea of remotely controlling cars via radio communication gave way to the concept of automated guided cars. The concept involved using buried wires in the road to steer and guide the vehicle. In early 1958, a demonstration of the first automatically guided vehicle was held by GM using a car fitted with two sensors in the front [5]. The sensors followed a buried wire in the road and allowed steering to be automatically adjusted. The work on automated guided cars continued in 1960s at GM as well as at many other organizations.

With the development of microelectronics, the idea of automated guided vehicles relying on buried wires got replaced by camera-based or vision-guided vehicles. A visually guided autonomous vehicle was demonstrated in Japan in 1977 [9]; it used two cameras to capture and process images of lateral guide rails on the road for guidance. A few years later a vision-guided vehicle successfully drove at a speed of about 60 miles/h on streets without traffic in Munich, Germany. This demonstration resulted in the joint European project, *PROMETHEUS*, for research and development of the autonomous vehicles [6]. In the same time frame, an experimental vehicle with onboard cameras, the *Stanford Cart*, was built to demonstrate autonomous navigation for short

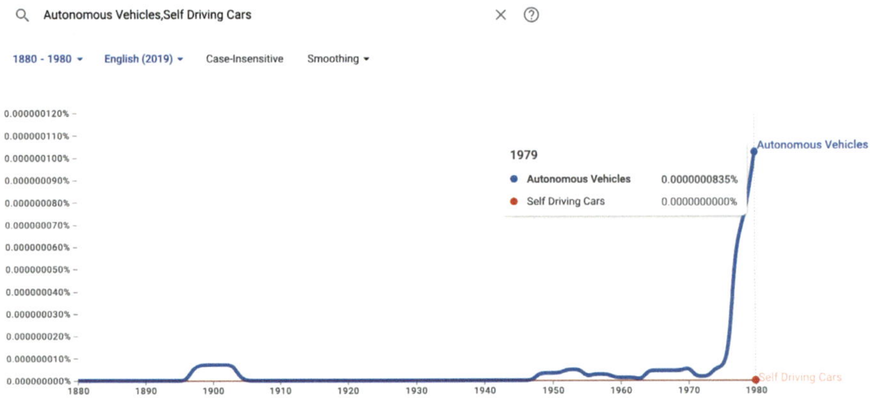

Figure 1.1 Usage of autonomous vehicles and self-driving car terms from 1880 to 1980.

runs [7]. With the successes of demonstrations in the USA and abroad, many partnerships between universities and industries began getting formed to give a boost to autonomous vehicle research. This increased level of activity is clearly evident in Figure 1.1.

The period beyond 1980 saw partnerships between universities and industry growing and showing impressive gains. In 1995, Carnegie Mellon University's NavLab drove a partially autonomous vehicle from Pittsburgh to San Diego in a demo termed "No Hands across America" [8]. The vehicle used vision-based sensing to steer autonomously while its braking and acceleration was under the control of a human driver. In 2004, the U.S. Department of Defense's research agency, DARPA, began a series of "Grand Challenges" to push the development of technologies related to autonomous vehicles. The earlier DARPA challenges were for off-road demonstrations of vehicles, and the winning team in 2005 completed the 150 miles long course in about seven hours. Later on, the challenge was set in a mock urban setting over a course of 60 miles obeying traffic rules and autonomously navigating alongside other road traffic [1]. With the push from DARPA resulting in the speedy development of sensor technologies, improved mapping, better software and systems, and the emergence of deep learning heralded by the performance of the Alexnet in 2012 [4], autonomous vehicles had arrived.

1.3 About the Book

The design, development, and deployment of autonomous vehicles present numerous technical and societal challenges. The chapters in this book are expected to provide an insight to some of such challenges. Other than the introductory chapter, there are 13 chapters in the book.

The advances in deep learning, which allows us to perceive and build three-dimensional model of the environment, has been a major factor in the growth and capabilities of the autonomous vehicles. The second chapter, Deep Learning for Autonomous Vehicles and Systems, reviews and summarizes the current deep learning approaches for the perception of the environment. The chapter looks at perception, localization, path planning, and motion control learning to illustrate the performance achieved by the current deep learning methods. The third chapter, Toward Autonomous Vehicles and Systems – Potential and Challenges of Artificial Intelligence, continues with the theme of using artificial intelligence for modeling for use in autonomous vehicles and systems. Its focus is on enormous data needs for building artificial intelligence models. It showcases an example of circumventing data needs via simulation and presents an example of an advanced driver assistance system design for an automated lane change system. The fourth chapter, Autonomous Vehicle Steering through Convolutional and Recurrent Deep Learning, describes work related to training a deep neural network to steer a vehicle on unmarked roads. Most of the prior work on this topic has dealt with steering on roads with lane markings. The chapter four presents the details of a study using various pretrained networks with no or little temporal context while generating steering control signals. The next chapter five address the leader-follower autonomy scenario. In this scenario, the movements of an autonomous vehicle designated as the leader are followed by one or more autonomous follower vehicles. It is not difficult to foresee applications of such a leader and follower autonomous vehicles in defense as well as in large scale emergency relief. The authors present an object detection and gesture recognition system that is used to map gestures to vehicle commands.

While a great deal of advances in autonomous vehicle systems can be attributed to the human-like recognition performance exhibited by deep convolutional networks, it is well-known that these networks can be fooled to make incorrect predictions by inserting adversarial noise to the input; the presence of natural noise in the scenes due to rain, fog, and other natural factors can also result in incorrect predictions. The chapter six, Adversarial Attacks and Defenses on Autonomous Vehicle Systems, looks into this aspect of deep networks and reviews different methods of adversarial attacks and the defenses being developed to counter such attacks.

The perception of the environment by an autonomous vehicle is done using many sensors of different modalities. These sensors, LiDAR, cameras,

Radar, etc., capture complimentary information at different rates and scale. The information captured by different sensors must be fused together to become useful for decision-making. Along with the information captured by the vehicle's sensors, there is a possibility of information being available from roadside sensors. The chapter seven, Multi-sensor Fusion for Multi-target Detection and Tracking, presents an overview of different sensors and their characteristics and describes the approaches for fusing the sensor data coming from multiple sensors.

Despite collecting millions of miles of driving data to create all possible scenarios for autonomous vehicles, it is likely that an autonomous vehicle will come across a situation where the built-in vehicle intelligence is unable to determine the next course of action. Using help from a remote operator via teleoperation is one potential solution for such situations. While the current remote assistants are humans, artificial intelligence-based agents are likely to be used in the future. The chapter eight, Teleoperation Technologies for Enhancing Connected and Autonomous Vehicles takes a look at how teleoperation technologies are being used and suggests evaluation criteria for teleoperation performance.

The use of hybrid edge/cloud solutions is attractive to efficiently process the large amounts of data produced by IoT devices. Such solutions are appealing in different application scenarios that require low latency, low energy scalable services. In particular, edge computing has been found useful in tasks such as driving assistance, collision avoidance, and traffic sign recognition. In chapter nine, Hybrid Edge/Cloud Solutions for Supporting Autonomous vehicles, the authors analyze how edge/cloud solutions can be exploited for efficiently managing tasks related to autonomous driving. Through simulation, the authors demonstrate that such solutions can yield significant performance benefits as the number of vehicles increases.

With an increasing number of autonomous vehicles on the road along with human-driven vehicles in the future, all kinds of interactions between these two types of vehicles are expected. In such a dynamic and diverse environment, decision-making will become challenging for autonomous vehicles. One possible approach to this issue is to employ deep reinforcement learning and game theory to develop agents through explorations with the help of a realistic simulator. The chapter ten, Game-Theoretic Decision Making for Autonomous Driving vehicles explores such a scenario using the Deep Q network so that vehicles can learn complex driving behaviors via self-play. The effectiveness of the proposed approach is demonstrated by testing the learned decision rules in different traffic scenarios.

The focus of the chapter eleven, State of Modeling and Simulation in Autonomous Vehicles, is to provide readers an in-depth understanding of challenges in developing realistic sensor simulations and complex driving scenarios. While the focus of the earlier chapters in the book has been on perception, localization and path planning, the control system design, and implementation is vital for a successful autonomous vehicle operation. In Path Following Control System Design and Analysis for Autonomous Vehicles, the author presents a design methodology to minimize the lateral deviation error from the reference path while the vehicle is traveling at the desired speed. The suggested methodology is demonstrated by simulating a high-degree-of-freedom vehicle model.

Chapter twelve, Public Transport Travel Time Prediction using Sequential Forward Floating Selection Algorithm and Stacked Autoencoderis, focused on mobility and suggests an algorithm for travel time prediction. The final chapter thirteen, A Look into the Ethics of Autonomous Vehicles Systems (AVS), summarizes the decision-making dilemmas faced by autonomous vehicles or their designers when faced with moral choices. The chapter explores and provides an in-depth discussion of the current and emerging ethical issues in the context of autonomous vehicles.

References

[1] Martin Buehler, Karl Iagnemma, and Sanjiv Singh. *The DARPA urban challenge: autonomous vehicles in city traffic*, volume 56. springer, 2009.

[2] Erik Eckermann. *World History of the Automobile*. Society of Automotive Engineers, 2001.

[3] KrÃűger Fabian. Automated driving in its social, historical and cultural contexts. In Barbara Lenz Markus Maurer, J. Christian Gerdes and Hermann Winner, editors, *Autonomous Driving: Technical, Legal and Social Aspects*. Springer, 2015.

[4] Alex Krizhevsky, Ilya Sutskever, and Geoffrey E Hinton. Imagenet classification with deep convolutional neural networks. In F. Pereira, C.J. Burges, L. Bottou, and K.Q. Weinberger, editors, *Advances in Neural Information Processing Systems*, volume 25. Curran Associates, Inc., 2012.

[5] General Motors. An automatically guided automobile cruised along a one-mile check road at general motors technical center today... *Press Release*, 1953.

[6] H.-H. Nagel. Eureka-projekt prometheus und pro-art (1986-1994). In B. Reuse and R. Vollmar, editors, *Informatikforschung in Deutschland*. Springer, 2008.
[7] Moravec Hans P. The stanford cart and the cmu rover. In Gordon T. Cox, Ingemar J. Wilfong, editor, *Autonomous Robot Vehicles*. Springer, 1990.
[8] Otis Port. Look ma, no hands. *Bloomberg*, 8-14-1995.
[9] T. Hirose S. Tsugawa, T. Yatabe and S. Matsumoto. An automobile with artificial intelligence. In *Proc. 6th IJCAI*, pages 893–895, 1979.

2

Deep Learning for Autonomous Vehicles and Systems

Guimin Dong[1], Mingyue Tang[2], Runze Yan[2], Zeyu Mu[2], Lihua Cai[3], and B. Brian Park[2]

[1]Amazon.com,
[2]University of Virginia, USA,
[3]South China Normal University, China

Abstract

Autonomous vehicles and systems (AVS) are becoming ubiquitous, aiming to approach human intelligence and ultimately replace human drivers. Modern AVS are developed to autonomously understand their surrounding environments and make intelligent decisions. The advancements in artificial intelligence (AI), especially in deep learning, have considerably enhanced the autonomy and intelligence of AVS. In this chapter, we review and summarize current research on the application of state-of-the-art deep learning (DL) approaches in the perception and decision-making of AVS. Perception, localization, path planning, and motion control learning, which are fundamental tasks in AVS, are reviewed to illustrate the state-of-the-art performance achieved by the surveyed DL methodologies. Examining the current practices of AI in AVS helps establish links between theoretical developments in AI and practical implementations in AVS and discovering challenges and issues in the path of achieving higher level autonomy. Finally, we suggest multiple promising research directions in order to improve AI-based AVS.

Keywords: Deep learning, autonomous vehicles and systems.

2.1 Review of Deep Learning Models for AVS

Recent attention to AI developments has been drawn to deep learning because of its state-of-the-art achievements in areas such as image classification and speech recognition [1]. This has led to a rise in the adoption of deep learning in autonomous vehicle applications, such as planning and decision-making, perception, mapping, and localization [49]. In this section, we discuss the fundamentals of deep learning technologies utilized in autonomous vehicles, as well as the capabilities of each paradigm. We concentrate on convolutional neural networks (CNNs), recurrent neural networks (RNNs), graph neural networks (GNNs), and deep Q-networks (DQNs).

2.1.1 Convolutional neural networks

CNNs are designed for processing grid-like data and are capable of extracting high-level features through convolutions. CNNs are typically composed of three types of layers including convolutions layers, pooling layers, and fully-connected layers. Among them is the core structure convolution layer that aims at learning feature representations from receptive fields. It contributes the majority of computation complexity in CNNs. This layer performs a dot product between an input matrix and a kernel matrix with learnable weights, and the bias term is added to accommodate richer data variations. An example is shown in Figure 2.1. As the kernel slides across the height and width of the input matrix and applies an element-wise nonlinear activation function on it, a feature map is generated and used as the input for the next layer. Commonly, Frobenius inner product is employed for dot product, and ReLU [70] for activation function. The activation step of a pooling layer is similar to

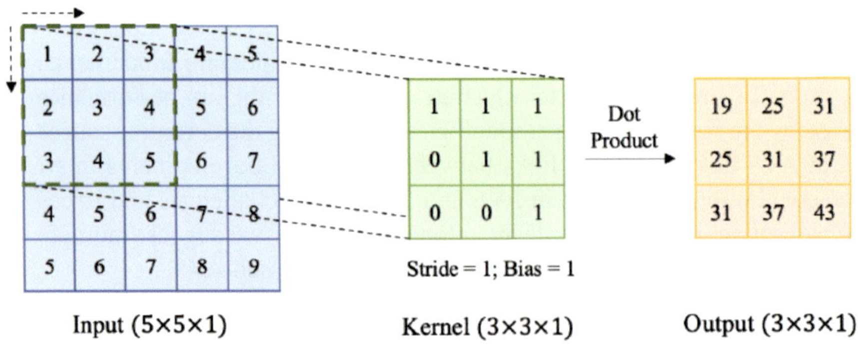

Figure 2.1 Procedure of convolution.

the convolution layer, while the main purpose of this layer is to reduce spatial size of convoluted features by extracting domain features, helping in reducing computation complexity. Max pooling and average pooling are two types of pooling, which return the maximum or average value from every slice of a given receptive field shown in Figure 2.2. The fully connected layer connects all neurons of two successive layers to generate global semantic information for classification where the Softmax function is commonly deployed.

With three types of layers, CNNs incorporate three architectural concepts, local representative fields, shared weights, and spatial or temporal subsampling, which results in a degree of shift, scale, and distortion invariance. Convolutional neural networks are designed to handle data in varying formats (e.g., picture, language, audio spectrogram, and video) and take advantage of signal qualities from local connections, shared weights, pooling, and the depth of layers. As a result, CNNs are most frequently used in analyzing visual images. Instead of using conventional fully connected networks, local connections, and shared weights are applied in the network to fully exploit the two-dimensional structure of the input data (e.g., an image signal). The approach produces a relatively small number of parameters, which speeds up and simplifies the training of the network. The process in the cells of the visual cortex is comparable to CNNs, and the smaller portions of a scene, as opposed to the whole area, are sensitive to these cells. In other words, the cells leverage the input as local filters and extract the spatially local correlation that already exists in the data.

Various CNN frameworks have been developed over the last decade or so for image classification because of their extraordinary capabilities of feature and classifier learning. ImageNet [18], a large-scale visual recognition challenge, makes a substantial contribution to the advancement of convolution neural networks and deep learning architectures [48, 105, 85]. Krizhevsky et al. [48] proposed AlexNet, which achieved the best result in ImageNet Large Scale Visual Recognition Challenge (ILSVRC) 2012, when

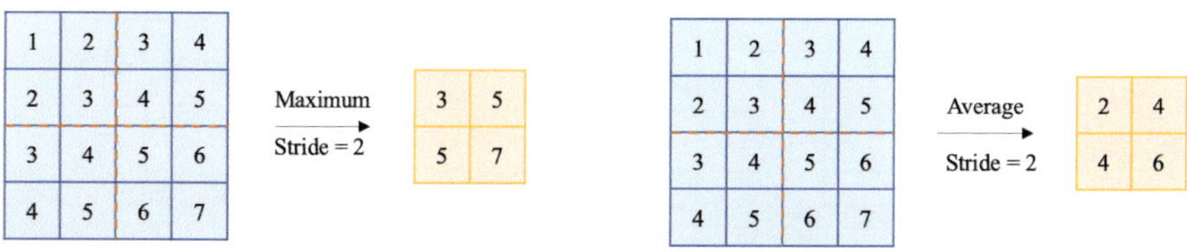

Figure 2.2 Procedure of pooling.

compared to results from earlier studies. AlexNet contains five convolution layers and three fully connected layers for effective learning abilities, and a highly optimized GPU implementation was adopted to accelerate the training process shown in Figure 2.3. With the success of deep architectures in classification accuracy, other important breakthroughs in CNN were made, including SqueezeNet [35], which reduced filter size for efficient distributed training with less overhead and dependence in hardwares; VGGNet [88], which was designed by VGG (visual geometry group) from University of Oxford; or GoogLeNet from Google [92] and residual neural network (ResNet) [32] from Microsoft Research, both of which increased network depth for higher accuracy in ILSVRC. Particularly, ResNet incorporated learning residual functions to the layer with respect to the inputs and increased depth to about 20 times deeper than AlexNet and 8 times deeper than VGGNet, wining the campaign of ILSVRC 2015. Because of its better capability in nonlinear function approximation and feature representations, various methods have been developed based on this network and applied in many other fields.

2.1.2 Recurrent neural networks

RNNs differ from feed-forward networks and CNNs in that they include connections that provide feedback into earlier layers. This is a key aspect of RNNs, using previously stored information in forward propagation through time. A simple RNN and its unfolded structure shown in Figure 2.4 have an input unit x, a hidden unit h, and an output unit y, all of which form a sequence through time t. The hidden state, referred to as "memory" of the system, is defined as

$$h_t = f_h(W_{XH}x_t + W_{HH}h_{t-1} + b_{ht}), \tag{2.1}$$

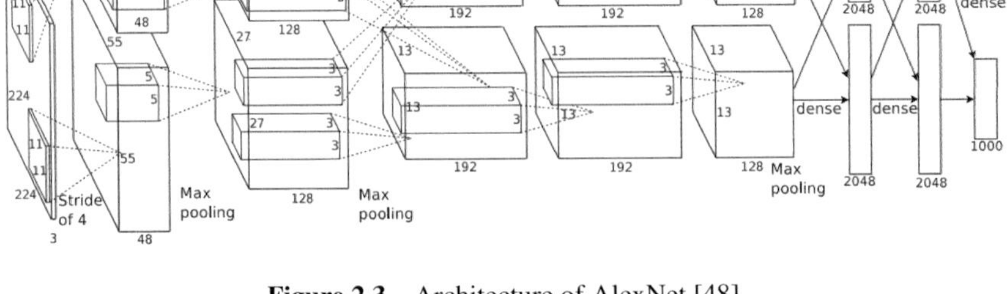

Figure 2.3 Architecture of AlexNet [48].

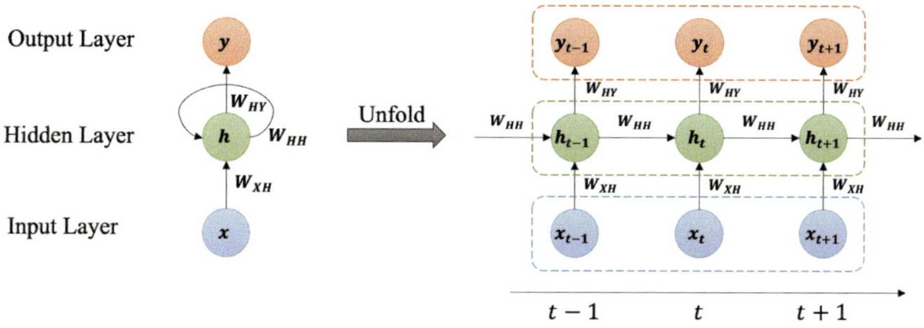

Figure 2.4 Architecture of a simple recurrent neural network and its unfolded structures.

where $f_h(\cdot)$ is the activation function of the hidden layer. Sigmoid, tanh, and specially ReLU have been widely employed as activation functions. W_{XH} is the weights connecting the input and the hidden layer, and W_{HH} is the recurrent weights connecting the hidden layers. b_{ht} is a bias term of the hidden layer at time t. The output state is calculated based on the hidden state h_t,

$$y_t = f_y(W_{HY}h_t + b_{yt}), \tag{2.2}$$

where $f_y(\cdot)$ is the activation function of the output layer, W_{Hy} contains the weights connecting the hidden layer and the output. b_{yt} is a bias term of the output at time t. With one input and one output, this simple RNN is referred to as one-to-one RNN. With respect to the number of input and output in the architecture, one-to-many, many-to-one, and many-to-many RNNs have been developed and applied in many fields (e.g., natural language processing and speech recognition).

With the use of back-propagation technique for unfolded network through time [41], RNNs can be efficiently and effectively trained. However, the capability of RNN learning long-term sequential dependency is limited by the training procedure. To address this issue, long short-term memory (LSTM) [104] was proposed by employing a unit of memory known as "cell", which can retain its value for a significant amount of time. This assists the neural network remembering the most recent computed value. A standard LSTM contains input, forget, and output gates and a cell with activation shown in Figure 2.5. The memory unit or cell consists of three gates that govern the flow of information into and out of the unit and the cell. Similar to LSTM units and gates, the gated recurrent unit (GRU) [15] is formed by two gates: an update gate and a reset gate. The duty of an update gate is to communicate

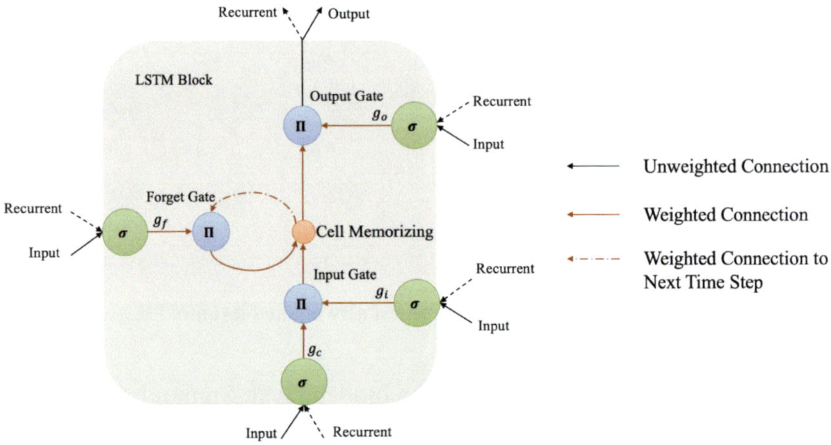

Figure 2.5 Architecture of a standard long short-term memory [28].

the prior cell's content maintenance requirements. The reset gate defines the process of transferring its previous cell's contents to the current input. GRU represents a conventional RNN by initializing the reset gate to 1 and the update gate to 0 and is recognized as a simplified LSTM.

2.1.3 Graph neural networks

Compared to CNNs, which perform convolution to deal with Euclidean data (e.g., image and text), GNNs have better capacity in performing convolution on non-Euclidean data like graphs shown in Figure 2.6, by capturing complicated interacting connections between nodes, and providing high-level representations of the graph inputs. Graph can be utilized to depict various objects, for example, transportation networks, social media networks, and molecules. A graph is generally defined as $G = (V, E)$, where $v_i \in V$ denotes a node in set V, and $e_{ij} = (v_i, v_j) \in E$ is an edge connecting v_i and v_j in set E. Edges can be divided into directed or undirected edges based on if directional dependencies exist between pairs of vertices. The neighborhood of a node v can be represented as $N_u = \{u \in V|(v, u) \in E\}$. The core intuition of GNNs is that the nodes in a graph can be defined by their neighborhoods.

The fundamental processes of GNNs include the iterative aggregation of neighborhood information for each node, followed by the incorporation of the aggregated information into the original node representation through neural information propagation. Differentiated by the aggregation and node update

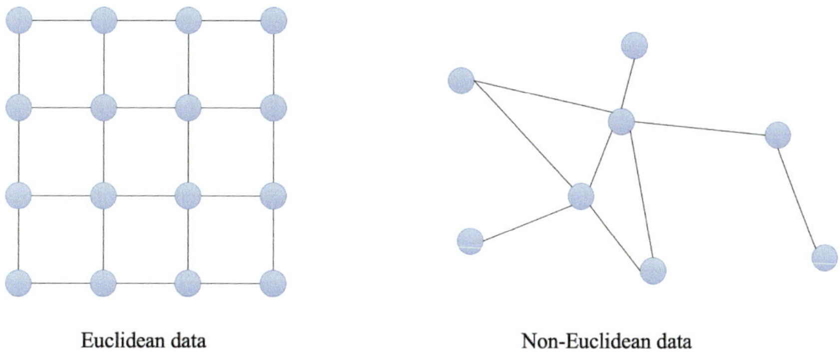

Figure 2.6 Euclidean data and non-Euclidean data.

operations, three representative GNN models are graph recurrent neural networks (GRNs), spectral graph convolutional neural networks (GCNs), and spatial GCNs. GRNs utilize a recurrent neural framework to learn node representations with shared function. The nodes in a graph are assumed to constantly exchange information with their neighbors until the equilibrium is achieved. For example, gated GNN (GGNN) [59] applies a gated recurrent unit (GRU) [15] as a shared parameterized function to generate node representation by learning previous nodes and neighbors' hidden states. GCNs with similar idea to CNN sum the neighboring pixels using a kernel with learnable weights, providing a technique for aggregating neighborhood information to update node representation and stack numerous graph convolutional layers to build high-level node embeddings. Spectral GCNs and spatial GCNs are two basic types of GCNs. Spectral-based GCNs conduct graph convolutions by processing incoming graph signals via a collection of learnable filters, relying on graph signal processing algorithms to aggregate information [107]. The geographical and temporal fluctuations in real-world graphs, such as transportation networks, social networks, and skeleton-based human behaviors, are commonplace. Nodes and edges are distributed differently in graphs throughout time, and their topological structures alter as a result. ST-GNNs are used to describe the spatial and temporal relationships in dynamic graphs. An ordered list or asynchronous collection of graphs may be both used to describe a dynamic graph. GCNs and RNNs are frequently incorporated in the design of ST-GNNs so that topological information can be recovered from graph inputs using GCNs, and the temporal relationships between graphs can be captured by RNNs.

2.1.4 Deep Q-networks

Through an agent interacting with its surrounding environments, the reinforcement learning (RL) paradigm is developed to learn skills that are necessary to perform better on a given task by maximizing the rewards illustrated in Figure 2.7. An RL problem can be seen as agents sensing their surroundings via sensors and acting upon that environment through actuators [71]. An expert does not tell RL agents what to do; instead, the reward function evaluates how well an agent performs. In each state s_t at time t, the agent takes an action a_t based on a policy $\pi(a_t|s_t)$ and gets a reward r_{t+1} and observations s_{t+1} from the environment depending on that decision's usefulness. For the agent, the ultimate goal is to maximize the total benefits it receives over the course of its whole existence. Typically, the Markov decision process (MDP) (S, A, P, R) is used to define an RL model where S is a set of states, A is the actions, $P(s_t, a_t, s_{t+1},) = \{s_{t+1}|s_t, a_t\}$ is the probability that an action a_t in a state s_t at time t leads to state s_{t+1} at time $t + 1$; $R(s_t, a_t, s_{t+1})$ is the reward gained from the action a_t transitioning from s_t to s_{t+1}; a discount factor, $\gamma \in [0, 1]$, determines the scale of distant future rewards to those from the immediate future. If $\gamma = 0$, the agent only cares about immediate rewards.

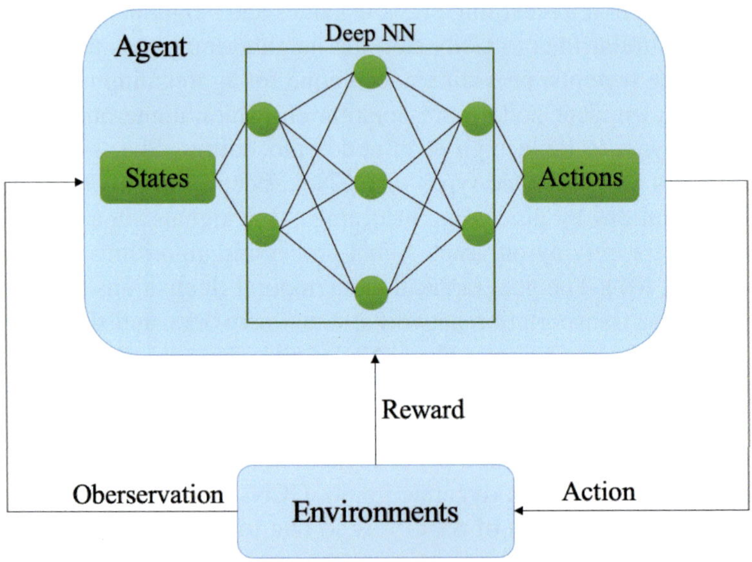

Figure 2.7 Framework of deep reinforcement learning.

Deep RL [66] combines reinforcement learning and deep neural networks that are capable of extracting features from high-dimensional data to realize human-like intelligence. A deep RL algorithm's training is shown to have distinct difficulties. Managing trade-offs between exploration and exploitation in RL is one of the most challenging aspects. To optimize its rewards, an agent should use its knowledge to pick actions that are known to yield the highest reward. However, in order to identify these beneficial activities, it can incur risks of attempting new actions that potentially lead to bigger rewards than the present best-valued actions for each state of the system. Q-learning is one of the most used RL algorithms. In this model-free temporal difference (TD) approach, the Q-functions are used to assess the utility of specific states and actions. As long as all actions in all states are sampled infinitely frequently and the state-action values are represented discretely, Q-learning has been demonstrated to converge to the optimal state-action values for a Markov decision process [65]. If a significant number of samples are collected for each state-action pair, Q-learning will be able to learn (near) optimum values for each state-action combination. In the event that a Q-learning agent has converged to the optimum Q-values for an MDP and then chooses actions greedily, it will get the same anticipated total of discounted rewards. To approximate the nonlinear Q-function across high-dimensional state spaces (such as the pixels in a frame of an Atari game), Deep Q-Networks (DQN) [103] use deep neural networks (DNNs) as a nonlinear Q-function approximator. Neural networks may reliably predict the value of each action without the requirement for any explicit domain-specific information or hand-crafted attributes. In practice, DQN enhances sampling efficiency by disassociating successive experience samples via experience replays.

2.2 Tasks in Autonomous Vehicles and Systems

The development and deployment of autonomous vehicles and systems (AVS) have been greatly facilitated by the growth of artificial intelligence (AI). AVS are becoming more dependent on AI to understand their surroundings and make appropriate decisions while in motion. For the ultimate objective of self-driving cars, understanding how AI functions in AVS is essential. With artificial intelligence, AVS will be able to accomplish the programmed work on their own, anywhere, and at any time. Self-driving cars, for example, are one of several areas in which AI-based technologies such as machine learning (ML) are rapidly being deployed. Techniques for machine learning (ML) may cope with uncertainty by utilizing a large quantity of data from the vehicle's sensors, even if the information is ambiguous. Camera and Lidar sensor data are two examples of data that are gathered. Sensor fusion methods are used to provide a semantic representation of the raw data. These representations are then used to create machine learning models for recognition of the surrounding objects including vehicles and pedestrians, and avoidance of collision with them. Given their remarkable outcomes in many other tasks such as image recognition and decision-making, deep learning models as summarized above have been widely adopted in self-driving algorithms. This section examines the use of AIs to assist the four core AVS applications: 1) perception, 2) localization, 3) path planing, and 4) motion control learning.

2.2.1 Perception

Figure 2.8 demonstrates how perception utilizes the computational algorithms of detection and localization to provide an enhanced understanding of the scenes mid-level representation of perception state. We organize the following three responsibilities of AVS that are related to perception conceptually: scene comprehension, decision-making, and planning. Perception builds on the computational algorithms of detection and localization to offer

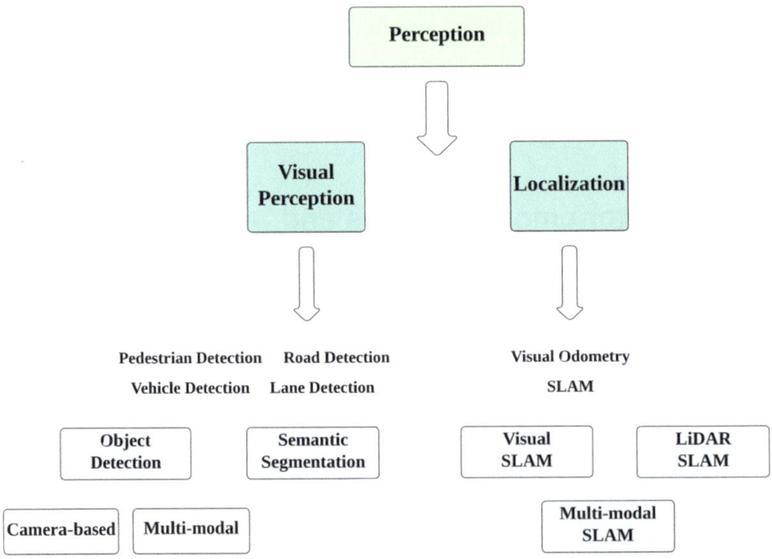

Figure 2.8 Flow chat of perception applications and tasks in AVS.

a higher level of comprehension of the scenes. It tries to reliably generalize to different scenarios as the AI-empowered processors become more abstract and high-level understanding of the surroundings by integrating disparate sensor inputs. Fusion simulates sensor noise and detection uncertainty across many modalities, such as LiDAR, camera, radar, and ultra-sound, and gives a sensor-agnostic representation of the environment [25].

Perception is how the AVS "see" the world. Usually, the perceptual data are incorporated into a learning model in order to make the best judgment possible. Perception of AVS has various important applications, such as pedestrian detection [83], road detection [13], vehicle detection [22], lane detection [7], visual odometry [97], and simultaneous localization and mapping (SLAM) [10]. Most of the applications are computer vision tasks (e.g., pedestrian detection, and road detection) and require two primary strategies utilized in the perception of autonomous cars, including semantic segmentation and object detection. These two tasks operate mostly with cameras: 1) Semantic segmentation is to assign each pixel in a picture to a certain object class. These class labels could represent a person, a bicycle, or a tree. At the pixel level, semantic segmentation is seen as an image classification problem. 2) Object detection is the process of recognizing an object of interest inside a picture, as well as generating a bounding box to localize it. Applications like visual odometry and SLAM involve distance measurement and localization, both coming from LiDAR, radar, and other radio-related sensors, and we cover this topic in Section 2.2.2.

Object Detection: Object detection is a task for determining which class instances an object belongs to. To detect a comprehensive 3D vision of the surroundings, self-driving vehicles must identify the types of objects included in an image and their specific placements [30]. To achieve the desired goal of identifying critical objects (e.g., pedestrians), we can either rely on single camera sensor (computer vision tasks) or attempt multiple sensors (multi-modal data fusion tasks) to have better performance in complex environment settings like fog [8].

For the single camera image processing cases – when applied to ImageNet [48] and Pascal VOC [24], various object detection architectures like YOLO [78] and SSD [63] have produced impressive accuracy and low error rates, but whether these architectures can produce low error in real-time self-driving scenarios remain a concern. The combination of region proposals and CNNs is referred to as "R-CNN", a visualization of which can be seen in Figure 2.9. "R-CNN" demonstrated effective performance for AlexNet and visual geometry group 16 (VGG-16). However, it was sluggish since each

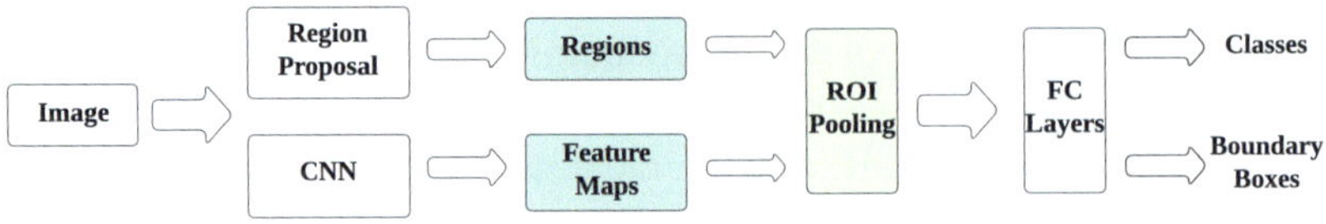

Figure 2.9 The fast R-CNN pipeline [2].

area inference required a complete forward pass of CNN [109]. CNN weights change as a result of the network's components pointing toward comparable areas for a large number of regions where it is necessary to analyze CNN for each region.

Vehicle detection in aerial data was accomplished using box annotations to train an R-CNN mask [31]. Even though CNN operation paired with image region proposal calculation is quicker, it is not a reliable method for concurrently training an end-to-end whole autonomous driving system. Fast R-CNN was a solution to the weight update issue. It implements region of interest (RoI) pooling by running an input picture via a convolution and pooling layer, followed by a Fourier convolutional neural networks (FCNN) layer that computes low-level features using back-propagation, max pooling, and instance segmentation [95]. Fast "R-CNN" addresses object identification as a variable size regression problem, which is proved ineffective for autonomous vehicle real-time object detection. The reported test time for fast R-CNN is near 0.2 s. In autonomous vehicles, the sliding window approach for extracting regions, followed by CNN results in a sluggish "R-CNN" during testing [12]. Faster "R-CNN" has been used for vehicle identification in traffic monitoring using unmanned aerial vehicle footage recorded above signalized junctions [58]. It is resistant to variations in light, and detection is indifferent to the number of vehicles in a frame. The use of Faster "R-CNN" to identify various types of transportation, such as buses, trucks, motorcycles, and bicycles, is still the subject of study.

Compare to other object detection approaches, YOLO ("you only look once") is a single-stage detector (one-step regression) rather than multistage classifiers. The special structure design makes YOLO becomes a very fast object detector (45 frames/s) [78]. The performance of YOLO on benchmark datasets was degraded despite the fact that it was quicker than the faster "R-CNN" [79]. Faster "R-CNN" variations on VGG [62], ResNet, and GoogLeNet (inception) all surpassed YOLO in terms of performance. However, they were all slower than YOLO [90]. When compared to other object recognition algorithms for intelligent visual surveillance systems, YOLO is orders of magnitude quicker; nonetheless, it has difficulty identifying tiny things inside an image. YOLO was able to identify adjacent automobiles in autonomous vehicles with a high degree of certainty; however, remote cars that seemed to be tiny were recognized and localized with a lower degree of certainty. This discrepancy may be related to the spatial limits imposed by the algorithm. In AVS, the ability to identify and locate objects in proximity using DL will fulfill the AI-empowered functions.

Nowadays, transformers demonstrate exemplary performance on various tasks, especially in text classification, machine translation, and other natural language processing (NLP)-related applications [60]. The superior performance on NLP tasks drives the development of transformers in computer vision community and therefore adoption of transformer models for vision and multimodal tasks [44]. In object detection task, vision transformer (ViT) achieves excellent results compared to state-of-the-art CNNs while requiring substantially fewer computational resources for pretraining [21]. The CNN-based methods use pixel arrays, whereas ViT splits the images into visual tokens. ViT splits the images into a series of image patches, like the series of word embeddings used when using transformers to text data. Then it inputs the series of visual tokens to transformer structure, then learns the self-attention map of hierarchies and alignments inside the input data (see Figure 2.10). A great advantage of transformer model is its ability to pretrain itself on a large dataset. Therefore, when compared to CNN-related approaches, ViTs

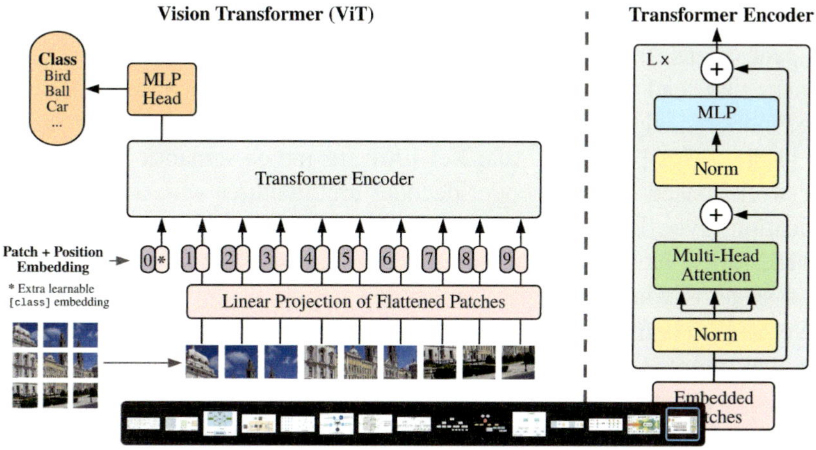

Figure 2.10 The ViT framework [21].

require lower computational resources and can achieve almost four times faster performance.

To handle real-world situations that are more complicated, researchers employed multimodal sensor data to assist the sole data coming from camera images. This brings us to multimodal fusion [25] and semi/self-supervised learning [17] in AVS deep learning fields. In the existing literature, most of the fusion methods used in object detection tasks are based on the deep-learning methods (e.g, R-CNN, fast R-CNN, and YOLO) which is discussed previously. There are three schemes of sensor fusion methods – 1) data fusion(early fusion), fusion takes place at raw data level, 2) feature fusion (half-way fusion), fusion happens after feature extraction and then fuses features, and 3) decision fusion (late fusion), combines multiple classifiers to form a final prediction result [86]. The multimodal fusion methods roughly follow the above three procedures. For instance, Wagner et al. [99] took the thermal cameras as additional input that helped the performance of the object recognition tasks. Their fusion method is based on R-CNN structure in both pixel-level fusion (early) stage and decision fusion (late) stage. The presented results showed the late fusion with pretrained R-CNN model achieved the state-of-the-art performance over all baseline function. Similar to sole camera tasks, though R-CNN-based models have obtained astonishing accuracy, they required around 47s for simple image recognition [25]. Therefore, the fusion methods that are based on fast R-CNN, YOLO, and ViT are more realistic and more popular [61, 56, 45, 3].

Semantic Segmentation: Semantic segmentation categorizes each pixel in an image to interpret driving scenes. Pixels may be labeled to indicate drivable space, pedestrians, traffic participants, buildings, etc. It is one of the high-level tasks that lead to total scene interpretation, employed in autonomous driving, indoor navigation, and virtual and augmented reality. SegNet [5], ICNet [108], ENet [73], and AdapNet [96] are major semantic segmentation networks based on the encoder-decoder architectures with a pixel-wise classification layer. In the same way, as bounding-box detectors have been improved, so too have attempts been made to enhance the calculation time of these systems when applied to embedded targets. Treml et al. [94] and Paszke et al. [73] demonstrated ways to accelerate embedded device data processing and inference for autonomous driving. Both models are lightweight networks that provide comparable outcomes to SegNet at a lower computational cost. AdapNet was used to optimize the robustness target for semantic segmentation. Adaptively learning features of expert networks depending on scene circumstances enables the model to perform robust segmentation in several situations. Using architectures, a combined bounding-box object detector and semantic segmentation result may be achieved. By adding a branch for object mask prediction in tandem with the current branch for bounding box identification, the approach enhances the efficiency of Faster R-CNN for instance segmentation. Recently, ViT-based models have lead the state-of-the-arts performance also in semantic segmentation tasks.

2.2.2 Mapping and localization

Accurately determining AVS's location and orientation are prerequisites for an autonomous driving system to function properly [46]. In order to help vehicles with their maneuvering responsibilities, avoid surrounding impediments, and conduct the appropriate driving actions, and it is necessary for vehicles to have localization algorithms that are accurate, dependable, and durable [75]. In addition, the localization system must be reliable such that it can operate in a variety of complicated situations and during harsh weather conditions. The process of mapping is an essential component of autonomous driving.

After an area has been mapped, the present location of the car may be pinpointed on the map. The initial demonstrations of reliable autonomous driving were carried out by Google and were mostly dependent on the vehicle's ability to localize itself within premapped locations [4]. Because of the scope of the issue, conventional mapping strategies have been supplemented with semantic object detection in order to achieve reliable disambiguation. In addition, it is possible to employ localized high-definition maps (HD maps) as a prior for object detection.

SLAM methods are being studied because they provide the generation of a map and the estimate of the position of sensors in an unknown environment simultaneously. Visual-based SLAM approaches play a crucial role in this sector since they are based on a low-cost and tiny sensor system, which ensures these benefits in comparison to other sensor-based SLAM techniques [64]. Deep-learning approaches have shown great improvements in accuracy, efficiency, reliability, and robustness compared to traditional SLAM algorithm [25]. LF-Net (local feature network) [72] makes a scale-space score map along with dense orientation estimates to choose the keypoints, and image patches around the chosen keypoints are cropped with a differentiable sampler and fed to the decoder network to make a descriptor. SuperPoint [19] is a fully convolutional neural network that is trained to find interest points and figure out their associated descriptors. In one forward pass,

SuperPoint infers both the pixel-level locations of interest points and their associated descriptors. VINNet [16] is an on-manifold sequence-to-sequence learning approach to motion estimation using visual and inertial sensors. It performs fusion of the data at an intermediate feature representation level in a two-stream LSTM network (camera and inertial measurement unit (IMU)). VINNet uses visual and inertial sensors to determine motion. Not only does DeepVO [101] automatically train an effective feature representation using CNN, but it also implicitly models sequential dynamics and relations using RNN. It is an end-to-end system for monocular visual object recognition that uses recurrent convolutional neural networks (R-CNNs). CNN-SLAM [93] combines naturally CNN-predicted dense depth maps with depth measurements collected directly through direct monocular SLAM. In a similar manner, Deep Virtual Stereo Odometry [102] integrates direct virtual stereo measurements into Direct Sparse Odometry (DSO) in the form of deep depth estimates. Another end-to-end deep-learning-based approach, UnDeepVO [57] uses sequential monocular pictures for testing while relying on stereo image pairs to restore the scale during training. Additionally, the loss function created for the networks' training takes advantage of spatial and temporal dense data. The system is able to accurately determine the attitude and field of view of a monocular camera.

In addition to visual SLAM methods, LiDAR-based SLAM approaches also play an important role in deep SLAM area. The advantage of using LiDAR is that it does not require a well-lit environment and can measure distances accurately. Li et al. [54] proposed LO-Net that learns feature representation for lidar odometry (LO) estimation with a new mask-weighted geometric constraint loss. In order to incorporate sparse but accurate depth measurements with visual, Li et al. designed a self-supervised visual-lidar odometry (self-VLO) framework that takes both visual and lidar input and generates depth measurement and poses in an end-to-end training manner. For other sensor-based/multisensor-based SLAM, deep-learning methods are often used to improve the sensing performance under adverse environmental conditions (e.g, smoky environments) that cameras and LiDAR cannot work well. For example, millimeter-wave (mmWave) radar is a lightweight and low-cost sensor used commonly in SLAM tasks. However, the data received by mmWave radar is noisy and sparse. Therefore, Huang et al. [33] proposed a cross-modal contrastive learning [36] for representation method, which provides the maximum agreement between mmWave Radar and LiDAR data, then generates dense mmWave presentation for extreme environments SLAM and navigation.

2.2.3 Path planning

Path planning, as one of the fundamental components in AVS, aims to find a path for a moving agent between a start and a goal point within a specified environment [29]. The objective of path planning is the creation of the optimal route for autonomous vehicles that are free of obstacles and collisions. The organizational structure graph of a path planning system is presented in Figure 2.11. The path planning system requires to take real-time data from multiple onboard sensors, including LiDAR, camera, odometry, GPS, sonar, etc. The location of both the start and the destination points remain the same or change while the vehicle is in motion. It is necessary to navigate around both stationary and moving objects. The used criterion for selecting routes comes from the passenger's perspective (e.g., safety, comfort, and the least journey time spent) or algorithm performance (e.g., feasibility, optimality, and computational complexity). The algorithms for the planning of routes can be either model-based or model-free approaches. It may entail a technique of local planning or involve a way of global planning for multiagents. In this section, we first discuss the conventional algorithms specifically designed for holonomic and nonholonomic agents. Then, we discuss the application of deep neural networks in path planning, as well as how these methods meet the criterion and solve the uncertainty existing in AVS. Finally, we

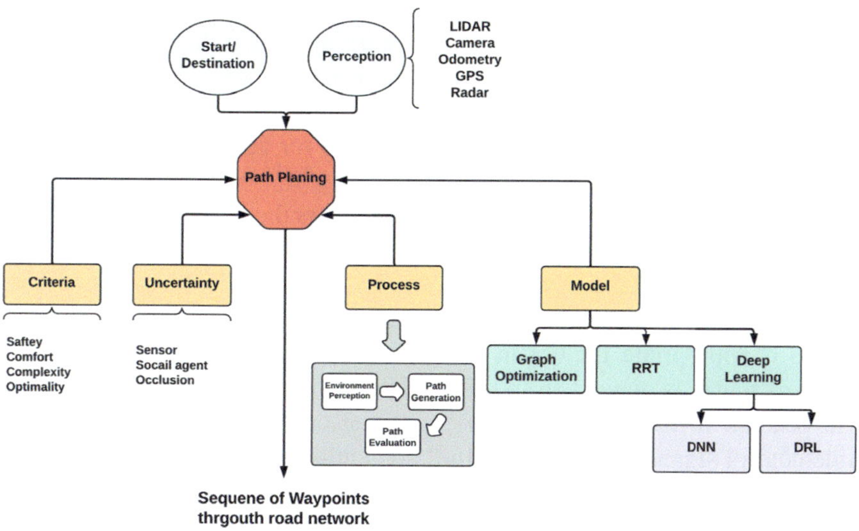

Figure 2.11 Architecture diagram of path planning process.

will summarize the challenges prior researchers encountered and future opportunities.

Based on the way robotic agents move, they can be classified as holonomic or nonholonomic drive types. A holonomic agent can control six degrees of freedom (DOF), while an agent is nonholonomic if it has less controllable DOFs [47]. Holonomic agents are easier to model when compared with nonholonomic agents. Most prior works on path planning yielded fast algorithms for holonomic cases, for example, Dijkstra [106] or A* algorithms [47]. Dijkatra's algorithm is a graph search algorithm which was proposed by Dutch scientist Edsger W. Dijkstra in 1959, mainly to solve the shortest path problem of weighted graphs [26]. This algorithm was later improved and applied to mobile robots to solve the path planning of mobile robots. Its basic principle is to select a point in a specific area as the starting point, then calculate all the paths from the starting point to a point in the area, and finally compare these paths to select the shortest path. The implementation of Dijkstra's algorithm is relatively simple, and the complexity is relatively low, but this algorithm needs to traverse all the distances from the origin to the node, which will lead to a long consumption time and a large calculation memory. The A* is considered an extension of Dijkstra's algorithm and has a better performance [47]. However, those algorithms tend to produce paths that are nonsmooth and do not generally satisfy the nonholonomic constraints of the vehicle. Rapidly exploring random trees [50], often known as RRTs, are a kind of nonholonomic algorithm that explores the configuration space via the use of random sampling and the construction of obstacle-free paths. There are several different implementations of RRT that are being utilized in the motion planning pipelines for autonomous vehicles right now.

We discuss the next important deep-learning methods applied in route planning approaches, including convolutional neural network, sequential neural network, graph neural network, and deep reinforcement learning methods. Convolutional neural network is a powerful tool to extract picture characteristics and is appropriate for use in categorization endeavors. However, the planning of routes is distinct from difficulties relating to picture categorization in general. It makes use of a variety of sensors, including cameras, to provide control signals. In this study [68], the authors made use of a camera that faces forward and mapped picture pixels to the angle of the steering. The activities associated with navigation, such as path planning, is included unconsciously in the mapping process. Mygapula et al. [69] outlined an end-to-end learning approach that uses CNN to forecast the steering angle that will allow a self-driving automobile to stay in its lane. The CNN architecture receives pictures as input and predicts the steering angles thereafter based on those visuals. This system is able to automatically learn the internal representations of the necessary procedures, such as detecting favorable road characteristics, by using the driving steering angle in conjunction with road photographs as the training input. Mygapula et al. demonstrate that convolutional neural networks have the potential to manage a self-driving car by analyzing the visuals that are input into the network. The authors of [52] propose a lightweight version of the U-Net for the purpose of end-to-end learning of lane detection and path prediction (PP) in autonomous driving. This version of the U-Net is inspired by the U-Net architecture of semantic image segmentation and uses depth-wise separable convolutions (D-UNet).

Sequential neural networks are deep-learning models that input or output sequences of data, which includes video clips, text sentences, and other time-series data. Long Short-Term Memory (LSTM) is a popular algorithm used in sequence models [91]. Additionally, the LSTM's scene context fusion module jointly captures past motion history, semantic scene contexts, and multiagent interactions. The ranking and refinement are iteratively performed using a feedback mechanism to improve the accuracy of predictions [34]. Bai et al. [6] proposed a deep-learning-based model for motion planning called a spatiotemporal LSTM network. This model creates a real-time reflection based on the extraction of spatiotemporal information. To be more specific, the spatiotemporal LSTM network model has three main structures. First, the convolutional LSTM (Conv-LSTM) is applied to find hidden features in sequential image data. Then, the 3D-CNN is used to take the multiframe feature information and pull out the information about space and time. Lastly, the fully connected neural network is used to build a control model for the steering angle of an autonomous vehicle. In [89], the authors integrated two traditional deep-learning models, a CNN and an LSTM. These models are used to extract spatial and temporary characteristics of the input pictures, respectively. For the purpose of performing end-to-end planning, the suggested model is able to accommodate nonlinear connections that exist between the input picture sequence and the output motion parameters. Transformer, as the latest sequential deep neural network, has some applications in trajectory planning for autonomous vehicles. Transformer's self-attention and multi-head attention mechanisms are well suited for context modeling and learning interaction behaviors for autonomous vehicles [81]. Prakash et al. proposed a novel multimodal transformer named TransFuser to fuse image and LiDAR data streams [77]. The proposed method achieved state-of-the-art driving performance while reducing collisions by 76% through validations in an urban environment with complex scenarios. To solve the problem that deep neural networks are usually hard to interpret, authors in [76] proposed a novel driving framework InterFuser based on the transformer model. The proposed framework is able to process multimodal sensor features and generate intermediate interpretable features.

Graph Neural Network (GNN) is a neural network that can directly be applied to graph structural data [20]. GNN provides a better way of analyzing the concept of relationships and interactions (e.g., the application of dealing with social network analysis). As such, there has been the use of GNN in path planning for the multiagent autonomous vehicle system. For example, Li et al. made use of GNN to communicate local extracted sensor features among multiple agents [55]. Khan et al. showed a GNN model could help improve traditional motion planning methods (e.g., RRTs) and outperform existing machine learning approaches [43]. Blumenkamp et al. designed a multirobot system based on GNN [9]. The system solved potential real-world problems for a GNN-based system, such as the poor telecommunication issues caused by noisy, delayed, or even lost messages. The system has been verified under a ROS2-based simulation environment.

The aforementioned deep-learning methods are usually supervised methods with the requirement of massive prelabeled road and map data. However, in some cases, an autonomous vehicle only knows part of or does not know any of the environmental information. To solve this challenge, we review deep reinforcement learning (DRL) concepts that are able to constantly update environmental information according to the obtained sensor data during the training process [53]. Kendall et al. [40] proposed a model based on DQN to solve the challenge of negotiating the right of way at an intersection. The proposed model is able to develop a strategy for lane following utilizing just a single monocular picture as input after only a few training events. They designed the reward signal as the distance driven by the car before the safety driver took charge of it, which can be easily obtained in a broad sense. The authors made use of a model-free deep reinforcement learning system to execute all explorations and optimizations on the vehicle itself [42]. For policy-based method, Silver et al. [87] demonstrated that there is a deterministic policy gradient for Markov decision process (MDP) that meet certain requirements, and that there is a simple model-free form for deterministic policy gradients that follows the gradient of the action-value function. Silver's research was published in the journal *Science*. As a consequence of this, stochastic policy gradients only integrate over the state space, while deterministic policy gradient (DPG) only integrates over the state space, which leads to fewer samples in situations with large action spaces. The authors in [76] proposed an inverse reinforcement learning system named DriveIRL, which generates driving trajectories based on the experiences collected from a self-vehicle, as well as from all the vehicles it observes. The system uses the behavior of other agents to create diverse driving scenarios without collecting additional data. They validated the system in read-world traffic conditions for more than 500 hours. Trajectory planning under uncertainty in the real-world is one of the main challenges in developing autonomous driving vehicles. Such uncertainties can come from 1) limitations of sensor hardware such as occlusion and limited range of perception, 2) difficulty of trajectory predictions for social agents (e.g., other vehicles and pedestrians), and 3) unknown environments in new areas. Traditional RL is unsafe and unrobust because it maximizes the average expected rewards in simulated or experimental environments. Rezaee et al. proposed a solution based on the distributional RL framework that handles two types of uncertainty, including detecting pedestrian crossing and winding roads with a limited field of view, through optimizing for worst-case outcomes [80]. This method has been evaluated against two challenging driving scenarios.

In order to develop efficient algorithms for planning trajectories, it is important to learn the optimality criterion each method has set for finding the optimal route. The most significant optimality criterion of algorithm performance involves minimum-time complexity, minimum energy, and minimum jerk. Minimum-time trajectory planning algorithms received a lot of consideration in autonomous vehicle work, because path planning, as a search problem, has low computational efficiency caused by too large a search space, which makes it difficult for many existing standard algorithms to be directly applied to practical situations.Other critera can also be suitable for different needs and requirements. Trajectory planning based on energetic criteria is interesting in many aspects, such as improving the fuel efficiency and decreasing the journey time.The degree of jerking such as the derivative of the vehicle acceleration is used to evaluate the comfort level when generating a trajectory. A hybrid of the above optimality criterion has also been proposed.Most existing work solved the trajectory planning problem from the following perspectives: 1) how to reduce the computational complexity of existing algorithms while ensuring the completeness (e.g., probabilistic completeness, resolution completeness, etc.) or optimality (e.g., globally optimal, locally optimal, asymptotically optimal) of a certain algorithm; 2)

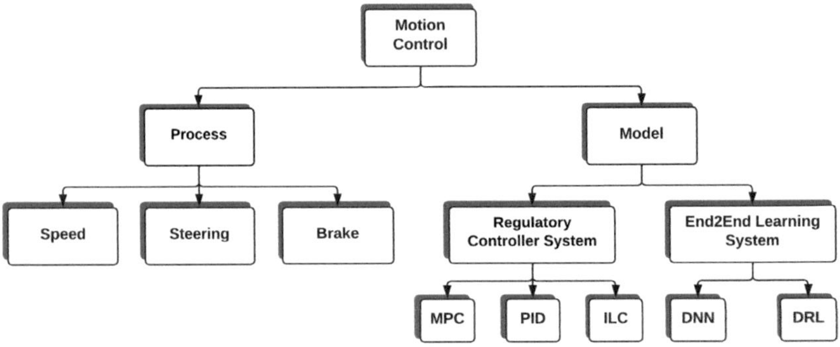

Figure 2.12 Structure of motion control learning.

how to solve the four types of uncertainties listed above, such as the occlusion of the field of view in the environment, the uncertainty of the behavior of other agents.

2.2.4 Motion control learning

Motion control is another crucial component of autonomous vehicle systems, following the path planning component. Figure 2.12 illustrates the idea of motion control, exploring it from two perspectives: the process and the model. Motion control process defines the actions autonomous vehicles need to execute and follow the planned path. For example, motion control learning algorithms are embedded within an ECU to manage speed, steering angle, and brake actions required at every point in a predetermined map such as Google Maps [38]. Furthermore, motion control of a vehicle requires building a temporal model of the vehicle dynamics and then displaying way points in chronological order over time. The established temporal models can be divided into a regulatory controller system and an End2End learning system. We discuss specific examples of the two types of controllers and summarize the advantages and limitations of using them.

Regulatory controller systems include traditional forms of closed-loop control strategies, such as proportional integral derivative (PID) controllers [51], model predictive control (MPC) [11], and iterative learning control (ILC) [74]. PID is one type of controller designed to minimize a cost function that consists of three terms: the effect of the current error, which is represented by the proportional term; the effect of errors in the past, which is represented by the integral term; and the effect of errors in the future, which is represented by the derivative term. PID has the advantages of simple structure and easy implementation. However, the fundamental PID controller can only control one basic process at a time and cannot meet the complex constraints of autonomous vehicles. MPC, a promising control technology, which computes the optimal control action based on the predicted agent's future behavior. While the MPC technique family's overarching goal is to keep the vehicle on the course, it was programmed while maintaining a stable behavior. Although MPC can achieve excellent results in complex scenes, but it is not suitable for situations where the trajectory needs to be repeated, such as racing or parking. In this case, the control performance cannot be improved by increasing the number of iterations. In contrast to MPC, ILC can improve the performance of the same task by learning from previous repeated executions [98]. ILC usually converges adequately in just several iterations [39]. However, regulatory controller systems suffer from a few limitations, including 1) precise dynamic modeling of autonomous systems is challenging, making it impossible to cope with complex application scenarios (e.g., high speeds or in a new environment); 2) they are not black-box solutions that can be simply extended to other application scenarios, and the differences between different application scenarios will affect their generalizability.

Another category of temporal models of vehicle dynamics is End2End learning systems, which usually refer to an integrated deep-learning-based black-box model mapping from raw sensor signals to final motion commands. End2End systems compress multiple components in traditional control systems into one neural network module to achieve better performance. For example, in the DARPA Autonomous VEhicle (DAVE) project, Muller et al. [68] proposed a neural network model for a scaled-down radio-controlled autonomous vehicle. The model was built using training data obtained from two forward-facing cameras while a person remained in charge of the car during the training process. The model was trained using a six-layer CNN to avoid collisions with objects while traveling at speeds of 2 meters per second. Building on the methodology of DAVE, NVIDIA used supervised learning with a CNN to design an end-to-end control system for steering a vehicle. In addition, Eraqi et al. [23] used a convolutional LSTM (C-LSTM) to learn how to drive a car based on visual and dynamic temporal dependencies. This was accomplished by introducing temporal contexts to a deep-learning steering model. The deep-learning model can be a CNN, RNN, and deep RL methods. After training the network to predict steering angles based on visual inputs, a basic CNN architecture was utilized to evaluate how well the network performed. When the C-LSTM network was used in experiments, the findings demonstrated increased accuracy as well as smoother steering fluctuation.

Reinforcement learning was employed by Wang et al. [100] in order to train an agent to perform lane change actions by using a Deep Q-Network (DQN). The network produces a continuous value for the intended yaw acceleration by using the host vehicle's speed, longitudinal acceleration, location, yaw angle, target lane, lane width, and road curvature. In order to ensure that Q-learning could be used to output continuous action values, a modified Q-learning approach was used to support continuous action values. In this approach, the Q-function was a quadratic function that was approximated by three single hidden layer feed-forward neural networks.

There is a close connection between optimal control and reinforcement learning. Optimal control can be viewed as a model-based reinforcement learning problem in which well-defined differential equations model the dynamics of the vehicle or environment. Reinforcement learning can be used to achieve optimal control. Approaches of reinforcement learning were created to handle stochastic control issues as well as ill-posed problems with uncertain rewards and state transition probabilities. These challenges were taken into consideration throughout the development of these methods. Although an End2End system is an excellent tool for solving the elaborate task of autonomous vehicle control, some barriers still affect its development: 1) an End2End system requires a massive amount of data for training, and a lack of data (e.g., data from different road types and weather) can weaken its performance. 2) An End2End system has a black-box structure with a complex architecture, making it difficult for interpretation and modification. If the system needs to be modified or improved (e.g., transfer to another application scenario or incorporate new sensor features), the old model will become useless. The whole neural network has to be replaced and trained all over again.

2.3 Future Trends and Challenges

2.3.1 Reliability

A major malfunction or failure in a safety-critical system, such as a vehicle, might result in death or severe injury to persons. Therefore, the safety of road users must be ensured before commercial deployments of such vehicles and systems. However, it might be difficult to ensure functional safety in deep-learning systems. As neural networks get more complicated, it becomes harder to comprehend the decisions they provide and how they arrive at them. The term for this is the black-box dilemma. The obscurity of these solutions is an impediment to their implementation in safety-critical applications. While it is possible to demonstrate that these systems provide good performance in our validation environment, it is impossible to test them in all of the real-world environments they might encounter. If we do not understand how they make decisions, it becomes progressively difficult to ensure that no harmful judgments are made in new contexts. It is much more difficult by the fact that online learning techniques to modify their rules during operation and, thus, safe policies might become harmful over time [82].

Any autonomous vehicle system must not only be capable of driving safely but also of responding safely to other cars or people who behave erratically. For example, if another motorist is driving carelessly or a previously unknown person rushes into the road, it might be impossible to ensure the safety of any vehicle operator. Therefore, it would be beneficial to include risky and aggressive driving behaviors of other cars in the vehicle controller's training data so that it may learn how to handle such circumstances. Utilizing a trauma memory that stores seldom occurring bad experiences (such as crashes) is one way to increase dependability and safety in such scenarios. These are then used in training to continually remind the agent of these occurrences and ensure that it maintains a safe demeanor.

Additionally, safety must be maintained throughout any real-world training or testing. During early training of a reinforcement learning agent, for instance, the agent is more likely to employ exploration than the exploitation of prior experiences, meaning that the agent would essentially learn by trial and error. Consequently, care must be taken to ensure that any exploration is conducted safely. This is particularly true in any setting with other motorists or pedestrians, since exploration might lead to hazardous decisions. An agent is encouraged to execute random behaviors during exploration, which might lead to disastrous outcomes if not anticipated beforehand. The use of demonstrations to provide examples of safe behavior could serve as a baseline policy. Simulated exploration, which occurs in a simulation environment, can limit exploration in state spaces that are deemed safe, and human oversight, which may be limited in scalability and infeasible in some real-time systems, are both potential solutions. The same holds true for any testing and assessment of the system; all safety measures must be followed until the system is assessed to work in a sufficiently safe way.

2.3.2 Computational and energy efficiency

Large amounts of data and time are necessary for adequate system training using deep-learning techniques, particularly deep reinforcement learning approaches. This may result in lengthy training periods, which can delay and increase the expense of designing an autonomous vehicle. Combining reinforcement learning with supervised learning is a typical method for reducing the amount of training data necessary or the amount of time required for training, which helps lower the training time while preserving flexibility.For a completely autonomous car, however, the quantity of training data necessary to construct a dependable and resilient system might be substantial. Due to the massive amount of data that must be gathered, it is difficult to train a vehicle to drive in all potential real-world conditions. Waymo, Zoox, and Toyota are studying autonomous driving using machine learning, and the quickest way from experimental to commercial systems would be to collaborate and share data [84].

The computing complexity of solutions due to high-dimensional vehicle control issues has a substantial impact on AVS. As demonstrated in the preceding cases, while discretization of the system might lower its complexity, it can result in a decline in system performance. In general, high computational burden of DNNs presents a challenge not only for the development and training of the networks but also for the deployment of such systems in vehicles. High computational overhead of deep-learning algorithms demand high onboard processing capabilities. This will drive up system costs as well as power consumption, which need to be kept in mind throughout the design phase of AVS.

2.3.3 Data sparsity

"Data is the new oil" has been one of the most common phrases in the car business during the last decade [37]. The availability of training data, quantity and quality of them have a strong correlation with the effectiveness of deep-learning systems. In general, current deep-learning algorithms are evaluated not only on the amount but also the quality of the training data. An algorithm's performance will increase according to the quality of the input data.

The amount of data acquired by AVS everyday is on the order of petabytes. Consequently, both parallelization of the training process and the storage infrastructure confront challenges. In recent years, simulation environments have been utilized to bridge the gap between the dearth of accessible data and deep learning's ravenous hunger for training examples. The degree of realism achievable in a virtual environment is not yet comparable to that of driving in the real-world. It is considered that traditional procedures can handle the majority of driving situations. However, the remaining challenges are edge circumstances that require the logic and intelligence of a human driver at this point. It is necessary to increase the generalization capabilities of deep-learning algorithms so that they can better handle edge scenarios. Generalization in deep learning is of the highest importance when it comes to comprehending potentially hazardous scenarios that might lead to mishaps. This is especially true when considering the scarcity of training data for such edge circumstances. This calls for the creation of one-shot and low-shot learning strategies, which can be taught with fewer training examples.

2.3.4 Human factors in AVS

Up to this point, we solely emphasize on the side of algorithms in AVS but fail to consider human factors both in the development process of AVS and integration of the human-in-the-loop design principle. Human controller plays a critical role in all phases of realizing an automatic driving vehicle or system. They can contribute to AVS's development while simultaneously being their primary users even when complete automation is achieved.

Specifically, in the development process of AVS, involvement of human controller can mitigate the data scarcity problem in model training, particularly in edge scenarios, where safety issue is placed above all other considerations. Edge scenarios are complex, rare, and when they occur, they may not be recorded due to safety considerations. Human controller is usually in control of the vehicle or system. When we adopt human-in-the-loop design principle, these data can be captured and applied in training through certain feedback mechanisms with proper postlabeling efforts.

Involving human users can also lead to better data quality due to more accurate labeling. The US National Highway Traffic Safety Administration (NHTSA) has published human factors design guidance for level 2 and level 3 automated driving systems [14]. In fact, combining active learning and annotation with effective UI design [67] can result in more accurate 'ground truth' information, more representative data, and more effective annotation of learning data in developing AVS. Additionally, the participation of human controller can provide direct feedback to system training, which can greatly expedite the learning process. These direct human feedback are generally more accurate and substantial in optimizing AVS. How to effectively incorporate them into deep-learning algorithms remains a largely uncharted area. Another substantial topic in development and adoption of AVS is related to how we can make them seamlessly collaborate with S human controller in all phases of their life cycles to fulfill all relevant goals, including but are not limited to safety, acceptance, trust, productivity, entertainment and comfort [27].

2.4 Conclusions

Over the past decades, there have been significant developments in the realm of AVS technology, particularly using artificial intelligence and deep learning. In today's world, various components of a self-driving automobile are designed using the most cutting-edge AI techniques. Not only have deep-learning strategies impacted the architecture of typical perception-planning-action pipelines, but they have also made it possible for end-to-end learning systems that can directly transfer sensory inputs into steering instructions. When it comes to the implementation of AI-based AVS on public roads, many challenging problems must be overcome. Given the prevalent formalism and explainability of neural networks, one of the most significant obstacles is how to demonstrate that these vehicles have enough functional safety. In addition to this, deep-learning systems are dependent on considerable computer hardware and vast training datasets.

References

[1] Laith Alzubaidi, Jinglan Zhang, Amjad J Humaidi, Ayad Al-Dujaili, Ye Duan, Omran Al-Shamma, José Santamaría, Mohammed A Fadhel, Muthana Al-Amidie, and Laith Farhan. Review of deep learning: Concepts, cnn architectures, challenges, applications, future directions. *Journal of big Data*, 8(1):1–74, 2021.

[2] Shilpa Ananth. Fast r-cnn for object detection.

[3] Alireza Asvadi, Luis Garrote, Cristiano Premebida, Paulo Peixoto, and Urbano J Nunes. Multimodal vehicle detection: fusing 3d-lidar and color camera data. *Pattern Recognition Letters*, 115:20–29, 2018.

[4] Mrinal R Bachute and Javed M Subhedar. Autonomous driving architectures: insights of machine learning and deep learning algorithms. *Machine Learning with Applications*, 6:100164, 2021.

[5] Vijay Badrinarayanan, Alex Kendall, and Roberto Cipolla. Segnet: A deep convolutional encoder-decoder architecture for image segmentation. *IEEE transactions on pattern analysis and machine intelligence*, 39(12):2481–2495, 2017.

[6] Zhengwei Bai, Baigen Cai, Wei ShangGuan, and Linguo Chai. Deep learning based motion planning for autonomous vehicle using spatiotemporal lstm network. In *2018 Chinese Automation Congress (CAC)*, pages 1610–1614, 2018.

[7] Massimo Bertozzi, Luca Bombini, Pietro Cerri, Paolo Medici, Pier Claudio Antonello, and Maurizio Miglietta. Obstacle detection and classification fusing radar and vision. In *2008 IEEE Intelligent Vehicles Symposium*, pages 608–613. IEEE, 2008.

[8] Mario Bijelic, Tobias Gruber, Fahim Mannan, Florian Kraus, Werner Ritter, Klaus Dietmayer, and Felix Heide. Seeing through fog without seeing fog: Deep multimodal sensor fusion in unseen adverse weather. In *Proceedings of the IEEE/CVF Conference on Computer Vision and Pattern Recognition*, pages 11682–11692, 2020.

[9] Jan Blumenkamp, Steven Morad, Jennifer Gielis, Qingbiao Li, and Amanda Prorok. A framework for real-world multi-robot systems running decentralized gnn-based policies. In *2022 International Conference on Robotics and Automation (ICRA)*, pages 8772–8778. IEEE, 2022.

[10] Guillaume Bresson, Zayed Alsayed, Li Yu, and Sébastien Glaser. Simultaneous localization and mapping: A survey of current trends in autonomous driving. *IEEE Transactions on Intelligent Vehicles*, 2(3):194–220, 2017.

[11] RH Byrne and CT Abdallah. Design of a model reference adaptive controller for vehicle road following. *Mathematical and computer modelling*, 22(4-7):343–354, 1995.

[12] Zhaowei Cai, Quanfu Fan, Rogerio S Feris, and Nuno Vasconcelos. A unified multi-scale deep convolutional neural network for fast object detection. In *European conference on computer vision*, pages 354–370. Springer, 2016.

[13] Luca Caltagirone, Mauro Bellone, Lennart Svensson, and Mattias Wahde. Lidar–camera fusion for road detection using fully convolutional neural networks. *Robotics and Autonomous Systems*, 111:125–131, 2019.

[14] John L Campbell, James L Brown, Justin S Graving, Christian M Richard, Monica G Lichty, L Paige Bacon, Justin F Morgan, Hong Li, Diane N Williams, and Thomas Sanquist. Human factors design guidance for level 2 and level 3 automated driving concepts. Technical report, 2018.

[15] Junyoung Chung, Caglar Gulcehre, KyungHyun Cho, and Yoshua Bengio. Empirical evaluation of gated recurrent neural networks on sequence modeling. *arXiv preprint arXiv:1412.3555*, 2014.

[16] Ronald Clark, Sen Wang, Hongkai Wen, Andrew Markham, and Niki Trigoni. Vinet: Visual-inertial odometry as a sequence-to-sequence learning problem. In *Proceedings of the AAAI Conference on Artificial Intelligence*, volume 31, 2017.

[17] Shohreh Deldari, Hao Xue, Aaqib Saeed, Jiayuan He, Daniel V Smith, and Flora D Salim. Beyond just vision: A review on self-supervised representation learning on multimodal and temporal data. *arXiv preprint arXiv:2206.02353*, 2022.

[18] Jia Deng, Wei Dong, Richard Socher, Li-Jia Li, Kai Li, and Li Fei-Fei. Imagenet: A large-scale hierarchical image database. In *2009 IEEE Conference on Computer Vision and Pattern Recognition*, pages 248–255, 2009.

[19] Daniel DeTone, Tomasz Malisiewicz, and Andrew Rabinovich. Superpoint: Self-supervised interest point detection and description. In *Proceedings of the IEEE conference on computer vision and pattern recognition workshops*, pages 224–236, 2018.

[20] Guimin Dong, Mingyue Tang, Zhiyuan Wang, Jiechao Gao, Sikun Guo, Lihua Cai, Robert Gutierrez, Bradford Campbell, Laura E Barnes, and Mehdi Boukhechba. Graph neural networks in iot: A survey. *arXiv preprint arXiv:2203.15935*, 2022.

[21] Alexey Dosovitskiy, Lucas Beyer, Alexander Kolesnikov, Dirk Weissenborn, Xiaohua Zhai, Thomas Unterthiner, Mostafa Dehghani, Matthias Minderer, Georg Heigold, Sylvain Gelly, et al. An image is worth 16x16 words: Transformers for image recognition at scale. *arXiv preprint arXiv:2010.11929*, 2020.

[22] Xinxin Du, Marcelo H Ang, and Daniela Rus. Car detection for autonomous vehicle: Lidar and vision fusion approach through deep learning framework. In *2017 IEEE/RSJ International Conference on Intelligent Robots and Systems (IROS)*, pages 749–754. IEEE, 2017.

[23] Hesham M Eraqi, Mohamed N Moustafa, and Jens Honer. End-to-end deep learning for steering autonomous vehicles considering temporal dependencies. *arXiv preprint arXiv:1710.03804*, 2017.

[24] Mark Everingham, Luc Van Gool, Christopher KI Williams, John Winn, and Andrew Zisserman. The pascal visual object classes (voc) challenge. *International journal of computer vision*, 88(2):303–338, 2010.

[25] Jamil Fayyad, Mohammad A Jaradat, Dominique Gruyer, and Homayoun Najjaran. Deep learning sensor fusion for autonomous vehicle perception and localization: A review. *Sensors*, 20(15):4220, 2020.

[26] Michael L Fredman and Robert Endre Tarjan. Fibonacci heaps and their uses in improved network optimization algorithms. *Journal of the ACM (JACM)*, 34(3):596–615, 1987.

[27] Anna-Katharina Frison, Yannick Forster, Philipp Wintersberger, Viktoria Geisel, and Andreas Riener. Where we come from and where we are going: A systematic review of human factors research in driving automation. *Applied Sciences*, 10(24):8914, 2020.

[28] F.A. Gers, J. Schmidhuber, and F. Cummins. Learning to forget: continual prediction with lstm. In *1999 Ninth International Conference on Artificial Neural Networks ICANN 99. (Conf. Publ. No. 470)*, volume 2, pages 850–855 vol.2, 1999.

[29] Sorin Grigorescu, Bogdan Trasnea, Tiberiu Cocias, and Gigel Macesanu. A survey of deep learning techniques for autonomous driving. *Journal of Field Robotics*, 37(3):362–386, 2020.

[30] Abhishek Gupta, Alagan Anpalagan, Ling Guan, and Ahmed Shaharyar Khwaja. Deep learning for object detection and scene perception in self-driving cars: Survey, challenges, and open issues. *Array*, 10:100057, 2021.

[31] Kaiming He, Georgia Gkioxari, Piotr Dollár, and Ross Girshick. Mask r-cnn. In *Proceedings of the IEEE international conference on computer vision*, pages 2961–2969, 2017.

[32] Kaiming He, Xiangyu Zhang, Shaoqing Ren, and Jian Sun. Deep residual learning for image recognition. In *Proceedings of the IEEE conference on computer vision and pattern recognition*, pages 770–778, 2016.

[33] Jui-Te Huang, Chen-Lung Lu, Po-Kai Chang, Ching-I Huang, Chao-Chun Hsu, Po-Jui Huang, Hsueh-Cheng Wang, et al. Cross-modal contrastive learning of representations for navigation using lightweight, low-cost millimeter wave radar for adverse environmental conditions. *IEEE Robotics and Automation Letters*, 6(2):3333–3340, 2021.

[34] Zhi Huang, Jun Wang, Lei Pi, Xiaolin Song, and Lingfang Yang. Lstm based trajectory prediction model for cyclist utilizing multiple interactions with environment. *Pattern Recognition*, 112:107800, 2021.

[35] Forrest N Iandola, Song Han, Matthew W Moskewicz, Khalid Ashraf, William J Dally, and Kurt Keutzer. Squeezenet: Alexnet-level accuracy

with 50x fewer parameters and< 0.5 mb model size. *arXiv preprint arXiv:1602.07360*, 2016.

[36] Ashish Jaiswal, Ashwin Ramesh Babu, Mohammad Zaki Zadeh, Debapriya Banerjee, and Fillia Makedon. A survey on contrastive self-supervised learning. *Technologies*, 9(1):2, 2020.

[37] Marko Javornik, Nives Nadoh, and Dustin Lange. Data is the new oil. In *Towards User-Centric Transport in Europe*, pages 295–308. Springer, 2019.

[38] Xuewu Ji, Xiangkun He, Chen Lv, Yahui Liu, and Jian Wu. Adaptive-neural-network-based robust lateral motion control for autonomous vehicle at driving limits. *Control Engineering Practice*, 76:41–53, 2018.

[39] Nitin R Kapania and J Christian Gerdes. Path tracking of highly dynamic autonomous vehicle trajectories via iterative learning control. In *2015 American control conference (ACC)*, pages 2753–2758. IEEE, 2015.

[40] Christos Katrakazas, Mohammed Quddus, Wen-Hua Chen, and Lipika Deka. Real-time motion planning methods for autonomous on-road driving: State-of-the-art and future research directions. *Transportation Research Part C: Emerging Technologies*, 60:416–442, 2015.

[41] Henry J Kelley. Gradient theory of optimal flight paths. *Ars Journal*, 30(10):947–954, 1960.

[42] Alex Kendall, Jeffrey Hawke, David Janz, Przemyslaw Mazur, Daniele Reda, John-Mark Allen, Vinh-Dieu Lam, Alex Bewley, and Amar Shah. Learning to drive in a day. In *2019 International Conference on Robotics and Automation (ICRA)*, pages 8248–8254. IEEE, 2019.

[43] Arbaaz Khan, Alejandro Ribeiro, Vijay Kumar, and Anthony G Francis. Graph neural networks for motion planning. *arXiv preprint arXiv:2006.06248*, 2020.

[44] Salman Khan, Muzammal Naseer, Munawar Hayat, Syed Waqas Zamir, Fahad Shahbaz Khan, and Mubarak Shah. Transformers in vision: A survey. *ACM Computing Surveys (CSUR)*, 2021.

[45] Jong Hyun Kim, Ganbayar Batchuluun, and Kang Ryoung Park. Pedestrian detection based on faster r-cnn in nighttime by fusing deep convolutional features of successive images. *Expert Systems with Applications*, 114:15–33, 2018.

[46] Ye-Hoon Kim, Jun-Ik Jang, and Sojung Yun. End-to-end deep learning for autonomous navigation of mobile robot. In *2018 IEEE*

International Conference on Consumer Electronics (ICCE), pages 1–6. IEEE, 2018.

[47] B Ravi Kiran, Ibrahim Sobh, Victor Talpaert, Patrick Mannion, Ahmad A Al Sallab, Senthil Yogamani, and Patrick Pérez. Deep reinforcement learning for autonomous driving: A survey. *IEEE Transactions on Intelligent Transportation Systems*, 2021.

[48] Alex Krizhevsky, Ilya Sutskever, and Geoffrey E Hinton. Imagenet classification with deep convolutional neural networks. *Advances in neural information processing systems*, 25, 2012.

[49] Sampo Kuutti, Richard Bowden, Yaochu Jin, Phil Barber, and Saber Fallah. A survey of deep learning applications to autonomous vehicle control. *IEEE Transactions on Intelligent Transportation Systems*, 22(2):712–733, 2020.

[50] Steven M LaValle et al. Rapidly-exploring random trees: A new tool for path planning. 1998.

[51] Tuan Le-Anh and MBM De Koster. A review of design and control of automated guided vehicle systems. *European Journal of Operational Research*, 171(1):1–23, 2006.

[52] Der-Hau Lee and Jinn-Liang Liu. End-to-end deep learning of lane detection and path prediction for real-time autonomous driving. *Signal, Image and Video Processing*, pages 1–7, 2022.

[53] Xiaoyun Lei, Zhian Zhang, and Peifang Dong. Dynamic path planning of unknown environment based on deep reinforcement learning. *Journal of Robotics*, 2018, 2018.

[54] Bin Li, Mu Hu, Shuling Wang, Lianghao Wang, and Xiaojin Gong. Self-supervised visual-lidar odometry with flip consistency. In *Proceedings of the IEEE/CVF Winter Conference on Applications of Computer Vision*, pages 3844–3852, 2021.

[55] Can Li, Lei Bai, Wei Liu, Lina Yao, and S Travis Waller. Graph neural network for robust public transit demand prediction. *IEEE Transactions on Intelligent Transportation Systems*, 2020.

[56] Chengyang Li, Dan Song, Ruofeng Tong, and Min Tang. Illumination-aware faster r-cnn for robust multispectral pedestrian detection. *Pattern Recognition*, 85:161–171, 2019.

[57] Ruihao Li, Sen Wang, Zhiqiang Long, and Dongbing Gu. Undeepvo: Monocular visual odometry through unsupervised deep learning. In *2018 IEEE international conference on robotics and automation (ICRA)*, pages 7286–7291. IEEE, 2018.

[58] Shulin Li, Weigang Zhang, Guorong Li, Li Su, and Qingming Huang. Vehicle detection in uav traffic video based on convolution neural network. In *2018 IEEE Conference on Multimedia Information Processing and Retrieval (MIPR)*, pages 1–6. IEEE, 2018.

[59] Yujia Li, Daniel Tarlow, Marc Brockschmidt, and Richard Zemel. Gated graph sequence neural networks. *arXiv preprint arXiv:1511.05493*, 2015.

[60] Tianyang Lin, Yuxin Wang, Xiangyang Liu, and Xipeng Qiu. A survey of transformers. *arXiv preprint arXiv:2106.04554*, 2021.

[61] Jingjing Liu, Shaoting Zhang, Shu Wang, and Dimitris N Metaxas. Multispectral deep neural networks for pedestrian detection. *arXiv preprint arXiv:1611.02644*, 2016.

[62] Shuying Liu and Weihong Deng. Very deep convolutional neural network based image classification using small training sample size. In *2015 3rd IAPR Asian conference on pattern recognition (ACPR)*, pages 730–734. IEEE, 2015.

[63] Wei Liu, Dragomir Anguelov, Dumitru Erhan, Christian Szegedy, Scott Reed, Cheng-Yang Fu, and Alexander C Berg. Ssd: Single shot multibox detector. In *European conference on computer vision*, pages 21–37. Springer, 2016.

[64] Andréa Macario Barros, Maugan Michel, Yoann Moline, Gwenolé Corre, and Frédérick Carrel. A comprehensive survey of visual slam algorithms. *Robotics*, 11(1):24, 2022.

[65] Patrick Mannion, Jim Duggan, and Enda Howley. An experimental review of reinforcement learning algorithms for adaptive traffic signal control. *Autonomic road transport support systems*, pages 47–66, 2016.

[66] Volodymyr Mnih, Koray Kavukcuoglu, David Silver, Andrei A. Rusu, Joel Veness, Marc G. Bellemare, Alex Graves, Martin Riedmiller, Andreas K. Fidjeland, Georg Ostrovski, Stig Petersen, Charles Beattie, Amir Sadik, Ioannis Antonoglou, Helen King, Dharshan Kumaran, Daan Wierstra, Shane Legg, and Demis Hassabis. Human-level control through deep reinforcement learning. *Nature*, 518(7540):529–533, February 2015.

[67] Robert Munro Monarch. *Human-in-the-Loop Machine Learning: Active learning and annotation for human-centered AI*. Simon and Schuster, 2021.

[68] Urs Muller, Jan Ben, Eric Cosatto, Beat Flepp, and Yann Cun. Off-road obstacle avoidance through end-to-end learning. *Advances in neural information processing systems*, 18, 2005.

[69] D V Prasad Mygapula, Adarsh S, Sajith Variyar V V, and Soman K P. Cnn based end to end learning steering angle prediction for autonomous electric vehicle. In *2021 Fourth International Conference on Electrical, Computer and Communication Technologies (ICECCT)*, pages 1–6, 2021.

[70] Vinod Nair and Geoffrey E. Hinton. Rectified linear units improve restricted boltzmann machines. In *Proceedings of the 27th International Conference on International Conference on Machine Learning*, ICML'10, page 807-814, Madison, WI, USA, 2010. Omnipress.

[71] Amiya Nayak and Ivan Stojmenovic. *Wireless sensor and actuator networks: algorithms and protocols for scalable coordination and data communication*. Wiley Online Library, 2010.

[72] Yuki Ono, Eduard Trulls, Pascal Fua, and Kwang Moo Yi. Lf-net: Learning local features from images. *Advances in neural information processing systems*, 31, 2018.

[73] Adam Paszke, Abhishek Chaurasia, Sangpil Kim, and Eugenio Culurciello. Enet: A deep neural network architecture for real-time semantic segmentation. *arXiv preprint arXiv:1606.02147*, 2016.

[74] Matthew W Percival, Howard Zisser, Lois Jovanovic, and Francis J Doyle III. Closed-loop control and advisory mode evaluation of an artificial pancreatic β cell: Use of proportional-integral-derivative equivalent model-based controllers. *Journal of diabetes science and technology*, 2(4):636–644, 2008.

[75] Mark Pfeiffer, Michael Schaeuble, Juan Nieto, Roland Siegwart, and Cesar Cadena. From perception to decision: A data-driven approach to end-to-end motion planning for autonomous ground robots. In *2017 ieee international conference on robotics and automation (icra)*, pages 1527–1533. IEEE, 2017.

[76] Tung Phan-Minh, Forbes Howington, Ting-Sheng Chu, Sang Uk Lee, Momchil S Tomov, Nanxiang Li, Caglayan Dicle, Samuel Findler, Francisco Suarez-Ruiz, Robert Beaudoin, et al. Driving in real life with inverse reinforcement learning. *arXiv preprint arXiv:2206.03004*, 2022.

[77] Aditya Prakash, Kashyap Chitta, and Andreas Geiger. Multi-modal fusion transformer for end-to-end autonomous driving. In *Proceedings of the IEEE/CVF Conference on Computer Vision and Pattern Recognition*, pages 7077–7087, 2021.

[78] Joseph Redmon, Santosh Divvala, Ross Girshick, and Ali Farhadi. You only look once: Unified, real-time object detection. In *Proceedings of*

the IEEE conference on computer vision and pattern recognition, pages 779–788, 2016.

[79] Shaoqing Ren, Kaiming He, Ross Girshick, and Jian Sun. Faster r-cnn: Towards real-time object detection with region proposal networks. *Advances in neural information processing systems*, 28, 2015.

[80] Kasra Rezaee, Peyman Yadmellat, and Simon Chamorro. Motion planning for autonomous vehicles in the presence of uncertainty using reinforcement learning. In *2021 IEEE/RSJ International Conference on Intelligent Robots and Systems (IROS)*, pages 3506–3511. IEEE, 2021.

[81] Khaled Saleh. Pedestrian trajectory prediction using context-augmented transformer networks. *arXiv preprint arXiv:2012.01757*, 2020.

[82] Matthew U Scherer. Regulating artificial intelligence systems: Risks, challenges, competencies, and strategies. *Harv. JL & Tech.*, 29:353, 2015.

[83] Joel Schlosser, Christopher K Chow, and Zsolt Kira. Fusing lidar and images for pedestrian detection using convolutional neural networks. In *2016 IEEE International Conference on Robotics and Automation (ICRA)*, pages 2198–2205. IEEE, 2016.

[84] Wilko Schwarting, Javier Alonso-Mora, and Daniela Rus. Planning and decision-making for autonomous vehicles. *Annual Review of Control, Robotics, and Autonomous Systems*, 1(1):187–210, 2018.

[85] Pierre Sermanet, David Eigen, Xiang Zhang, Michael Mathieu, Rob Fergus, and Yann LeCun. Overfeat: Integrated recognition, localization and detection using convolutional networks. 2014. Publisher Copyright: © 2014 International Conference on Learning Representations, ICLR. All rights reserved.; 2nd International Conference on Learning Representations, ICLR 2014 ; Conference date: 14-04-2014 Through 16-04-2014.

[86] Xian Shuai, Yulin Shen, Yi Tang, Shuyao Shi, Luping Ji, and Guoliang Xing. millieye: A lightweight mmwave radar and camera fusion system for robust object detection. In *Proceedings of the International Conference on Internet-of-Things Design and Implementation*, pages 145–157, 2021.

[87] David Silver, Guy Lever, Nicolas Heess, Thomas Degris, Daan Wierstra, and Martin Riedmiller. Deterministic policy gradient algorithms. In *International conference on machine learning*, pages 387–395. PMLR, 2014.

[88] Karen Simonyan and Andrew Zisserman. Very deep convolutional networks for large-scale image recognition. *arXiv preprint arXiv:1409.1556*, 2014.

[89] Sheng Song, Xuemin Hu, Jin Yu, Liyun Bai, and Long Chen. Learning a deep motion planning model for autonomous driving. In *2018 IEEE Intelligent Vehicles Symposium (IV)*, pages 1137–1142, 2018.

[90] Shrey Srivastava, Amit Vishvas Divekar, Chandu Anilkumar, Ishika Naik, Ved Kulkarni, and V Pattabiraman. Comparative analysis of deep learning image detection algorithms. *Journal of Big Data*, 8(1):1–27, 2021.

[91] Ilya Sutskever, Oriol Vinyals, and Quoc V Le. Sequence to sequence learning with neural networks. *Advances in neural information processing systems*, 27, 2014.

[92] Christian Szegedy, Wei Liu, Yangqing Jia, Pierre Sermanet, Scott Reed, Dragomir Anguelov, Dumitru Erhan, Vincent Vanhoucke, and Andrew Rabinovich. Going deeper with convolutions. In *Proceedings of the IEEE conference on computer vision and pattern recognition*, pages 1–9, 2015.

[93] Keisuke Tateno, Federico Tombari, Iro Laina, and Nassir Navab. Cnn-slam: Real-time dense monocular slam with learned depth prediction. In *Proceedings of the IEEE conference on computer vision and pattern recognition*, pages 6243–6252, 2017.

[94] Michael Treml, José Arjona-Medina, Thomas Unterthiner, Rupesh Durgesh, Felix Friedmann, Peter Schuberth, Andreas Mayr, Martin Heusel, Markus Hofmarcher, Michael Widrich, et al. Speeding up semantic segmentation for autonomous driving. 2016.

[95] Ayşegül Uçar, Yakup Demir, and Cüneyt Güzeliş. Object recognition and detection with deep learning for autonomous driving applications. *Simulation*, 93(9):759–769, 2017.

[96] Abhinav Valada, Johan Vertens, Ankit Dhall, and Wolfram Burgard. Adapnet: Adaptive semantic segmentation in adverse environmental conditions. In *2017 IEEE International Conference on Robotics and Automation (ICRA)*, pages 4644–4651. IEEE, 2017.

[97] Michelle Valente, Cyril Joly, and Atnaud de La Fortelle. Deep sensor fusion for real-time odometry estimation. In *2019 IEEE/RSJ International Conference on Intelligent Robots and Systems (IROS)*, pages 6679–6685. IEEE, 2019.

[98] Lanh Van Nguyen et al. Iterative learning control for autonomous driving vehicles. In *2019 International Conference on Advanced*

Technologies for Communications (ATC), pages 243–247. IEEE, 2019.

[99] Jörg Wagner, Volker Fischer, Michael Herman, Sven Behnke, et al. Multispectral pedestrian detection using deep fusion convolutional neural networks. In *ESANN*, volume 587, pages 509–514, 2016.

[100] Pin Wang, Ching-Yao Chan, and Arnaud de La Fortelle. A reinforcement learning based approach for automated lane change maneuvers. In *2018 IEEE Intelligent Vehicles Symposium (IV)*, pages 1379–1384. IEEE, 2018.

[101] Sen Wang, Ronald Clark, Hongkai Wen, and Niki Trigoni. Deepvo: Towards end-to-end visual odometry with deep recurrent convolutional neural networks. In *2017 IEEE international conference on robotics and automation (ICRA)*, pages 2043–2050. IEEE, 2017.

[102] Nan Yang, Rui Wang, Jorg Stuckler, and Daniel Cremers. Deep virtual stereo odometry: Leveraging deep depth prediction for monocular direct sparse odometry. In *Proceedings of the European Conference on Computer Vision (ECCV)*, pages 817–833, 2018.

[103] Seonghun Yoon and Kyung-Joong Kim. Deep q networks for visual fighting game ai. In *2017 IEEE conference on computational intelligence and games (CIG)*, pages 306–308. IEEE, 2017.

[104] Yong Yu, Xiaosheng Si, Changhua Hu, and Jianxun Zhang. A review of recurrent neural networks: Lstm cells and network architectures. *Neural computation*, 31(7):1235–1270, 2019.

[105] Matthew D. Zeiler and Rob Fergus. Visualizing and understanding convolutional networks. In David Fleet, Tomas Pajdla, Bernt Schiele, and Tinne Tuytelaars, editors, *Computer Vision – ECCV 2014*, pages 818–833, Cham, 2014. Springer International Publishing.

[106] F Benjamin Zhan and Charles E Noon. Shortest path algorithms: an evaluation using real road networks. *Transportation science*, 32(1):65–73, 1998.

[107] Ziwei Zhang, Peng Cui, and Wenwu Zhu. Deep learning on graphs: A survey. *IEEE Transactions on Knowledge and Data Engineering*, 2020.

[108] Hengshuang Zhao, Xiaojuan Qi, Xiaoyong Shen, Jianping Shi, and Jiaya Jia. Icnet for real-time semantic segmentation on high-resolution images. In *Proceedings of the European conference on computer vision (ECCV)*, pages 405–420, 2018.

[109] Zhong-Qiu Zhao, Peng Zheng, Shou-tao Xu, and Xindong Wu. Object detection with deep learning: A review. *IEEE transactions on neural networks and learning systems*, 30(11):3212–3232, 2019.

Authors' Biography

Guimin Dong received his Doctor of Philosophy (PhD) in Systems Engineering at the University of Virginia in 2022. His research interest includes Cyber-human Systems, Deep Graph Learning, Ubiquitous Computing, and IoT. Now, he works as a research scientist at Amazon.

Mingyue Tang is a PhD student at the University of Virginia majoring in Systems Engineering. Her research interests include Mobile Sensing, Data Mining, Pervasive Computing, and IoT Healthcare Systems.

Runze Yan is a PhD student of Systems and Information Engineering at the University of Virginia. His research topics include mobile sensing, ubiquitous computing, and machine learning.

Zeyu Mu is a PhD student of Systems and Information Engineering at the University of Virginia. Her research topics include control, optimization, and simulation of autonomous vehicles.

Lihua Cai is an Assistant Professor at the Aberdeen Institute of Data Science and Artificial Intelligence at South China Normal University. He obtained his PhD degree in System Engineering at the University of Virginia in 2020. His research interests lie in the intersection of sensing technologies and artificial intelligence, focusing on the integration of mobile sensing with IoT systems to solve real-world problems through innovative AI approaches. These include the application of GNN methods in heterogeneous sensor fusion problems, solutions to multiagent task scenarios, and systems with automatic learning via reinforcement learning and active learning frameworks.

B. Brian Park is a Professor in the Engineering Systems and Environment Department and a member of Link Lab at the University of Virginia. Before joining the University of Virginia, he worked as a Research Fellow at the National Institute of Statistical Sciences and a Post-Doctoral Research Associate at the North Carolina State University. Dr. Park received the BS and the MS from Hanyang University, Seoul, Korea, respectively, and the PhD from Texas A&M University. He is an Associate Editor of the *American Society of Civil Engineers Journal of Transportation Engineering*, *Journal of Intelligent Transportation Systems*, and the *KSCE Journal of Civil Engineering*, and an editorial board member of the *International Journal of Sustainable Transportation*. Furthermore, he is a member of the TRB (a division of the National Academies) Artificial Intelligence and Advanced Computing Applications Committee.

3

Toward Autonomous Vehicles and Systems – Potentials and Challenges of Artificial Intelligence

C. Hürten[1], A. Stockem Novo[2], Dieter Schramm[1], and P. M. Sieberg[3]

[1]University of Duisburg-Essen, Germany
[2]Institute of Computer Science, University of Applied Sciences, Germany
[3]Schotte Automotive GmbH & Co. KG, Germany
E-mail: christian.huerten@uni-due.de, dieter.schramm@uni-due.de, anne.stockem-novo@hs-ruhrwest.de, philipp.sieberg@schotteautomotive.de

Abstract

Autonomous vehicles are considered as an important part of future transportation systems. However, several challenges must be overcome on the journey toward autonomous vehicles and systems. In this context, the application of artificial intelligence is a promising approach.

The utilization of artificial intelligence offers great potential and provides advantages in multiple respects. Compared to the classical approach of theoretical modeling, artificial intelligence, and machine learning methods do not necessarily require expert knowledge, and the explicit modeling effort is reduced. Moreover, the use of artificial intelligence-based models can improve accuracy and reduce computational effort. In contrast to the potential of artificial intelligence, however, some challenges are involved in its use. One of these challenges is the availability of sufficient amounts of adequate quality data used for training. Simulations are one way of generating such data in large amounts. Depending on the complexity of the simulation models used, however, creating such data also requires an enormous amount of time. The selection of the right complexity of simulation models is considered in this contribution. The focus is on reducing computation time and the associated impact on the performance of trained artificial intelligence.

Keywords: Advanced driver assistance systems, vehicle models, automated lane change, support vector machine, model quality, machine learning, multifidelity modeling.

3.1 Introduction

In the context of advancing automation, more and more automated systems are also being used in the automotive sector. Such advanced driver assistance systems (ADAS) support the driver in a wide range of tasks and usually aim to increase comfort and safety in the vehicle. By supporting drivers with a wide range of advanced driver assistance systems, the number of personal injuries in road traffic has already been reduced in recent years. Such active safety systems include the electronic stability program (ESP) and adaptive cruise control (ACC) [1].

However, despite a downward trend in the number of personal injuries, more than 18,800 people still died in road traffic accidents in Europe in 2020 [2]. About 3.7% of all personal injuries in road traffic are caused by driver misconduct when changing lanes [3]. These accidents could be partly prevented by appropriate advanced driver assistance systems that support lane changes. Most passenger cars are already equipped with systems that warn of possible collisions when changing lanes. There are also systems available on the market that automatically perform the lane change after initiation by the driver [4].

A further development of these systems toward autonomous driving would be the fully automated lane change, in which the initiation of the driver is no longer required, but the decision about the execution of the lane change is made autonomously by the system.

Bringing such fully automated systems to the market is safety-critical and costly since functional safety can only be guaranteed if all possible situations have been tested properly. Conducting real driving tests would no longer be economical due to a large number of test cases to be performed, which is why such systems are developed with the help of simulations – real driving tests are only used in the final test phase. However, simulating a vehicle can also be computationally intensive and, therefore, expensive and time-consuming – the effort involved here depends on the complexity of the vehicle model used, with the most complex models usually producing the most realistic results. Fortunately, not every simulation requires the most complex model for realistic results, so determining the complexity of the simulation model to use is critical to the development phase to ensure that the results are accurate enough but not too many resources are needed to compute them [5].

The following sections deal with estimating the required complexity of the simulation model. For this purpose, existing vehicle models are revised in Section 2 and applied to a simple lane change scenario in a simulation model. Finally, Section 3 presents the results of the model, which is discussed in Section 4.

3.2 Modeling and Simulation

3.2.1 Vehicle models

Over the course of time, a wide variety of models have been developed to describe the behavior of vehicles. Examples of these are the single-track model, the twin-track model, and advanced multibody simulations. These models differ in the complexity of the system description; single-track models have the fewest degrees of freedom, usually 2–7 – compared to multibody simulations with usually more than 20 degrees of freedom or even more than 500 when using finite element methods. Within the mentioned model types, there are further gradations in the complexity. For example, single-track models can be simplified with idealized linear relationships (linear single-track model) and modeling of nonlinear behavior (nonlinear single-track model). In the following section, these single-track models are presented, with the modeling being limited to pure front-axle steering [6].

3.2.1.1 Linear single-track model

The linear single-track model was introduced in 1940 by P. Rieckert and T. Schunk [7]. This point model considers the total mass m of the vehicle at the center of gravity S being projected onto the road surface. This means that no wheel load differences occur at the wheels of an axle – these properties can thus be regarded as identical. This results in the simplification of combining an axle's wheels in the vehicle's center plane [8, 9].

Furthermore, it is assumed that rotations around the longitudinal vehicle axis (roll) as well as around the lateral vehicle axis (pitch) and movements of the vehicle along the vertical axis (lift) do not occur. These simplifications result in three of the six degrees of freedom that a rigid body exhibits in space. What remains is rotation about the vehicle's vertical axis (yaw) and the directions of motion along the vehicle's longitudinal and lateral axis, respectively. In addition, the assumption that the vehicle moves with the constant center of gravity velocity eliminates another degree of freedom, so that the vehicle behavior can be described with the simplifications made via the yaw angle ψ_V and the direction of the center of gravity velocity. The latter is specified with the aid of the side slip angle β, which sets the velocity direction in relation to the longitudinal axis of the vehicle. With the knowledge of the yaw angle, these quantities can be expressed in inertial coordinates [9, 10].

Furthermore, aligning torques and tire caster resulting from the tires' slip angle, as well as longitudinal tire forces occurring due to maintaining constant speed, are neglected. The wheel load distribution between the two axles is also considered constant [6].

3.2.1.2 Nonlinear single-track model

The nonlinear single-track model is also based on the combination of both tires of the respective axle to one tire in the center plane. Also, the consideration of the vehicle mass m being concentrated in the center of mass S remains valid. However, a constant center of gravity velocity v is no longer assumed. Therefore, the circumferential forces on the tires are not neglected, and the drive and braking torques applied to the tires are modeled. However, the rotations around the vehicle's lateral and longitudinal axis are still omitted [6].

The linear single-track model uses a linear approximation of the tire forces, which limits the model to cases of low slip angles. In the nonlinear single-track model, tire models represent the nonlinearities of tire behavior. Examples include the Dugoff et al. tire model [12] and the "Magic Formula" model of Pacejka [13]. The latter is briefly presented below.

This tire model describes the course of the forces acting on the wheels as a function of the occurring kinematic tire sizes with the aid of mathematical approximations. For this purpose, data is collected on the basis of quasi-static driving tests in order to represent these measured relationships by means of mathematical functions. The Magic Formula model is thus an empirical model [14, 14, 15].

The kinematic tire quantities required for estimating the tire forces are necessary to describe this model. The first of these quantities is the lateral slip. This is equal to the tangent of the slip angle, but for normal driving conditions, it can be set approximately equal to the slip angle α and describes the lateral sliding of a tire during cornering. The other kinematic tire parameter is the longitudinal slip s, which describes the sliding of a tire in a longitudinal direction.

3.2.1.3 Linear roll and pitch model

The nonlinear single-track model presented in the previous section represents the driving behavior of a vehicle under the restriction of the nonexistence of rolling and pitching motions. However, these rotations of the vehicle body occurring, in reality, have an influence on the wheel contact forces F_z – and thus also on the horizontal tire forces F_x and F_y [11, 16].

To model this behavior, the nonlinear single-track model can be extended with a roll model as well as a pitch model, which approximates the wheel load distribution. These approximated wheel loads can then be provided to the nonlinear single-track model as well as its tire model to more accurately determine the horizontal wheel forces [6, 16].

A linear roll model is presented first, followed by a brief outline of a linear pitch model.

Body roll is the rotation of the vehicle body relative to the chassis about the longitudinal axis of the vehicle, which is mainly caused by centrifugal force acting on the body. This movement can be modeled as a rotation of the body around a "virtual joint" [16], which connects the body to the chassis. This joint is the roll axis of a vehicle, which passes through one roll pole W per vehicle axle – the instantaneous pole of the roll motion in the respective vehicle axle plane. The position of the roll pole depends to a large extent on the design of the suspension system [4, 16].

The linear roll model is based on some simplifications, which are explained below: [16]

1. The front and rear axles are combined, resulting in only one roll pole. This roll pole is simplified and positioned at the vehicle's axle height.
2. Only small roll angles φ are considered, with $|\varphi| \ll 1$. This allows linearization of the equations of motion without large errors.
3. The chassis as well as the body are considered as rigid bodies.
4. The suspension components – springs, dampers, and anti-roll bars – are described via linear relationships.
5. The total vehicle mass m is assigned to the body and considered to be concentrated in its center of gravity S, while the chassis is assumed to be massless. It follows that lateral acceleration acts solely

on the body's center of gravity S and generates a centrifugal force there.

Furthermore, pitching describes the rotation of the vehicle body relative to the chassis about its lateral axis. These movements mainly occur during braking and acceleration [14]. Analogous to the linear roll model, some simplifications are also made for the linear pitch model:

1. Only small pitch angles θ with $|\theta| \ll 1$ are considered. On this basis, the equations are again linearized.
2. The tires or wheel suspensions of each axle are combined centrally.
3. The rotation is represented analogously to the roll pole as a rotation about a pitch pole N on the straight line connecting the two-vehicle axles at the height of the body center of gravity S.
4. The chassis and the body are regarded as rigid bodies which do not deform.
5. The relevant suspension components – springs and dampers – are described using linear relationships.
6. The total vehicle mass m is assigned to the body and considered to be concentrated in its center of gravity S, while the chassis is assumed to be massless. It follows that the longitudinal acceleration acts solely on the center of gravity S of the body.

Corresponding to the linear roll model, the wheel load distribution or the individual wheel loads can be calculated by determining the pitch angle θ, which is then used in the calculations of the nonlinear single-track model and its tire model, respectively.

3.2.2 Simulation setup

This work aims to test the behavior of traffic agents in a lane change event. A lane change is triggered due to a slower vehicle in front of the ego while the ACC of the ego vehicle issues deceleration due to the negative relative speed. A following vehicle on the neighbor lane complicates the situation and must be taken into account.

Figure 3.1 shows the situation that is simulated for this section. Here, a two-lane roadway is considered, with a straight route without curves as a simplification. The situation involves a total of three vehicles, which are referred to in the following the ego vehicle, the front vehicle, and the following vehicle.

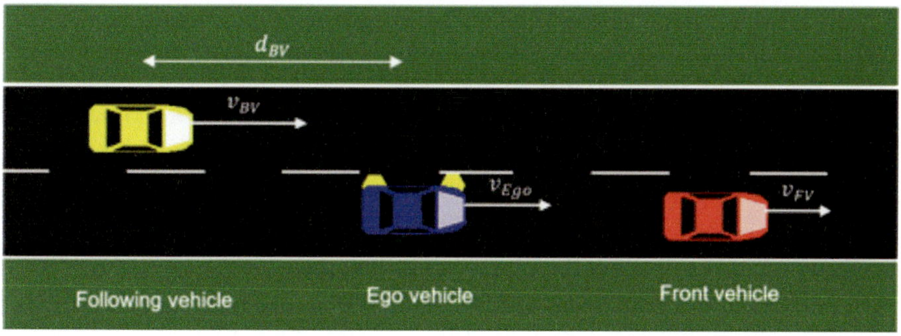

Figure 3.1 Simulated scenario, according to [17].

The main focus here is on the ego vehicle. This vehicle moves at a constant speed v_{Ego} on the right lane. The vehicle in front also travels in the same lane. Here, the ego vehicle has a higher speed than the front vehicle, that is, it approaches the vehicle in front, which is moving at a constant speed of v_{FV}. The ego vehicle is equipped with the ADAS ACC, an assistance system for speed and distance control during longitudinal driving.

In the simulation, a lane change of the ego vehicle is triggered as soon as the ACC – based on the distance of the ego vehicle to the vehicle in front as well as the speed difference of both vehicles – issues a decelerating signal. In this case, the following vehicle, which is traveling at a higher speed v_{BV} than the ego vehicle and is behind the ego vehicle by d_{BV} at the time of the lane change must be taken into account.

The ego vehicle is equipped with appropriate sensor technology to detect obstacles and determine their distance – for example, via LIDAR or radar sensors. This obstacle detection is capable of working both at the front (to supply information to the ACC) and at the rear and side of the vehicle (detection of the following vehicle).

Thus, an automatic lane change, triggered by the ACC, to a lane being used by another faster vehicle is simulated. It is examined whether the lane change would be feasible in the given situation. No collision between the vehicles should occur. On the other hand, the following vehicle should always be able to keep the minimum legal distance from the ego vehicle. However, this distance is not specifically defined, for example, in the StVO, the German road traffic regulations. In the corresponding paragraph of Section 4, it is stated that it should only be large enough to be able to stop behind the vehicle in front without collision. In this work, therefore, a widespread rule is used: In towns, one-second distance should be kept; out of town, 2 s. These time intervals multiplied by the vehicle speed result in the distances to be maintained. This rule is also recommended by the German Road Safety Council.

Furthermore, the lane change is investigated to determine whether the following vehicle had to brake excessively. In such cases, the lane change is considered to be infeasible. For this purpose, the maximum braking deceleration of the following vehicle was set to $b_{\mathrm{max}} = 0.5\frac{\mathrm{m}}{\mathrm{s}^2}$. This value is recommended by the ITE (Institution of Transportation Engineers), [18]. In [19], such braking deceleration is determined by the authors as a moderate value.

In summary, the considered situation depends on four parameters: the velocities of the vehicles v_{Ego}, v_{FV}, and v_{BV} and the longitudinal distance of the following vehicle to the ego vehicle at the time of the lane change start, d_{BV}. Simulations are performed for these parameters, and the obtained data are evaluated.

Two different simulation programs – MATLAB/Simulink as well as Eclipse SUMO – are used in the simulation. For the simulation, SUMO uses an update of the position of the vehicles at user-defined time steps with the help of an integration by the Euler method. In principle, the vehicles are modeled for simulation as a point mass whose velocity is constant over a time step. Between these time steps, the velocity can jump based on the vehicle acceleration. This acceleration is computed by a car-following model based on a model by S. Krauß [20], a driver model for modeling human driving behavior on the road considering various factors such as reaction time, traffic density, and driver experience, but also interactions with other road users [21].

By default, SUMO only calculates the longitudinal movement of a vehicle using the presented method, a lateral movement on the lane is not modeled. Also, a lane change of a vehicle is modeled only rudimentarily by default, in that the vehicle jumps between two simulation time steps from one lane to the adjacent one. Alternatively, the "Sublane Model" can be activated to simulate the lateral movements of vehicles. Here, the same method of integration is used for longitudinal motion. When using this model for lateral movement, vehicles can then be localized to several lanes. Even several vehicles in one lane next to each other are allowed. The vehicle width is included for the internal collision calculation and the underlying driver model, which is disregarded in the standard lane change model [21, 22].

In this work, SUMO is used for the most part only to simulate the traffic participants – that is, the vehicle in front and behind, and MATLAB/Simulink is used to model the behavior of the ego vehicle. For this purpose, both programs are coupled in a co-simulation. The two programs communicate via TraCI (Traffic Control Interface) [23], integrated with SUMO as a communication interface by default. On the MATLAB side, TraCI4Matlab [24] is used to implement the TraCI interface in MATLAB using Java libraries.

Due to the different complexity of the vehicle models used, according to Section 2.1, the step size varies during the simulations. While the nonlinear single-track models, as the most complex models, must be simulated with a step size of 0.001 s, otherwise errors occur in the integrations, the linear single-track model can be simulated with 0.01 s without the integrations becoming problematic. The SUMO internal vehicle model is even running with step sizes of 1 s by default, but in this work, the SUMO step size is set to 0.1 s to improve the update rate between Simulink and SUMO. The two less complex vehicle models could have also been run with the step size of 0.001 s. However, this would significantly increase the computation time without

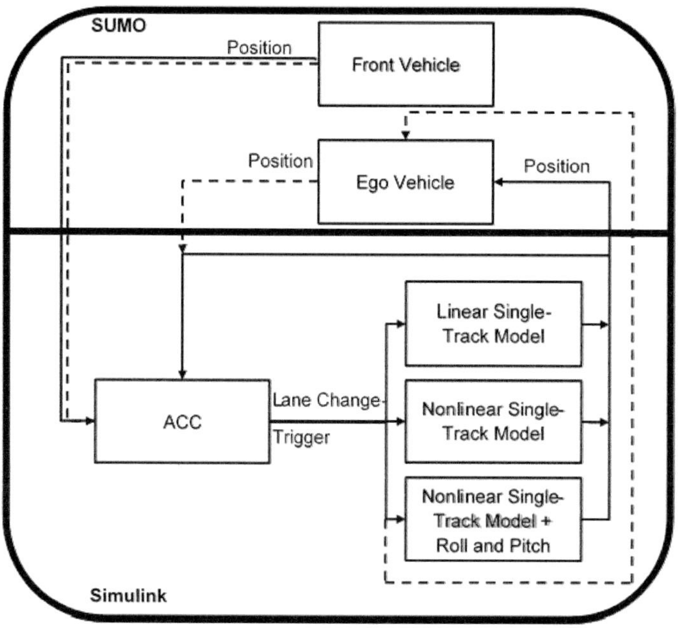

Figure 3.2 Structure and dataflow of the co-simulation, according to [17].

making the obtained results more accurate. The SUMO step size was chosen for the frequency of communication between the two simulation programs, since the information stored in SUMO is only processed and updated at this rate.

Figure 3.2 shows the structure and data flow of the co-simulation. The solid connections in the figure represent the case where the simulation is operated with a Simulink vehicle model – the dashed connections correspond to the communication flow when using the SUMO vehicle model. The following vehicle has been excluded from the graphic to ensure clarity – this vehicle has no relevant interfaces between the two simulation programs anyway.

SUMO is generally responsible for the simulation of the road users and sends the current position of the front vehicle to Simulink for the calculations of the ACC. Additionally, the ACC receives as input the current ego vehicle position from the corresponding vehicle model used. Since both vehicles are assumed to have constant speeds, these are known at the beginning of the simulation and do not need to be sent to the ACC at each time step.

The decision on whether to initiate a lane change, obtained from the ACC calculations, is subsequently transmitted to the corresponding vehicle models.

In the figure, the controllers related to the steering angle, pedal positions, and gear selection are also integrated into the respective vehicle model blocks. Furthermore, additional information is transferred from SUMO to Simulink, which is stored there during the simulation and is of interest for the evaluation of the driving maneuver. These are, for example, the position of the following vehicle and its speed.

The simulation then runs as follows:

1. The vehicle in front advances with v_{BV} and at a given distance on the right lane in front of the ego vehicle.
2. The ego vehicle advances with v_{Ego} on the right lane.
3. The following vehicle travels at a distance d_{BV} on the left lane, initially at the speed v_{Ego}.
4. As soon as the ACC of the ego vehicle gives to brake due to approaching the slower vehicle in front, the lane change is started.
5. At this moment, the speed of the catching vehicle is increased abruptly to v_{BV}.
6. The simulation is terminated 5 s after the lane change is completed. Additionally, the simulation is terminated prematurely if a collision occurs between the ego vehicle and the closing vehicle.

3.2.3 Data generation

Using the setup described in the previous section, lane change maneuvers with different parameters were simulated for each of the four vehicle models (SUMO, linear single-track model, nonlinear single-track model, and nonlinear single-tack model with pitching and rolling). The parameters are the speed of the ego vehicle v_{Ego}, the speed of the vehicle in front v_{FV}, and that of the following vehicle v_{BV} as well as the distance between the ego vehicle and the following vehicle at the time of lane change initiation d_{BV}.
Table 3.1 shows for which parameter values the simulation was performed in each case. When selecting the parameters, a maximum speed of 70 km/h was assumed so that rural roads with speed limits are also covered in addition to urban traffic. For safety reasons, it was also considered that the following vehicle does not adhere to the applicable speed limit.

Overall, it was considered that the interaction of the parameters makes sense with respect to the lane change maneuver under investigation. Thus, parameter combinations where $v_{\mathrm{FV}} > v_{\mathrm{Ego}}$ or $v_{\mathrm{BV}} \leq v_{\mathrm{Ego}}$ were discarded. In the first case, there would be no lane change triggered by the ACC at all; in the second case, a collision during the lane change would be ruled out – the vehicle behind would not catch up at all.

Considering these cases, a total of 1505 parameter combinations remain, for each of which four simulations were performed with corresponding different vehicle models. During the simulations, variables characterizing the lane change were stored in MATLAB/Simulink, namely the distances between the road users and the ego vehicle d_{BV} and d_{FV}, the lateral acceleration a_y, and the moving average over half a second of the lateral jerk j_y. The latter two variables could only be determined for the Simulink models, and SUMO does not explicitly calculate these variables – due to the simple modeling with velocity jumps and constant lateral velocity during lane changes, the subsequent calculation of these variables by deriving the position would also not yield meaningful results.

Table 3.1 Parameter variation for simulation data acquisition.

Parameter	Start value	Interval size	End value
v_{Ego}	30 km/h	5 km/h	70 km/h
v_{FV}	25 km/h	10 km/h	65 km/h
v_{BV}	40 km/h	10 km/h	140 km/h
d_{BV}	50 m	25 m	200 m

Following the simulation of the lane changes, these maneuvers are checked for feasibility. Here, it is first considered whether a collision has occurred due to the maneuver. Subsequently, whether the distance between the ego vehicle and the following vehicle corresponds to the required minimum distance is examined. In the Simulink models, the lateral acceleration and the jerk are additionally checked for compliance with the limits according to ISO 21202 – this was given for all lane changes, so the parameterization of the steering angle controller can be considered successful in this respect. An investigation of the braking deceleration of the following vehicle could be dispensed with since this was limited in SUMO anyway.

Finally, the findings were combined into a data set for each vehicle model, which is structured as follows:

$$\begin{aligned} \Omega &= \left\{\left(\boldsymbol{x}^1, y^1\right), \ldots, \left(\boldsymbol{x}^M, y^M\right)\right\} \\ \boldsymbol{x}^m &= \left[v_{\text{Ego}}^m, v_{\text{VF}}^m, v_{\text{HF}}^m, d_{\text{HF}}^m\right]^T \\ y^m &\epsilon \ \{0,1\} \\ m &= 1 \ldots\ M \text{ with } M = 1505 \end{aligned} \tag{3.1}$$

Here y describes the feasibility of the lane change: 1 (positive) means that the lane change can be initiated, and 0 (negative) means that there would be a hazard when executing the lane change.

3.2.4 Model training for the automated lane change

For the mapping of the correlations in the data set Ω, a support vector machine (SVM) is used in this work. An SVM is to be found in the training process, which can best represent the data set for the corresponding vehicle model. For this purpose, hyperparameter optimization is performed based on the grid search and the method of cross-validation.

For the training, the data set Ω is first divided into a part for the cross-validation $\Omega_{T,V}$ and a part for the final test Ω_{Test} – a share of 30% is chosen here for the test data. The cross-validation is performed with $k = 5$, that is, the partial data set $\Omega_{T,V}$ is divided into five equal parts, each of which is used once as validation part – the others correspond to the training data, respectively. Figure 3.3 shows the division of the data set for each of the five training sets. Before splitting the data, they were randomly mixed in the data set; the splitting and mixing are identical for each vehicle model, so for instance, the test data set each contains the same maneuvers in terms of simulation parameters. The training and validation data

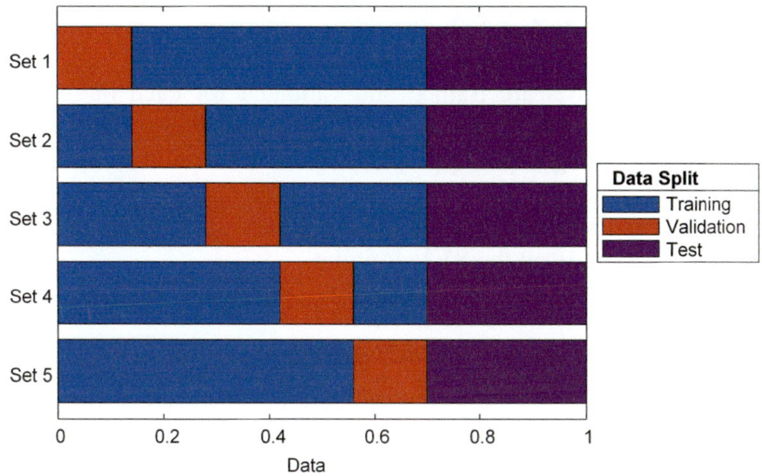

Figure 3.3 Split the data set into training, validation, and test data.

comprise 1053 driving maneuvers for each vehicle model used, the test data correspondingly 452.

The test data are not considered in the training process and are only of interest for the model validation.

In this work, the data are standardized. The standardization transformation is determined using only the training data, as is common in machine learning but is applied to the entire data set.

The hyperparameters of an SVM are, on the one hand, the regularization parameter C, and on the other hand, the kernel function used, which in turn has different hyperparameters. This work examines four different kernel functions: the linear, the polynomial, the sigmoid, and the RBF kernel function. Table 3.2 breaks down the hyperparameter combinations considered. Here, the recommendations of the authors in [25] were followed.

In addition, another hyperparameter is considered for each kernel: a weighting of the training data. The foremost goal of the lane change maneuver should be safety, and thus any training example with hazard in case of lane change ($y = 0$) according to the definition in Section 3.2 should be classified correctly in any case. Therefore, the influence of weighting in hyperparameter optimization is also investigated, and the training data expressing hazard in case of lane change is therefore weighted five times as well as ten times. In addition, a run without weighting is performed for comparison. Thus, a total of three variations are tested with respect to weighting. All hyperparameter combinations together result in 3402 different possibilities.

Five different SVMs are trained and validated for each of these possibilities according to cross-validation. The result of these five training or validation runs is averaged per hyperparameter combination and considered in the evaluation.

The Python module Scikit-learn [26], a library specialized in machine learning, is used to perform the SVM training.

3.2.5 Model validation for the automated lane change

In the following, the best parameter configuration for each of the different vehicle models is selected based on the results of training with cross-validation and hyperparameter optimization. For this purpose, the training and validation results of the five cross-validation sets are averaged for each hyperparameter combination.

As mentioned before, security should play the most important role in the development of an ADAS function. For this reason, only hyperparameter configurations are considered in the model selection, which did not give any lane change permission in both training and validation despite possible hazards. In this work, this case is present when the SVM outputs $y_{\text{pred}} = 1$, although the actual class membership of the considered data point is $y_{\text{True}} = 0$. Such misclassification is also called false positive (FP). The counterpart to this is the classification as true positive (TP) – in the context of this work, this designates a lane change initiation, with which the road users were not exposed to any danger.

A crucial metric to study the performance of a classifier with respect to FP classification is precision, which is defined by the number of TP classifications N_{TP} and FP classifications N_{FP} as follows:

$$\text{precison} = \frac{N_{\text{TP}}}{N_{\text{TP}} + N_{\text{FP}}} \tag{3.2}$$

Table 3.2 Hyperparameter variation.

Kernel	**Regularization**	**Kernel parameters**			**Hyperparameter**
	C	r	γ	d	**combinations**
Linear	$2^{-5}, 2^{-3}, \ldots, 2^{15}$	–	–	–	12
Polynomial	$2^{-5}, 2^{-3}, \ldots, 2^{9}$	$-10, -1, 0, 1, 10$	$2^{-15}, 2^{-13}, \ldots, 2^{-1}$	2, 5	640
RBF	$2^{-5}, 2^{-3}, \ldots, 2^{15}$	–	$2^{-15}, 2^{-13}, \ldots, 2^{5}$	–	132
Sigmoid	$2^{-5}, 2^{-3}, \ldots, 2^{7}$	$-10, -1, 0, 1, 10$	$2^{-15}, 2^{-13}, \ldots, 2^{1}$	–	350

Hence, the goal is that the precision is one – because then $N_{\text{FP}} = 0$ applies. Therefore, all parameter combinations are discarded for which the SVMs have achieved a precision of less than one in training and validation. The decision criterion between the remaining parameter combinations is then the accuracy on the validation data. Here, accuracy describes the proportion of correct classifications – thus, the classifier reaching the highest accuracy is desired to be selected.

In order to avoid the classification as FP, the influence of a possible weighting of the data points leading to a hazard was also examined in the hyperparameter optimization. As it turned out in the evaluation, a precision of one is only achieved if data weighting is used. All hyperparameter combinations without weighting always led to misclassifications with respect to FPs.

The selected kernel function has a great influence on the accuracy of an SVM. Since linear separability can be assumed in the fewest real situations, the transformation into a higher-dimensional space is crucial. Nevertheless, by means of the linear kernel function the case of not performing a transformation of the input data is considered.

Here it becomes clear that the linear kernel function – apart from only a small number of models resulting in full precision – does not reach the accuracy of the other kernel functions.

The best results are obtained for all vehicle models using an RBF kernel function; this is consistent with the findings of the authors in [25], who generally recommend the RBF kernel function. However, it should be noted that in this work, the polynomial and the sigmoid kernel functions also achieve quite similar results. Overall, the SVMs with these three kernel functions seem to be similarly well-suited to represent the respective data correlations. A lower scatter in accuracy for the hyperparameter evaluations using the RBF kernel functions is also noticeable; on the one hand, this could speak for better representability of the correlations by this function – on the other hand, the number of parameter combinations considered here is significantly lower than for polynomial or sigmoid functions.

The best regularization parameters C depend strongly on the considered kernel functions and their parameters so that a closer consideration is omitted here. Table 3.3 gives the best hyperparameter combination for the vehicle models in each case. The average accuracy achieved on the validation data is also noted here. In general, a decrease in the accuracy of the vehicle model complexity can be observed here.

So far, only the data for the cross-validation $\Omega_{T,V}$ have been considered. Now that these data have been used to select the best SVM models for

Table 3.3 Best SVM parameterization found per vehicle model.

Vehicle model	SVM parameters	Validation accuracy
SUMO	RBF kernel, $C = 128,\ \gamma = 2^{-5}$ Fivefold weighting	0.9867
Linear single-track model	RBF kernel, $C = 8,\ \gamma = 2^{-5}$ Tenfold weighting	0.9383
Nonlinear single-track model	RBF kernel, $C = 1,\ \gamma = 2^{-3}$ Fivefold weighting	0.9288
Nonlinear single-track model with pitch and roll	RBF kernel, $C = 1,\ \gamma = 2^{-3}$ Fivefold weighting	0.9288

the individual vehicle model simulations, the performance of these can be evaluated using the previously unseen test data Ω_{Test}. For this purpose, the SVM models with the parameters found are now trained using the entire partial data set $\Omega_{T,V}$, instead of only four of five equally sized parts as before, precision and accuracy are calculated on the test data as final evaluation criteria.

The results of the evaluation of the test data are shown in Table 3.4.

Again, all four SVMs show a precision of one, so they do not misclassify, which would lead to a hazard for the ego vehicle and road users. Furthermore, the accuracy on the test data is better than the averaged accuracy from the cross-validation for most of the vehicle models; only for the linear single-track model, it remains comparable. This phenomenon can be explained by the amount of training data used, which is increased by 25% in the case of the test evaluation. This increased amount of training apparently positively influences the classification of the SVMs; with even more data, even better results could probably be achieved – although in machine learning, it is generally true that models become more accurate with an increasing amount of training data.

Table 3.4 Best SVM parameterization found per vehicle model.

	SUMO	Linear single-track model	Nonlinear single-track model	Nonlinear single-track model with pitch and roll
Precision	1,0	1,0	1,0	1,0
Accuracy	0.9889	0.9381	0.9447	0.9425

3.3 Results

The different vehicle models are compared to each other in order to address the central question of estimating the required complexity of the simulation model. It is assumed that the nonlinear single-track model with pitch and roll represents reality and is therefore considered as "ground truth."

In the previous section, the SVM models were trained and validated on the data which was generated with a specific vehicle model. No considerations were made regarding the model quality and the differences between the individual vehicle models. This will be done in the following by evaluating the SVM models on the most realistic simulated data. Subsequently, the limits of each vehicle model are defined, in which these models deliver realistic results in the simulation, and their complexity therefore is sufficient to model real-world behavior.

3.3.1 Differences in vehicle model quality

A total of four vehicle models with varying complexity were used. While the standard SUMO model is the simplest, the linear single-track model is considerably more complex. Furthermore, the nonlinear single-track model

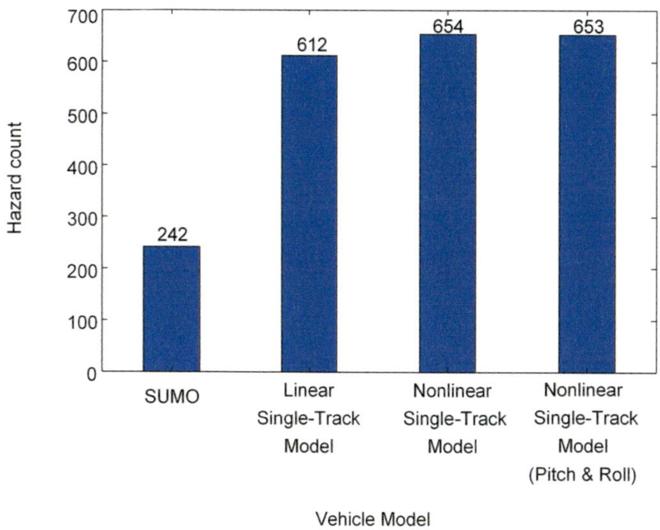

Figure 3.4 Number of hazardous lane changes per vehicle model used in the simulation.

without pitch and roll model and the one with the pitch and roll model are considered. These models have even higher complexity; the latter model is the most complex vehicle model in this work. Due to this high complexity, it is assumed that the results of that model also most accurately represent reality. The simulation results which were generated using the nonlinear single-track model with pitch and roll are therefore considered as "ground truth," thus representing reality due to lack of actual real-world data. The results of the other models are measured against this.

If only the results from Table 3.4 were considered, the conclusion would be that the SUMO vehicle model is best used to develop an ADAS for automated lane changing by simulation. However, this would lead to very inaccurate ADAS systems. For example, although the trained SVMs can best reproduce the simulation results using the SUMO model and achieve a test accuracy of 98.89%, if the SVM trained on SUMO simulation data is now tested against the ground truth data, the accuracy drops to 73.01%. Even worse, the precision of this model drops to 67.03% – thus, a large number of misclassifications are performed, which leads to a hazard for the ADAS-controlled vehicle and road users.

In fact, the data obtained in the simulation regarding a possible lane change differ between the vehicle models used, in some cases considerably. Figure 3.4 shows the number of simulated hazards and collisions per vehicle model. It can be seen that significantly fewer hazards occurred in the simulation with the SUMO vehicle model than with the more complex vehicle models – between which, however, differences also exist.

These differences can be explained with the help of Figure 3.5, in which the progressions of a lane change simulated at a speed of 50 km/h with the different vehicle models are shown exemplarily. It can be seen that the lane change simulated by the SUMO model is completed much faster, and hence the distances to the vehicle behind remain larger. The course between the other vehicle models differs less – accordingly, the number of hazards is similar (see Figure 3.4). In Figure 3.5, for the sake of clarity, the nonlinear single-track model is not further divided with respect to pitch and roll modeling, since the differences in the curves were not recognizable here.

Table 3.5 shows the test evaluation results on the ground truth data for all vehicle models used in the simulation. In addition, the development of precision and accuracy is shown in comparison to the results from Table 3.4, which are based on the simulation data of the respective own vehicle models.

Based on these results, it can be stated that both the SVM trained on the simulation results of the SUMO vehicle model and the SVM trained on the results of the linear single-track model do not realistically represent the

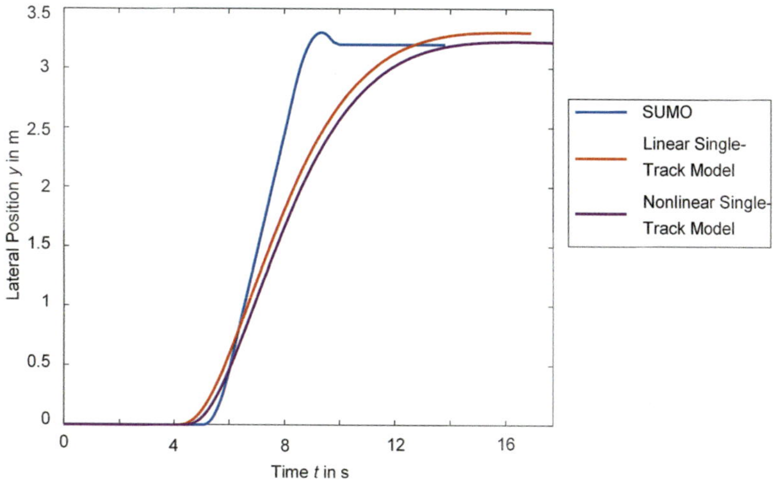

Figure 3.5 Comparison of the lane change progressions of the different vehicle models at 50 km/h.

Table 3.5 Test evaluation of the SVM models on ground truth data with a difference to the test results on the respective vehicle model data.

	SUMO	Linear single-track model	Nonlinear single-track model	Nonlinear single-track model with pitch and roll
Precision	0.6703	0.9790	1.0	1.0
Δ Precision	−0.3297	−0.021	0	0
Accuracy	0.7301	0.9558	0.9425	0.9425
Δ Accuracy	−0.2588	+0.0177	−0.0022	0

classification with respect to a possible lane change. Furthermore, for both SVMs, the precision on the test data decreases – dangerous misclassifications would thus be accepted as a consequence. Regardless of this, the separation plane of the SVM trained on the linear single-track model seems to be able to better represent the ground truth data overall; the accuracy increases by 1.77%, but since safety should have a higher priority, this result is not further considered.

It can be seen that only the nonlinear single-track models can maintain the full precision on the test data – although this should not be surprising for the nonlinear single-track model with pitch and roll modeling as the vehicle model data used in training is the ground truth data. However, the modeling of the pitch and roll behavior does not seem to have a major impact on the simulation data overall; the results of both SVMs based on the nonlinear single-track model data are ultimately identical on the ground truth test data considered. Figure 3.4 also confirms this assessment; there is only one more dangerous simulated lane change maneuver using the nonlinear single-track model without pitch and roll motions.

Based on these results, the conclusion could now be that for the development of an ADAS for an automatic lane change, an SVM should be trained whose training data is based on a simulation with the nonlinear single-track model – ideally with consideration of the pitch and roll behavior, because this model has the highest complexity and most accurately represents reality.

The disadvantage of such an approach is the computation time required for such a simulation as it is rapidly growing with increased model complexity. Figure 3.6 shows the computation time required for the simulation of the 1505 lane change maneuvers. Here, it can be seen that the simulation with the nonlinear single-track models required considerably more time than with the other vehicle models. The fastest simulation was using the SUMO model, where this was faster by a factor of 35.75 than the slowest model, which was the nonlinear single-track model with linear pitch and roll model.

In practice, however, an ADAS based on machine learning is not developed on about 1500 data points but on a considerably larger database – so that the required computing time plays an important role. It would therefore be advantageous to minimize the computing time. This would be possible if the less complex models were used to generate the data in trivial driving maneuvers. The results in Table 3.5 certainly allow for some data points to be correctly represented with the simpler models – otherwise, the SVMs would show significantly worse performance. Also, the results in Figure 3.4 concerning the simulated hazards show that the simulation with less complex vehicle models led to hazards too. However, in the borderline areas where the complexity of the models is important, more complex simulation models would have to be used to obtain reliable data.

3.3.2 Definition of the model boundaries

In order to connect the different models in a way that the simple vehicle models are used for trivial driving maneuvers – but outside their model capacities, more complex vehicle models are used – the limits must be known in which a vehicle model is sufficiently accurate. Thus, a multi-fidelity model is to be set up using a hierarchical approach. Therefore, selection criteria are assigned to the various vehicle models.

The goal here is to learn the relationships, that is, the selection criteria, in the data with the help of machine learning methods. For this purpose, various machine learning methods can be used, such as artificial neural networks, clustering algorithms, or SVMs.

In this work, again an SVM is chosen to learn which vehicle model is applicable in the simulation and when. The decision to use the SVM was made because the task is ultimately a binary classification – the vehicle model is applicable, or it is not – and an SVM seeks a separation plane maximizing the distance between both classes. Many other machine learning models aim only to find an arbitrary separation plane.

In the first step, the simulation results of the different vehicle models are compared with the ground truth data. This leads to an evaluation of the feasibility of the vehicle models at each considered data point – and ultimately to a binary distinction of whether a model is sufficient or not at a given data point. In a second step, the SVM shall be trained to interpolate these data points to produce a mapping of the models' feasibility in the simulation parameter space. The training of the SVMs has performed analogously to Section 2.4. The results of the comparison are then standardized; 30% of these data are considered as the test data set Ω_{Test} and with the remaining data, a hyperparameter optimization with cross-validation is performed. Thus, after training, the SVM should be able to decide whether a

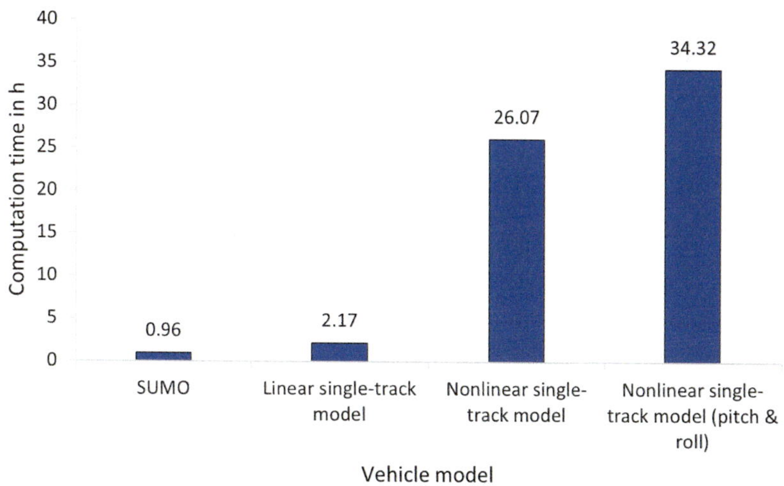

Figure 3.6 Total computation time for the simulation of the maneuvers per vehicle model.

vehicle model can represent reality or not for a certain simulation parameter configuration.

Additionally, the influence of a weighting of the data is considered again. The goal of the training is that the SVM is able to make a statement about whether the respective vehicle model is able to generate the same simulation result as the ground truth data considering an input vector $x = [v_{\mathrm{Ego}}, v_{\mathrm{FV}}, v_{\mathrm{BV}}, d_{\mathrm{BV}}]^T$. Since a safety-relevant ADAS is to be trained on these simulated data afterward, it is crucial to avoid misclassifications, meaning false-positive classifications that result in collisions. A vehicle model should thus only be used if it is suitable for the specific maneuver – otherwise, a more complex model should be used. The requirement that the precision of the trained SVM must be one should again ensure that such FP classifications are avoided.

A detailed evaluation of the hyperparameter optimization will be omitted in the following; only the best SVM in each case will be briefly presented.

For learning the SUMO vehicle model boundaries, 3402 hyperparameter combinations were tested in combination with the investigation of data weighting. Three cases were considered again for weighting: no weighting, fivefold weighting of cases where the simulation results do not match the ground truth data, and tenfold weighting of these. The best hyperparameter combination considered with precision results in an averaged validation accuracy of 89.74%. On the test data, this SVM also achieves 100% precision, with an accuracy of 88.94%. The parameter configuration of this SVM is described in Table 3.6.

A total of 5670 hyperparameter combinations were validated for the approximation of the model boundaries of the linear single-track model. This increased number is due to the fact that the weighting of the data had to be considered in more detail. The standard weightings considered in this work (none, fivefold, and tenfold) did not produce an SVM with a precision of one. A subsequent examination of the underlying data set showed that it consists of 1454 positive and only 51 negative data points – since the negative ones should be classified correctly in any case, they have to be weighted significantly more. Thus, a 30-fold and a fiftyfold weighting of the data points of the negative class were considered.

With the new weights considered, some of the trained SVMs achieved a precision of one – the best parameter combination found is again given in Table 3.6. The SVM with these parameters was also able to show perfect precision on the test data set; the accuracy averaged 62.49% on the validation and reached 60.18% on the test data, which is significantly lower than for the SUMO vehicle model boundaries.

For the nonlinear single-track model without pitch and roll, no SVM could be trained on the model boundaries, since only a single data point was different when compared with the ground truth data. Therefore, training a binary classifier on a data set consisting of 1505 data points where one of the classes consists of only on point would not provide usable results, respectively, would not be reasonable to validate.

For this reason, the single deviating data point is instead cut out of the scope of the vehicle model.

For the nonlinear single-track model with pitch and roll consideration, no application boundary can be defined because these data are considered ground truth data.

3.3.3 Development of a driver assistance system based on model quality

After the model quality of the four different vehicle models was examined in the previous section and the application limits of these models were determined, an SVM has trained again in this section as the decision maker of an ADAS for automated lane changing. Opposite to Section 2, however, the vehicle models are combined in consideration of their area of application. As shown in Figure 3.6, the required computation time for the different vehicle models is very different. In this section the training of the SVM is done on a larger data set than before and the influence on the required computation time and the achieved accuracy is investigated for the case that the generated multi-fidelity model is used – and hence the vehicle models are used all together for simulation according to their application suitability. The comparison is made against an SVM trained exclusively on data simulated with the nonlinear single-track model with pitch and roll modeling.

3.3.3.1 Data generation and model training

The training of the SVMs in this section is fundamentally similar to that in Section 3.2, with one difference being the size of the underlying data set – a

Table 3.6 Best-found parameter combinations for the model boundary approximation.

Vehicle model	SVM parameters	Test accuracy
SUMO	RBF kernel, $C = 8,\ \gamma = 2^{-3}$ Fivefold weighting	0.8894
Linear single-track model	Polynomial kernel of second degree, $C = 512,\ \gamma = 2^{-7},\ r = 1$ Fiftyfold weighting	0.6018

more extensive one is used this time. Instead of 1505 simulation runs, 5447 are examined. For this purpose, the grid of the speed of the closing vehicle v_{BV} and the initial distance from it to the ego vehicle d_{BV} is halved.

Again, only those parameter combinations were used which produce the desired lane change maneuver – accordingly, the front vehicle must be slower than the ego vehicle, and the following vehicle must be faster than this.

Another difference is the use of vehicle models. While in Section 3.2, these simulation runs were performed individually for each of the four vehicle models considered, the findings from Sections 3.3.1 and 3.3.2 will now be incorporated here. Basically, this means that the data set no longer consists of the simulation results of one vehicle model in each case but of a combination of the simulation results of different vehicle models.

For each simulation run, the vehicle model that is least complex – but still suitable for use according to the defined application limits of the vehicle model – is selected. Thus, every maneuver is simulated using the SUMO model if possible. If a maneuver lies outside the SUMO model limits, the linear single-track model is used. Only if this cannot be used for the maneuver either, the nonlinear single-track model without pitch and roll is used. As the most complex vehicle model, the nonlinear single-track model with pitch and roll modeling is only used if the other three vehicle models would operate outside their application limits.

The described procedure of generating a data set with corresponding model boundaries, which the trained SVMs define, is validated in the following. To evaluate the data set and the obtained results on their basis, a second data set is generated in which only the nonlinear single-track model with linear pitch and roll model is used for simulation. The distribution of the individual vehicle models in the data sets is shown in Figure 3.7.

For each of these two data sets, an SVM is to be trained. The procedure known from the previous sections is used. For each SVM, a hyperparameter optimization with fivefold cross-validation is performed. Again, the weighting of the data leading to lane change hazards is additionally investigated. This results in an investigation of 3402 differently parameterized SVMs for the data sets.

However, in contrast to the previous investigations, 30% of the data sets are not randomly declared as test data sets Ω_{Test}. To achieve comparability with the results from Section 3.2, the same simulation parameterizations are used for the test data as they were assigned to the test data set in Section 3.2; all remaining simulation runs are assigned to the Ω_{TV} cross-validation data. These are used for hyperparameter optimization and are also standardized again.

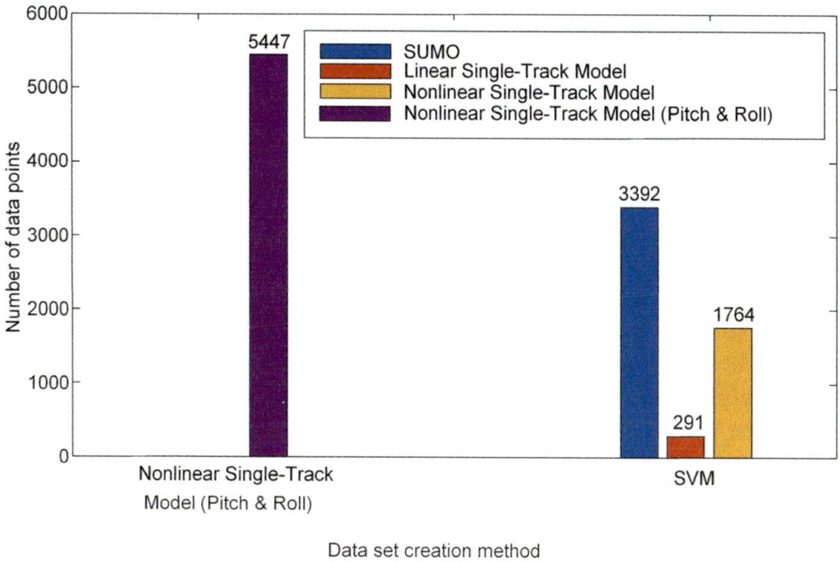

Figure 3.7 Distribution of vehicle models used for simulation per data set.

Table 3.7 Best hyperparameters found for the data sets considered.

Data set	SVM parameter	Validation accuracy
Comparative data set	RBF kernel, $C = 0.5$, $\gamma = 1$ Tenfold weighting	0.9401
Boundary finding using SVM	RBF kernel, $C = 1$, $\gamma = 0.5$ Fivefold weighting	0.9602

3.3.3.2 Evaluation of the trained support vector machines

Since the task of the SVMs is again the decision-making for an automated lane change, which in case of misclassifications can lead to hazards for road users, the condition that the achieved precision of the found SVMs in training and validation must correspond to one is considered again. Following the selection of the best SVM hyperparameters found in each case, an SVM with

Table 3.8 Precision and accuracy on training (Ω_{TV}) and test (Ω_{Test}) data of the best found ADAS-SVMs with respect to their own data set (left) and ground truth data (right).

	Comparative data set	Boundary finding using SVM
Precision (Ω_{TV})	1.0 I 1.0	1.0 I 0.9996
Precision (Ω_{Test})	1.0 I 1.0	1.0 I 1.0
Accuracy (Ω_{TV})	0.9465 I 0.9465	0.9648 I 0.9642
Accuracy (Ω_{Test})	0.9558 I 0.9558	0.9690 I 0.9668

these parameters is trained on the entire data set Ω_{TV} in order to evaluate the performance on the test data set Ω_{Test}.

First, the comparison data set based on the nonlinear single-track model with pitch and roll consideration is considered. The best hyperparameter combination for this is shown in Table 3.7; here, a validation accuracy of 94.01% is achieved on average in the cross-validation, and the precision achieved is 100%.

After re-training on the complete cross-validation data set Ω_{TV}, an accuracy of 95.58% was achieved on the test data – again with 100% precision. These test results as well as the corresponding ones on the training data, are noted in Table 3.8. This table also shows the accuracy and precision with respect to the ground truth data, which is the same data for the comparison data set, so the results are also the same.

In cross-validation, based on the data set created by model boundary determination using SVM, the ADAS-SVMs with these parameters achieved a mean validation accuracy of 96.02% – the precision is 100% in both training and validation.

After training an SVM parameterized in this way on the entire training data, a precision of one on the test data was achieved – when considering only the simulation results with vehicle models combined from the SVMs. The precision of the test data is 96.90%. These data can be found in Table 3.8.

Furthermore, the evaluation on the ground truth data is also summarized in this table. Here, in addition to a slight decrease in accuracy on both the training and test data, there is, above all, a decrease in precision on the training data. This decrease in precision can be attributed to the false-positive misclassification of one data point. Thus, this data point describes a lane change maneuver that resulted in a hazard in the ground truth data, while it was classified as executable in the simulation results combined by SVMs.

3.3.3.3 Evaluation of the results

The goal of combining the vehicle models based on their application limits is that the obtained data sets are identical to the ground truth data set – the comparison data set from the simulation with the nonlinear single-track model with pitch and roll. However, this goal was not fully achieved. On the one hand, this can be seen from the results in Table 3.8. Here, accuracy and partly also precision decrease slightly when compared with the ground truth data.

When directly comparing the combined data set with the comparison data set, it turns out that at eight data points of the data set compiled by means of SVMs, the simulation results do not yield the same decision for

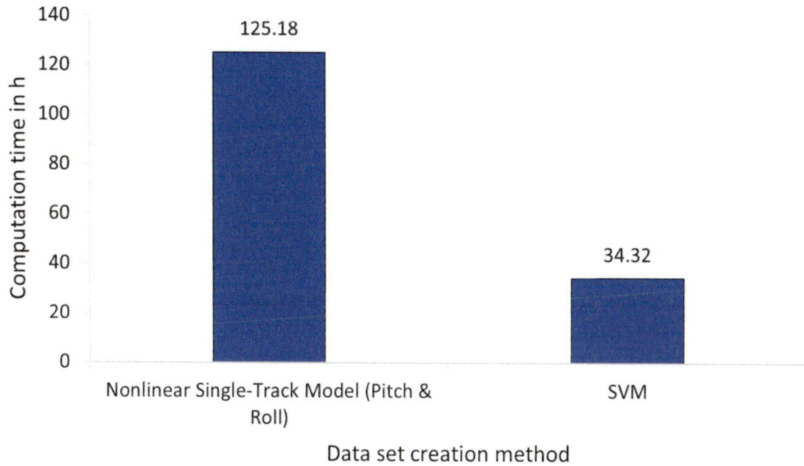

Figure 3.8 Required computation time for the simulation of the maneuvers per data set.

the lane change option as the ground truth data. Accordingly, 0.147% of the simulation runs used an inappropriate vehicle model. Thereby, the precision is also reduced in the data set combined using SVMs – which in turn affects the safety of the ADAS for an automated lane change.

In an SVM training process, the objective is to find a hyperplane separating the two classes with a maximum distance of each class to this decision hyperplane. Therefore, the optimization algorithm spans a margin between both classes, ideally permitting no data points lying inside. The SVM's decision boundary is placed in the middle of the margin. In the model boundary definition via SVMs, it can be shown that the data points for which an unsuitable vehicle model was chosen are relatively close to the decision hyperplane and in fact are located inside the margin. In the present cases of using an unsuitable vehicle model for the simulation of the lane change maneuver, a safety-oriented boundary determination would thus have avoided using an insufficient vehicle model.

Another crucial question is whether the additional effort for the model boundary finding as well as the training on the enlarged data set, is worthwhile. Here, the focus is on the one hand on the achieved accuracy on the test data, on the other hand, on the computing time required for the simulation.

The SVMs trained on the smaller data set could achieve a test accuracy of 94.25% under the condition of a precision of 100%. In contrast, the machine models trained on the larger data set in this section – ignoring the ADAS-SVM trained on the data set combined with SVMs due to lack of precision – were able to achieve a test accuracy of 95.58%. Using more training data thus increased the accuracy of the SVMs by 1.33%.

The declared goal of combining the simpler vehicle models according to their areas of application is primarily to reduce the computation time required in maneuver simulation. As shown in Figure 3.8, this goal is met. Due to the use of simpler vehicle models where possible, the time required for simulation could be reduced significantly. Using SVMs to define the application boundaries resulted in a 72.6% reduction in the time required, although it is important to note here that the precision was reduced with this method.

3.4 Conclusion

This chapter investigated the required complexity of a simulation model for a safety-critical traffic scenario, and therefore the influence of the quality of training data for a system based on an artificial intelligence classification scheme was analyzed. As an application example, the decision-making for a driver assistance system for automated lane changing was presented. The simulation used four vehicle models of different physical complexity to generate data for lane change maneuvers with three road users. As the first step, four SVM classifications were trained based on the generated data for each vehicle model, in order to classify the lane changes as feasible or not feasible. A lane change was annotated as "not feasible" in case road users were endangered during the maneuver according to a guideline for safety distances in German traffic.

Different support vector machines were trained on the generated data sets using hyperparameter optimization with cross-validation. Special focus was placed on ensuring that the trained SVMs do not make decisions that endanger road users at the expense of accuracy. The results showed that the least complex vehicle model simulation could be reproduced best by the machine learning model and with the highest test accuracy.

In the second step, the data generated with the most complex simulation model was taken as ground truth data and compared against the three less complex simulation models. The computational time required for the simulation of the most accurate model is a multiple of the simpler models. The difference between the simplest and the most complex model is about a factor of 35. Therefore, it was investigated in which cases the simpler models show good performance and are thus sufficient in terms of physics representations. As expected, the analysis showed that the SVMs trained on the results of the simpler vehicle models had to accept losses in accuracy in some cases. Most importantly, some decisions of the SVMs trained on the results of the simpler models would have been dangerous for road users who were not classified as such.

In order to keep the computation time as low as possible without loss of accuracy, especially when creating large data sets, which will lead to higher model accuracy when used for training machine learning algorithms, in the third step, a method has been developed in a way that aims to combine the different vehicle models for optimized accuracy with respect to computation time. The goal is to choose the simplest possible vehicle model for the simulation of a specific maneuver, which at the same time achieves a similar result compared to the most realistic data.

With this objective, a method for determining the required model complexity was developed, which works with the aid of machine learning methods, in this case also using SVMs. The application limits of the different vehicle models for the simulation of the lane change maneuver were determined by comparing the simulation results of the vehicle models with the realistic data and narrowing down areas of equal performances of either simulation.

Based on these model limits, a larger data set than before was created for each method variant, in which the vehicle models used were combined in each case according to the defined limits. The data set with the most complex vehicle model represented the ground truth annotations. After training an SVM with one of each of these larger data sets, the results obtained were evaluated against the annotations. Here, it was first shown that the larger data set actually has a positive influence on the test accuracy; an increase of 1.33%–95.58% test accuracy could be achieved. The ADAS-SVM trained on the data set created by the model boundary finding method using SVMs resulted in decisions with hazards for the road users in the evaluation despite even higher accuracy in comparison. In these combined data sets, however, vehicle models were also used, which were not suitable in the corresponding simulation runs. In [17], it could be shown that a considerably better result can be achieved by more restrictive model boundaries. This is shown by the boundary definition using SVMs, where a shift of the decision boundary to the edge of the separation band prevented incorrect assignments of usable vehicle models.

By combining the vehicle models, the goal of reducing the computation time required for simulation was achieved. The creation of the combined data sets required only about 35% of the time compared to the most complex model alone. With a deviation of the simulation results of ultimately only

0.018% of the data points in the boundary finding using SVM with shifted decision boundaries published in [17] and the achievement of the same accuracy of the ADAS-SVM in the lane change decisions, this approach can be considered successful. Overall, however, the limitation must be considered that the definition of the model boundaries itself also requires a certain effort, and this procedure is therefore only worthwhile if a large number of maneuvers are to be simulated subsequently.

In the following, we briefly describe how the results of this work can be used to conduct further investigations. The performance of the advanced driver assistance system can be improved by training on more comprehensive data sets, including larger data sets, but above all, also the consideration of more road users and higher traffic density in the simulation.

Furthermore, we emphasize the generalizability of our approach. The methodology for model boundary finding developed in this work is neither bound to the application for an advanced driver assistance system nor to the use on vehicle models. This methodology can be applied to all topics where simulation models of different complexity are available.

Additionally, it can be investigated to what extent an iterative approach influences the accuracy as well as the effort of the method. For example, a rough grid of the parameter space could be used in the beginning to estimate the limits of the investigated models. Along these estimated boundaries, a finer grid could be used to define the boundaries more precisely. An investigation concerning the required gridding of the search space for sufficient accuracy of the defined model boundaries would also be conceivable. As also shown in this work, the application is only limited if a model is very similar to the comparison data.

A further investigation concerning the similarity of the database and the applicability of the methodology could find solutions concerning the problem or appropriate decision support, when a difference between the data to be compared is too small and no trustworthy boundary can be found.

References

[1] M. Maurer, J. C. Gerdes, B. Lenz and H. Winner, Autonomous Driving. Berlin, Heidelberg: Springer Berlin Heidelberg, 2016.

[2] EU Commission. Mobility & Transport - Road Safety, Annual statistical report https://road-safety.transport.ec.europa.eu/statistics-and-analysis/data-and-analysis/annual-statistical-report_en, 2020.

[3] German Federal Statistical Office, Fehlverhalten der Fahrzeugführer bei Unfällen mit Personenschaden. https://www.destatis.de/DE/Themen/Gesellschaft-Umwelt/Verkehrsunfaelle/Tabellen/fehlverhalten-fahrzeugfuehrer.html
[4] M. Ersoy and S. Gies, Chassis handbook: fundamentals, driving dynamics, components, mechatronics, perspectives. Springer Science & Business Media, 2010.
[5] H. Winner, S. Hakuli, F. Lotz, and C. Singer (eds.). Handbook of driver assistance systems. Amsterdam, Netherlands: Springer International Publishing, 2014.
[6] D. Schramm, M. Hiller and R. Bardini, Vehicle dynamics. Modeling and Simulation. Berlin, Heidelberg 151, 2018.
[7] P. Rieckert and T.-E. Schunck, "Zur Fahrmechanik des gummibereiften Kraftfahrzeugs", 1940.
[8] P. Pfeffer and M. Harrer, Steering handbook. Cham, Switzerland: Springer International Publishing, 2017.
[9] P.M. Sieberg, and D. Schramm. Ensuring the Reliability of Virtual Sensors Based on Artificial Intelligence within Vehicle Dynamics Control Systems. Sensors 2022, 22, 3513. https://doi.org/10.3390/s22093513
[10] R. Rajamani, Vehicle Dynamics and Control (Mechanical Engineering Series).
[11] R. N. Jazar, Vehicle Dynamics. Cham: Springer International Publishing, 2017.
[12] H. Dugoff, P. S. Fancher and L. Segel, "An Analysis of Tire Traction Properties and Their Influence on Vehicle Dynamic Performance" in International Automobile Safety Conference, 1970, doi: 10.4271/700377.
[13] H. B. Pacejka, ed., Tyre and Vehicle Dynamics (Second Edition). Oxford: Butterworth-Heinemann, 2006.
[14] P.M. Sieberg, Hybrid Methods in Vehicle Dynamics State Estimation and Control–Exploiting Potentials and Ensuring Reliability of Artificial Intelligence. 2021. Dissertation, Duisburg, University of Duisburg-Essen, 2021.
[15] G. Genta and L. Morello, The automotive chassis. Dordrecht: Springer, 2009.
[16] D. Schramm, B. Hesse, N. Maas, and M. Unterreiner. Vehicle Technology: Technical foundations of current and future motor vehicles. Walter de Gruyter GmbH & Co KG, 2020.
[17] C. Hürten, P. M. Sieberg, and D. Schramm. Generating a Multi-fidelity Simulation Model Estimating the Models' Applicability with Machine

Learning Algorithms. In Proceedings of the 12th International Conference on Simulation and Modeling Methodologies, Technologies and Applications - SIMULTECH, 2022. ISBN 978-989-758-578-4; ISSN 2184-2841, pages 131-141. DOI: 10.5220/0011318100003274
[18] T. J. Gates, D. A. Noyce, L. Laracuente and E. V. Nordheim, "Analysis of Driver Behavior in Dilemma Zones at Signalized Intersections", Transportation Research Record 2030, No. 1, pp. 29–39, 2007, doi: 10.3141/2030-05.
[19] H. Saptoadi, "Suitable Deceleration Rates for Environmental Friendly City Driving", IJRCMCE, 4.1, 2017, doi: 10.15242/IJRCMCE.ER1216002.
[20] S. Krauß, "Microscopic Modeling of Traffic Flow: Investigation of Collision Free Vehicle Dynamics", 1998. https://elib.dlr.de/8380/.
[21] J. Erdmann, "SUMO's Lane-Changing Model" in Lecture Notes in Mobility, Modeling Mobility with Open Data, M. Behrisch and M. Weber, Hg., Cham: Springer International Publishing, 2015, pp. 105–123, doi: 10.1007/978-3-319-15024-6_7.
[22] P.A. Lopez, M. Behrisch, L. Bieker-Walz, J. Erdmann, Y.-P. Flötteröd, R. Hilbrich, L. Lücken, J. Rummel, P. Wagner, and E. Wiessner. Microscopic traffic simulation using sumo. In: 2018 21st international conference on intelligent transportation systems (ITSC). IEEE, 2018. pp. 2575-2582.
[23] A. Wegener, M. Piorkowski, M. Raya, H. Hellbrück, S. Fischer and J.-P. Hubaux, "TraCI: An Interface for Coupling Road Traffic and Network Simulators", Proceedings of the 11th Communications and Networking Simulation Symposium, CNS'08, 2008, doi: 10.1145/1400713.1400740.
[24] A. F. Acosta, J. E. Espinosa and J. Espinosa, "TraCI4Matlab: Enabling the Integration of the SUMO Road Traffic Simulator and Matlab® Through a Software Re-engineering Process" in Lecture Notes in Mobility, Modeling Mobility with Open Data, M. Behrisch and M. Weber, Hg., Cham: Springer International Publishing, 2015, pp. 155–170, doi: 10.1007/978-3-319-15024-6_9.
[25] C. Hsu, C. Chang and C.-J. Lin, "A Practical Guide to Support Vector Classification Chih-Wei Hsu, Chih-Chung Chang, and Chih-Jen Lin", 2003.
[26] F. Pedregosa et al., "Scikit-learn: Machine Learning in Python", Journal of Machine Learning Research 12, pp. 2825–2830, 2011.

Authors' Biography

Christian Hürten joined the Chair of Mechatronics at the University of Duisburg-Essen in 2021 after completing his master's degree studies in Mechanical Engineering with a specialization in Mechatronics. At the chair, he is employed as a research assistant and is working toward his doctorate. His research interests mainly focus on the use of machine learning methods for application in dynamic system control and prediction, particularly in the context of automated vehicles and inland vessels.

Prof. Dr. rer. nat. Anne Stockem Novo holds a PhD in theoretical physics from Ruhr University Bochum. She is a Professor of Applied Artificial Intelligence at the University of Applied Sciences Ruhr West. At ZF Automotive Germany GmbH, she worked more than five years in R&D for Automated Driving. Her area of research is deep learning in the context of automated driving.

Prof. Dr.-Ing. Dr. h.c. Dieter Schramm has headed the Chair of Mechatronics at the University of Duisburg-Essen since 2004. Before that, he worked in leading positions at Bosch GmbH and Tyco Electronics Ltd. In addition to his university position, he is director of Mercator Science & Education Ltd. in Kuala Lumpur, co-founder of Automotive Executive Education GmbH (AEE), Duisburg, and member resp. chairman of the advisory boards of university-related research institutes. In 2015, he was awarded an honorary doctorate by the University of Miskolc. His research focuses on manipulators, electromobility, vehicle dynamics, driver assistance, and automated driving of motor vehicles and ships.

Dr.-Ing. Philipp M. Sieberg is a Researcher and Lecturer at the Chair of Mechatronics of the University of Duisburg-Essen, Germany, and Managing Director of Schotte Automotive GmbH & Co. KG. His research interests are primarily in applied artificial intelligence, especially for intelligent transportation systems. He is a member of the Executive Committee of the IEEE Germany and the IEEE ITSS German Chapter. He has received several awards for his research, including from the IEEE ITSS and the University of Duisburg-Essen.

4

Autonomous Vehicle Steering through Convolutional and Recurrent Deep Learning

Mark Kocherovsky, Guiseppe DeRose Jr., Nicholas Paul, Ian Timmis, and Chan-Jin Chung

Department of Math and Computer Science,
Lawrence Technological University, USA

Abstract

Research into autonomous vehicles has made great strides in the past two decades; some of the literature utilizes deep learning to train a model to steer a vehicle. However, autonomous steering algorithms have previously focused on roads with lane markings. This chapter outlines work on training a neural network to steer a vehicle on entirely unmarked roads. These models were trained on data collected by a simple webcam mounted on our research vehicle, a Polaris Gem 2 modified with a drive-by-wire system and sensor components. Training data – images of the roads – were collected by driving the vehicle along several routes with a webcam recording and saving frames labeled with the steering angle, which was used to train convolutional neural networks. Our preliminary real-world testing with the ACTor vehicle shows that a CNN using a fine-tuned prebuilt network is capable of navigating unmarked roads. After collecting more training data, we trained various fine-tuned pretrained networks as well as configurations including different recurrent structures using temporal contextualization. Pruning techniques were also experimented with to achieve a better balance of training data. Testing the models on unseen data demonstrated the ability to predict steering wheel angles to within approximately 0.2 rad on average.

Keywords: Autonomous vehicles, deep learning, convolutional neural networks, recurrent neural networks, long short-term memory, gated recurrent unit, autonomous steering.

4.1 Introduction

The study of autonomous vehicles (AVs) has seen an explosion of interest in the last decade. Researchers and engineers have been working to develop vehicles that can autonomously navigate on road networks. It is believed that AVs are presumed to be safer than human drivers and more convenient for passengers. Shifting the responsibility of navigation and defensive driving to computers - which have faster reaction times than humans - is thought to decrease accidents and increase navigational efficiency, though significant issues remain in the current level of AV technology [31].

However, the task of actually steering on the road can be very challenging for computer systems. AVs need to be able to detect lane markings in order to follow the road safely. Traditional computer vision methods used to detect lane markings include several filtering, sampling, smoothing, and extraction techniques to detect expected colors and shapes (such as a yellow line for a lane boundary) and extrapolate the road's immediate layout. Some algorithms model lanes as mathematical functions to determine a line of best fit. Once lane markings are found, the steering wheel angle can be estimated mathematically. However, image processing tasks can be quite complex, given the variety of environmental conditions. A road might be unmaintained, leading to markings that are not sufficiently clear for the computer to detect correctly. The road might be intentionally unmarked or even entirely unpaved, further confusing computer vision models [32, 24, 33, 2, 19].

Deep learning, a subfield of machine learning, has been presented as an alternative [24]. Deep learning is the study of artificial neural networks (ANN); algorithmic structures inspired by - but not directly modeled on - the human brain. ANNs are usually trained in a supervised manner: The model is given a set of input data and the correct output data. The network iteratively tweaks its own parameters to minimize incorrect output, thus "learning" on the data. Once the model is trained, it can then be used on unseen input [8]. In 1988, the autonomous land vehicle in a neural network (ALVINN) demonstrated that it was possible to use artificial neural networks to automatically steer a vehicle using a video feed of the road in front of it [23]. Since then, advances in deep learning have enabled experiments with convolutional neural networks — neural networks designed for image-based tasks – and recurrent neural networks — which enable models to learn using time-sequential data [11, 18].

However, previous research focused exclusively on roads **with lane markings**. Though the problem of steering on **unmarked roads** has been acknowledged [32, 33], it has been ignored by experimentation. Timmis et al. were the first to demonstrate a deep learning-based system for autonomous steering on unmarked roads [30]. This work has since been continued by Paul [20] and the original research described in this chapter.

Our work was performed on the Autonomous Campus Transport (ACTor 1) vehicle, our research platform, shown in Figure 4.1. The ACTor is a Polaris GEM 2 modified with a DataSpeed drive-by-wire system and several sensors, the most relevant to this work being a webcam mounted to the front of the vehicle [21]. The ACTor's software is based on the Robot Operating System (ROS) platform, which allows programmers to enable connections and interactions between different hardware and software components — assuming those components' compatibility with ROS — in a decentralized and modular way using messages [27].

Figure 4.1 The ACTor 1 research platform [30, 21].

The original work by Timmis et al. used 4600 images collected using the ACTor labeled with the steering wheel angle at the time of recording in radians to train their model. The final CNN+DNN (**CD**) model (using the InceptionV3 pretrained model with fine-tuning on randomly initialized weights) had an average mean squared error of 0.265 rad [30, 28]. Trials ran with a small CNN+DNN+RNN (**CDR**) model made it apparent that RNN structures were nearly twice as effective as the baseline CD model [20].

Data was taken using a camera mounted on the front of the vehicle (our ACTor 1 platform), which was used to train neural network models that would report a predicted steering angle based on the image given to the model as input. These models can then be used to steer the vehicle.

New data were collected with the ACTor in order to broaden the scope of the training [10], which was used to measure the effectiveness of different pretrained models on the dataset [16]. Kocherovsky et al. then modified the previous work by adding RNNs (creating a CDR), where the RNN could be one of several different recurrent layers (RNN, LSTM, and GRU) to test whether using time-series layers would be effective compared to a pure CD solution [17].

We present DeepSteer (deep learning-based steering) as an alternative to traditional machine vision methods, especially in environments which lack lane markings. Though the obvious application of DeepSteer is in autonomous vehicles meant to navigate roads, more specifically-trained models could be used for robotic navigation in specialized zones, such as within factories or hospitals.

Section 4.2 describes how data was collected and preprocessed, as well as how our software was integrated with vehicle hardware. Section 4.3 explains the fundamentals of deep learning, the structure of each model, and techniques used to train each ANN. The results from each set of trials is shown in Section 4.4, and the chapter concludes in Section 4.5 with a summary and discussion of future research topics.

4.2 Data Collection

For each set of trials, we used neighborhood roads on and close to the LTU campus. These paths were chosen because they represent possible real-world scenarios where a vehicle would need to be able to steer without lane markings providing guidance. While driving down the paths, shown in Figure 4.2, our program would take frames from the camera at a set interval. The steering wheel angles would be recorded at the same time and would be saved as part of each frame's filename. Figure 4.3 gives a basic overview of this process. The car was driven down each path both ways to reduce possible training biases and provide more data to learn from [30].

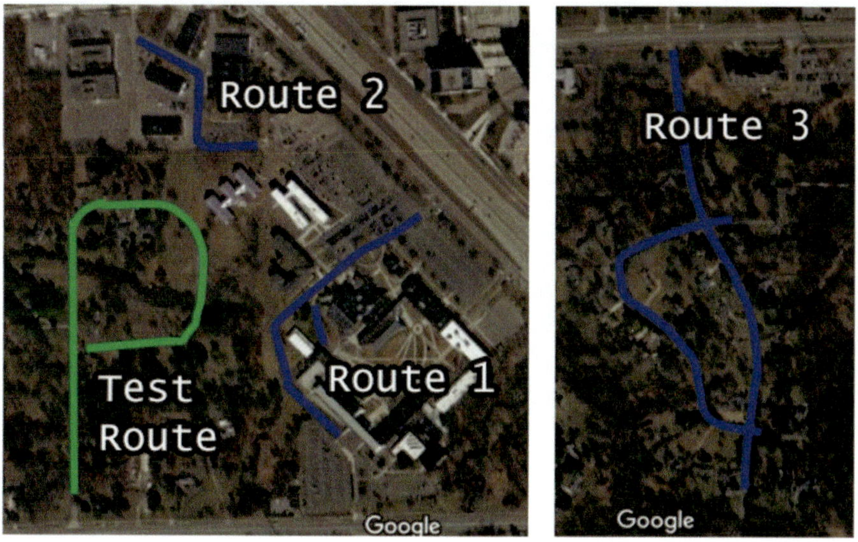

Figure 4.2 Our data collection paths. The ACTor would be driven down each path twice (both ways) to collect as much data as possible. Routes 1, 2, and 3 (route) comprised the training dataset, while the Test Route (green) made up the testing dataset. **Route 3 was only used in the latter two sets of trials** [10].

Figure 4.3 Data collection overview. Images of the road are saved together with the steering angle at the time of recording [30].

Figure 4.4 An example of training data. The name of the file is 151_-1.1275.jpg, indicating that the file is the 151st picture in that set and that it depicts a turning angle of 1.1275 rad to the right. Parts of the hood, windshield, and left side-view mirror are visible [17].

After collection, but before training, each image was resized to 320x240 pixels and the steering wheel angle was extracted from the filename. Figure 4.4 shows an example data image where the steering wheel is turned about 1.1 rad to the right. Timmis et al. [30] used 4600 images, which was also used for SimpleRNN testing [20]. The dataset collected in [10] contains approximately 6200 images.

4.2.1 Integrating software modules with the vehicle

The Robotics Operating System [27], despite its name, is not an operating system but a middleware platform designed for building decentralized robotics control systems and is available for a variety of programming languages. We primarily use C++ and Python ROS libraries. Each software component is built as its own **node**. Nodes communicate using message passing through **topics**. Nodes may **publish** messages (broadcast data) to topics or **subscribe** to topics (receive input data). Note that there is no explicit connection between nodes: individual nodes do not care which nodes, if any, subscribe to the topics they publish to or publish to the topics they subscribe to. This makes it easy to develop modular systems, which improves performance and maintenance.

ROS is especially useful since hardware components (sensors, drive-by-wire, etc.) can support ROS interaction, depending on the manufacturer's choices. On ACTor1, we only use hardware with ROS support: each component has its own nodes and topics they subscribe to. In this work, we make use of a webcam, connected using USB, and a Dataspeed Drive-By-Wire system with ROS support. In our programs, the camera node broadcasts the images taken by the camera on a certain topic, and the vehicle control

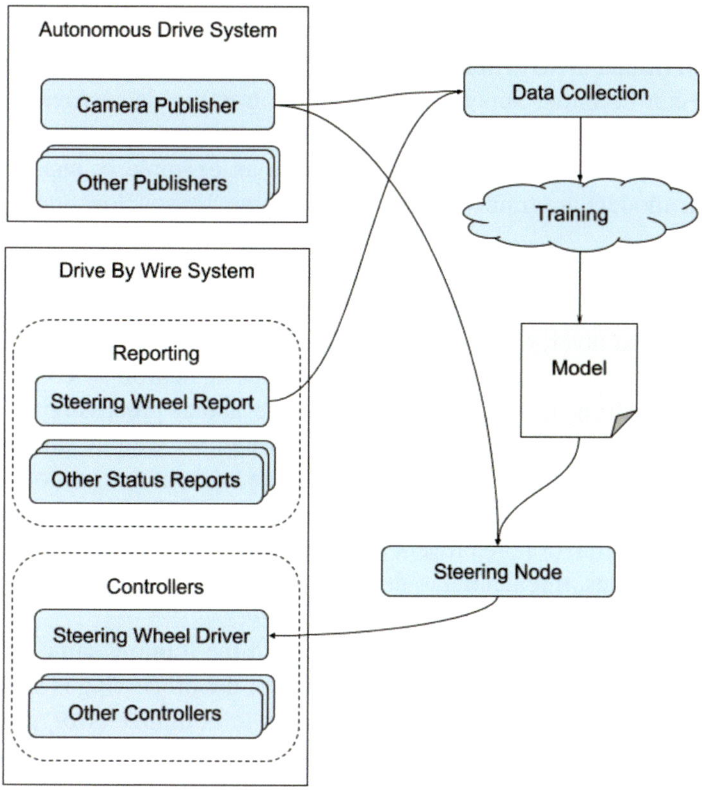

Figure 4.5 A diagram explaining our software pipeline. The Autonomous drive and drive by wire systems collect data, which is then used to train a model outside of the vehicle. This model is then used to control the steering wheel [30].

nodes broadcast the steering wheel angle. Another node subscribes to these topics and performs the necessary preprocessing on the images. Once the images are preprocessed, they are used to train models using TensorFlow and Keras [1, 6]. If we need to change any component, it can be easily done without major change to the other one. Our pipeline can be seen graphically in Figure 4.5 [30].

4.3 Model Training

4.3.1 Deep learning fundamentals

Deep learning (DL) is a subfield of machine learning (ML), itself a subfield of the study of artificial intelligence (AI). ML techniques focus on training models using previously known training data, which are then used to make predictions based on the analysis of new, unseen, data. Deep learning algorithms use **artificial neural network** (ANN) structures to create versatile models that can be either supervised, where the training data provides the desired output, or unsupervised, where the desired output is not provided, if it is known at all. The work described here is an example of supervised learning. Our models are trained in Python using the TensorFlow and Keras libraries [1, 6].

ANNs are made up of **layers** of processing units called **neurons**. Each layer takes in input from the previous layers and returns output which is taken as input by the next layer, where applicable. **Dense neural networks** (DNNs) are the most basic form of ANNs. In a DNN, each neuron is connected to every other neuron in the adjacent layers. The actual processing comes from the connections between neurons, called **weights**. Weight values are multiplied by the corresponding input values to the neuron; the sum of these operations is then fed into an **activation function**, such as a sigmoid or hyperbolic tangent function, to prevent the values from "exploding" (reaching very high magnitudes, threatening to destabilize the model). The result of the activation function is the input given to the next layer.

The training itself consists of iterating through the training samples and adjusting the weight values so that the final result of the processing is as close to the desired output as possible. The main goal is to minimize error, usually called **loss**. The structure and hyperparameters of the ANN, including the amount of training iterations (**epochs**), are configured by the programmer.

Convolutional neural networks (CNNs) are a type of ANN often used for image recognition and analysis. When given the training dataset, the convolutional layers work to identify abstract features common to the samples.

In the context of autonomous vehicles, this might be the general shape of other vehicles, human beings, common road signs, etc. In this work, powerful **pretrained** networks are **fine-tuned** (partially retrained as described in 4.3.2) and applied on the images of the road immediately in front of the ACTor.

Finally, we also use **recurrent neural networks** (RNNs), which have the ability to "remember" previous samples in the short-term. This allows us to teach networks with temporal contexts, which we use since our dataset consists of sequences of images from the same courses. RNN variants include **long short-term memory** (LSTM), which can carry sequential information from longer intervals through training than RNNs, and **gated recurrent units** (GRUs), which have the ability to reset and forget irrelevant patterns within sequences, allowing for more versatile learning over longer durations [8, 5, 25].

4.3.2 Model structure

Each model follows the same basic structure. First, the input data (an image) is fed into a CNN (usually the InceptionV3 pretrained network). Then, there is a dense layer, which in CDR models would be followed by a recurrent layer, and then end in a linear neuron, which provides output. In a CD model, it immediately ends with the linear neuron.

The CD model shown in Timmis et al. consisted of a fine-tuned InceptionV3 model, a dense layer of 1024 neurons, and a final output neuron. This can be seen graphically in Figure 4.6. It is upon this model which all subsequent work is based [30].

The SimpleRNN model did not use InceptionV3, as it was merely meant as a small demonstration of the CDR pipeline on the dataset. SimpleRNN merely used two convolutional layers, a dense layer of 500 neurons, a

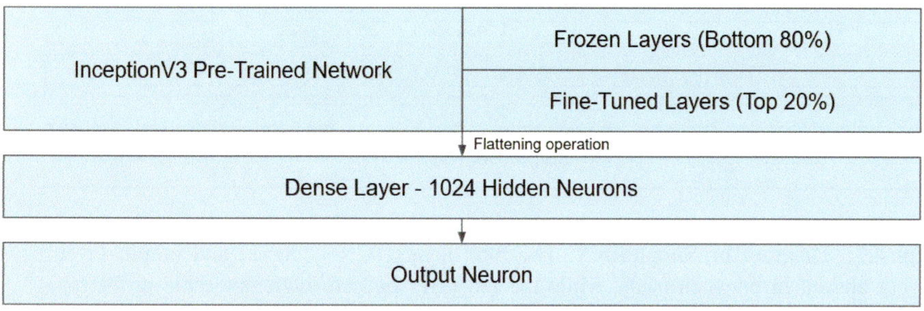

Figure 4.6 Original CD model used by Timmis et al. [30].

recurrent layer with 500 hidden neurons, and the final output neuron. The recurrent layer was interchangeable: it could be an RNN, LSTM, or GRU layer. The layer could also be disabled for baseline tests to compare the CDR with a CD structure more closely resembling the original work. A diagram depicting SimpleRNN is shown in Figure 4.7 [20].

The CD model shown in [16] closely followed Timmis et al. [30] with two key differences. First, the InceptionV3 network at the top of the model could be exchanged for other pretrained networks. Other than InceptionV3, this program also tested DenseNet, EfficientNet, MobileNet, ResNet50, Xception, VGG16, and VGG19 [15, 29, 14, 13, 7, 26, 28]. This is shown in Figure 4.8. This was accomplished by keeping references to each network in the program and calling on the necessary ones when needed. Each network had its top 20% layers unfrozen for fine-tuning, and each network was trained on the ImageNet dataset [9]. The second difference was the addition of extra data. New runs on each course were tested, and Route 3 (in Figure 4.2) was added. This work was also built on structures used in [10].

Kocherovsky et al.'s second set of trials [17] used a CDR model combining aspects of the previous CD model with SimpleRNN. The InceptionV3 pretrained network was fine-tuned with the top 20% of layers unfrozen. The output would be given to a layer of 1024 densely connected neurons. The next layer was a recurrent layer. Like with SimpleRNN, this recurrent layer

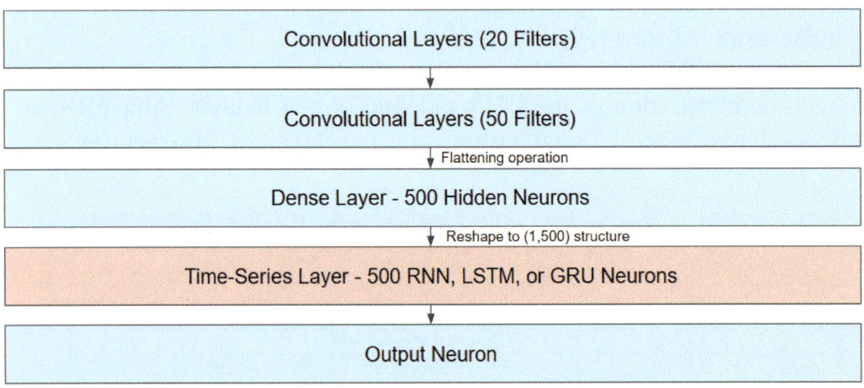

Figure 4.7 Diagram of SimpleRNN. The blue layers (CNN, dense, and output layers) were not altered in between trials, while the red layer had an interchangeable property: it could either contain an RNN, LSTM, or GRU layer. It could also be disabled for baseline comparison [20].

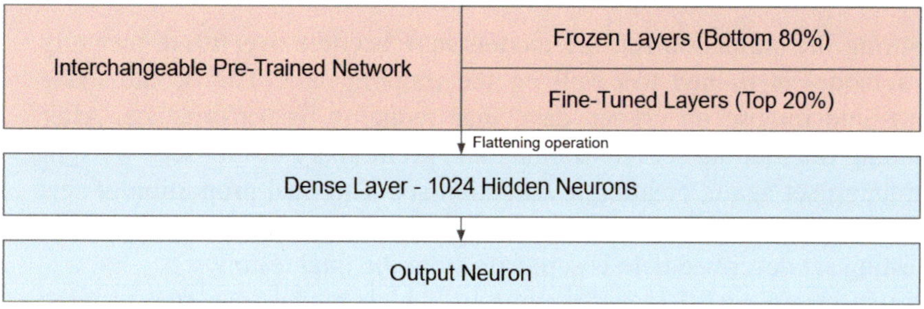

Figure 4.8 CD model used by Kocherovsky et al. The pretrained network, marked in Red, was interchangeable between eight different structures, each trained on the ImageNet dataset [9]. Each structure also had the top 20% of layers unfrozen for fine-tuning. The layers shown in blue remained constant [16].

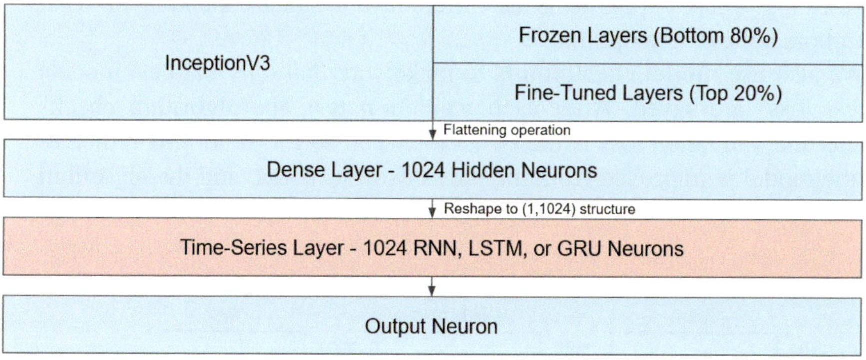

Figure 4.9 CDR model used by Kocherovsky et al. The time-series layer, shown in red, could consist of different recurrent structures or be disabled for comparison. The layers shown in blue remained constant [17].

could either consist of an RNN, LSTM, a GRU, or be entirely disabled for to compare with a CD baseline. This is shown in Figure 4.9.

4.3.3 Training techniques

During training, we used multiple techniques in addition to fine-tuning, the most important of which was **validation**. Validation testing involves creating a small test dataset from the training data to test the models during training rather than after. After each epoch, the model will test itself on the validation data, which it treats as unseen, and reports the validation loss. Without measuring the validation set, the models may become **overfitted**, meaning that a model is trained too well on the training data. This would make it inflexible for use on unseen data, thus reducing its performance. After separating the training set into training and preliminary testing sets, we split the training set again, creating a validation set. The final proportion of sets was 60% training, 20% validation, and 20% (preliminary) testing. Note that the testing set described here is **separate** from the final testing set.

Early stopping [8] is a technique by which models can stop training before the set number of epochs based on analysis of the validation data. Before training, the programmer sets the "patience" of the procedure as some integer value x. During training, a record of validation loss from the previous x epochs is kept. If the model's validation loss does not improve (closer to zero) after x epochs, then the model will stop training to prevent exacerbation of overfitting. These experiments use different values of x depending on what the authors believe was optimal.

We also use **model checkpoints** to make sure that only the best models (lowest loss) are saved. After each validation test, the algorithm checks whether the validation loss is the best value seen so far. If so, this indicates that the model is improved from the previous best model, and the algorithm saves the model, overwriting the previous record-holder. This ensures that the

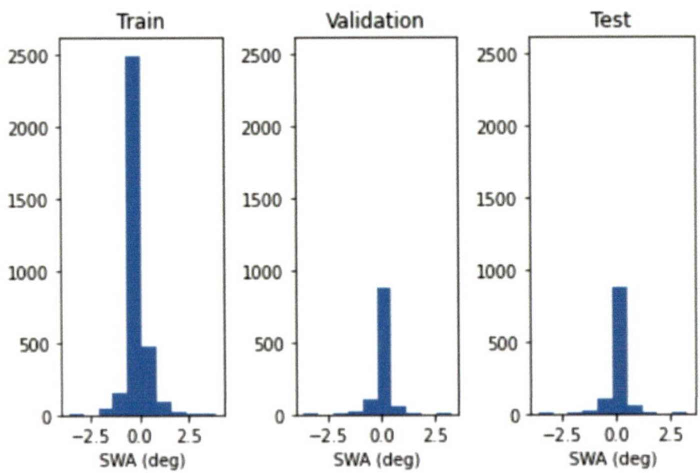

Figure 4.10 Histograms of the training dataset's content before pruning, sorted by steering wheel angle (SWA) in degrees. It can clearly be seen that low-angle input data overwhelmingly outnumbers high-angle input data, thus proving that the model is biased.

best model is saved for future use. Though this does not necessarily ensure that overfitting will not occur, the checkpoint technique is a powerful tool to prevent the phenomenon [8].

During the course of the DeRose [10] experiment, we noticed that our models reported much smaller loss values on low-angle data (-0.1 rad $\leq \theta \leq 0.1$ rad) than on high-angle data ($\theta < -0.1$ rad, $\theta > 0.1$ rad), indicating a possible bias toward low-angle inputs. Bias in our results can be seen graphically in Section 4.4. **Bias** occurs when a model is trained in such a way that it performs better on one set of inputs than others, even if this was not intended by the programmer. Correcting for bias is an ongoing problem in deep learning. For example, a model might learn to replicate human patterns of racial and/or gender discrimination if these prejudices are reflected in the training data [3]. Analysis of our training dataset reveals that the model trains on far more low-angle data than high-angle data. This is shown in Figure 4.9. Therefore, it became necessary to perform pruning.

Pruning is performed by identifying input data that might introduce bias or encourage overfitting and removing them before training. First, some input images were removed because they showed a curved road while the car was driving relatively straight. We believe that this may have confused the model by offering contradictory data. Next, we randomly removed 1550

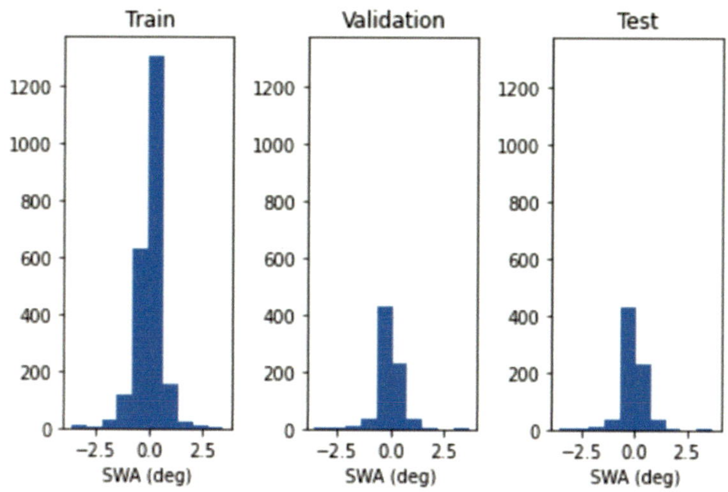

Figure 4.11 Histograms of the training dataset's content after pruning, sorted by steering wheel angle (SWA) in degrees. Low-angle input data still outnumbers high-angle input data, but the balance is definitely improved when compared to Figure 4.10 [17].

low-angle images, which made up half of that subset. As shown in Figure 4.11, our training dataset is better balanced. There are now approximately 3800 input images in the dataset rather than 7000. Even though low-angle data still dominates, there is more equality between the input data and thus reduced risk of bias. In Section 4.4.4, we demonstrate that this approach was successful [17].

4.4 Results

4.4.1 Original CD (CNN+DNN)

The original work by Timmis et al. [30] found a mean squared error of 0.2653 rad. Further testing reveals that the model can be confused by intersections: it is not sure if it should continue straight or turn. Otherwise, the team's results show a strong performance. A diagram of their validation loss in mean square error is shown in Figure 4.12. Figure 4.13 shares a frame captured from a YouTube video [22, 30] recorded during a test using the original CD integrated with ACTor 1. The vehicle is being driven down one of the training paths, and the webcam (rectangular object in the top-right) takes pictures, which is then combined with the steering wheel angle to be used as training data. Timmis et al. also performed real-world testing on the test path shown

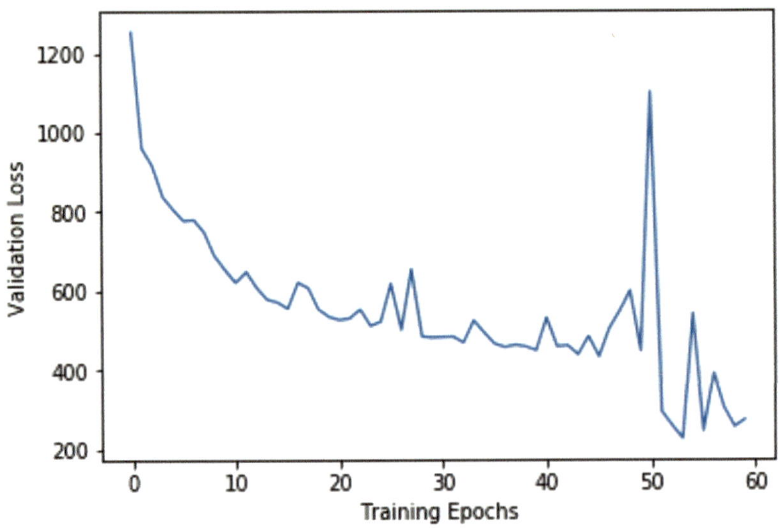

Figure 4.12 Validation loss of the original CD model during training [30].

Figure 4.13 A frame captured from a video [22, 30] demonstrating a test of the original CD model integrated with the ACTor vehicle. The webcam (top-right rectangle) reads in frames that are sent to the program. The program preprocesses the frames and send them to the ANN model, which determines a steering angle.

in Figure 4.14. Areas where the vehicle went the wrong way are marked in dashed circles. All three of these locations are private driveway intersections, indicating that the models were not sufficiently trained on driveway scenarios.

4.4.2 Simple RNN

Experiments documented in [20] after normalization show that the simple CD model returns an error of 0.721 rad (The error is higher than the original CD model since it does not use any pretrained networks; the purpose was to compare the baseline with recurrent structures.), the CD+RNN model had an error of 0.397, the CD+LSTM model had an error of 0.403, and the CD+GRU model had an error of 0.399. In Figure A.1, the training and validation losses during each epoch are displayed. Testing errors are shown in Figure 4.15, where the CD+RNN is clearly the best performer, and the CD is clearly the worst. Evaluation of the input does not require a long short-term memory, and unlike LSTM and GRU layers, and RNN layers easily forget old information, thus making RNNs more flexible in this context.

Figure 4.14 Detailed view of the Test Route (see Figure 4.2). In [30], real-world testing was performed on this path. Locations marked in dashed circles are approximately where the model steered in the wrong direction. Notice that these locations are all intersections with driveways!

4.4.3 CD with new data and interchangeable pretrained networks

Out of the eight networks tested on the data collected in [10], InceptionV3 had the lowest mean absolute error (0.2799 rad) while MobileNet had the highest (0.4738 rad). Full results are shown in Table 4.1 and in Figure A.2. Figure 4.16 shows a comparison between predictions and ground truths of the test course data [16].

From these results, we can see that EfficientNet did not learn from the data at all, whereas MobileNet, which at least learned some patterns, had

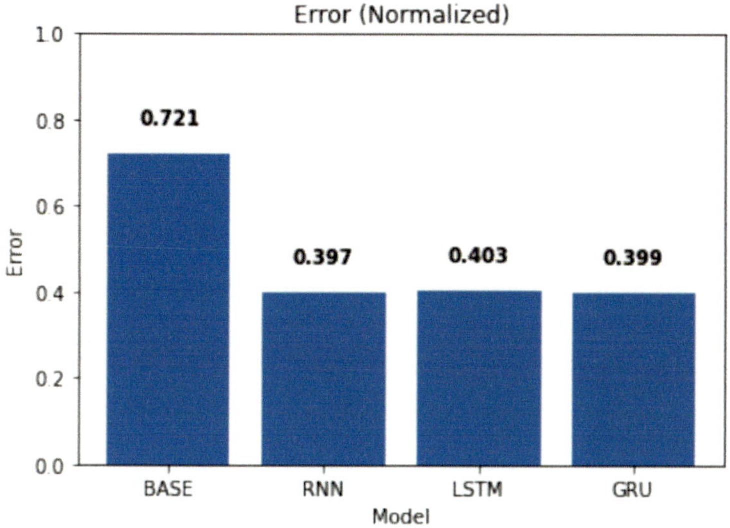

Figure 4.15 Error of CDR models using the simple RNN architecture.

worse results on average. This is likely because there were enough low-angle turns for EfficientNet's guesses of 0.0 to be more accurate than MobileNet. ResNet ended up overcompensating for the data and recommended angles that were usually too far left. DenseNet and Xception are closer, but their agreement with the true outputs were not as close as VGG16, VGG19, and InceptionV3. Despite VGG19 and VGG16 apparently being closer to the data than the InceptionV3 model, certain areas of disagreement contributed to their lower performance.

Table 4.1 Comparison of test loss (mean squared error) and test error (mean absolute error) for CD models with interchangeable pretrained networks. It is clearly shown that InceptionV3 maintained the best performance [16].

Pretrained network	Test loss (MSE)	Test error (MAE, rad)
DenseNet	0.3532	0.3836
EfficientNet	0.5170	0.4497
MobileNet	0.5217	0.4738
ResNet	0.3607	0.4425
Xception	0.2453	0.3106
VGG16	0.2356	0.3003
VGG19	0.2835	0.3211
InceptionV3	0.1794	0.2799

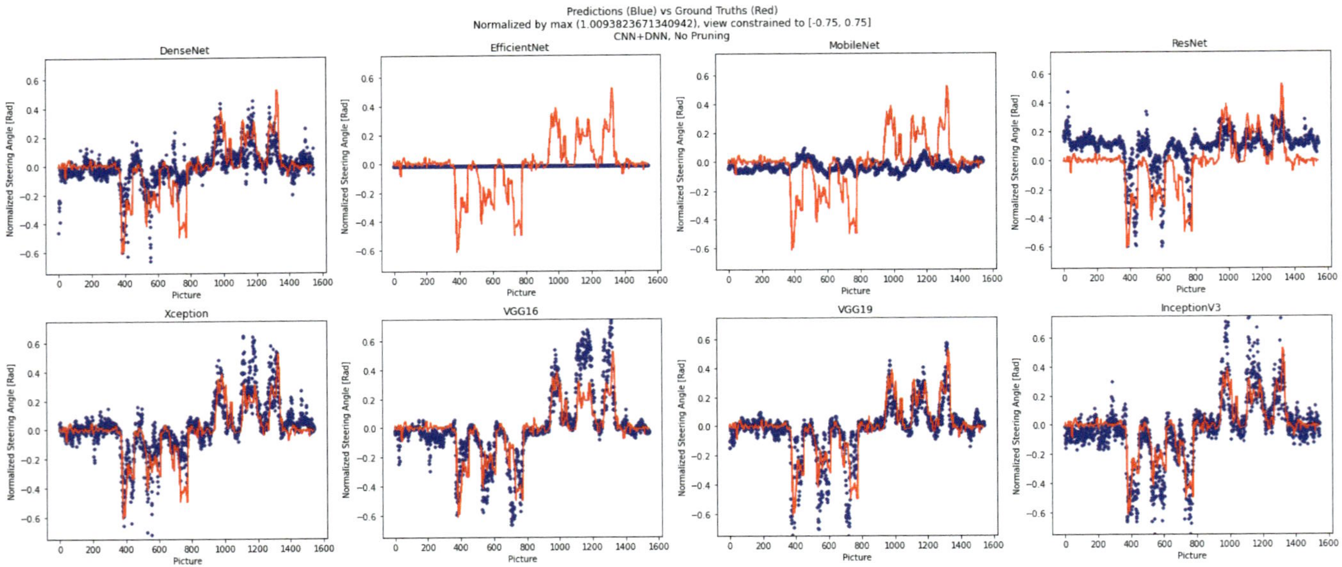

Figure 4.16 Comparison of model predictions (blue) with the true output values (red) for the test course. The view is constrained to ± 0.75 for better detail, as there are outlying predictions outside this range. Whereas EfficientNet, MobileNet, and ResNet failed to sufficiently learn from the training data, InceptionV3, VGG16, and VGG19 were close to the true data [16].

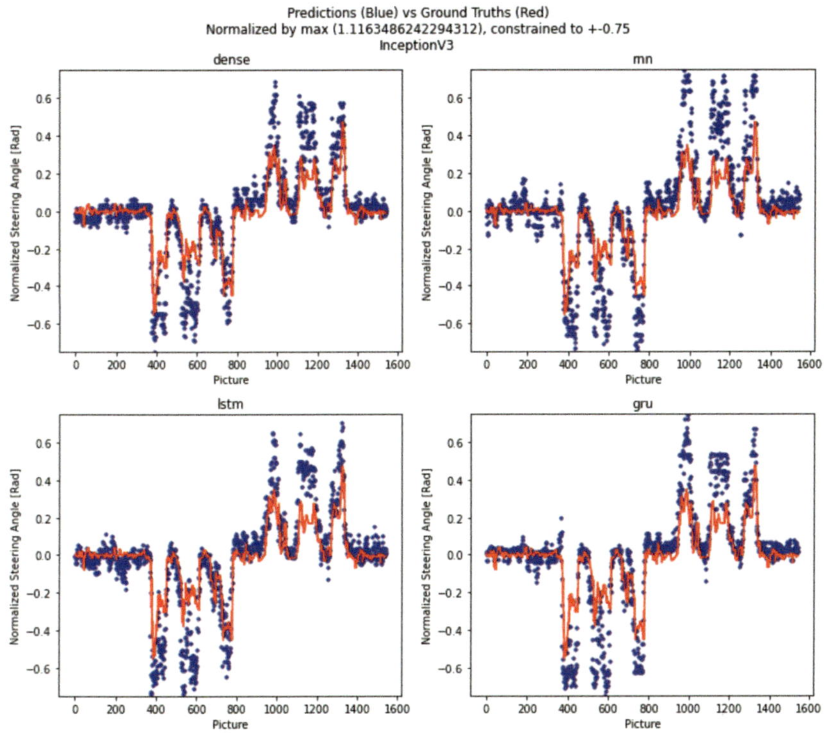

Figure 4.17 Comparison of model predictions (blue) with the true output values (red) for the test course. Bias toward low-angle input is present, as those predictions are closer to the ground truth than high-angle predictions.

4.4.4 CDR with new data

The introduction of interchangeable recurrent structures and pruning into the work done in [10] and [17] saw a marked improvement on the part of both techniques. Without pruning, the GRU performed best, with a mean absolute error of 0.2028 rad, while the CD network without pruning performed worst, with a mean absolute error of 0.2248 rad. The MAE range was 0.022, which is a very small amount. This required further analysis. A plot of each models' predictions on the test data compared to the ground truth (true value) showed that the model still performs better on low-angle input than on high-angle input. Figure 4.17 shows comparisons of the models' performance on a view constrained to ± 0.75, as outlying predictions are not shown. It can clearly be seen that the model makes more correct predictions on low-angle output than on high-angle output, indicating the presence of bias [17]. These

Table 4.2 Comparison of test loss (mean squared error) and test error for CDR models with interchangeable recurrent layers, both with and without pruning. It is clearly shown that the models trained on pruned data performed better than those that did not [17]. It is also important to note that the models shown in the literature are trained on roads **with** lane markings, whereas our models are trained for use on roads **without** lane markings [4], and that the model in [12] was tested in a simulated environment.

Model	Test loss (MSE, rad)	Test error (MAE, rad)
Dense (no pruning)	0.1198	0.2248
Dense (pruning)	0.0829	0.1916
RNN (no pruning)	0.1095	0.2211
RNN (pruning)	**0.0721**	**0.1799**
LSTM (no pruning)	0.1166	0.2243
LSTM (pruning)	0.0875	0.1846
GRU (no pruning)	0.0996	0.2028
GRU (pruning)	0.0791	0.1865
[4] (w/ lane markings)	-	0.0422
[12] (sim. env. w/ lane markings)	-	0.022

results influenced our decision to prune the training data as described in Section 4.3.3.

After training on the pruned data, the RNN (pruned) model gave the best performance with a mean absolute error of 0.1799 rad, and the CD (pruned) network still had the worst performance with a mean absolute error of 0.1916 rad. **Each model trained on the pruned dataset outperformed each nonpruned model**. Analysis of each prediction shows, despite the higher maximum value, somewhat better agreement with the true values. This is shown in Figure 4.18.

The combined set of test results and comparisons with other models is shown in Table 4.2. It shows that models trained on the pruned dataset perform better than those trained on the full dataset, indicating that the procedure has worked to reduce, but not eliminate, bias. These results are shown graphically in Figure A.3. While models described in the literature are trained and tested on marked roads, two examples are included here for reference.

Figure 4.19a shows a comparison of predictions between each CDR structure on a single example image. Each model attempted to predict the angle on an image whose ground truth was −0.2845 rad. This image was chosen because of its moderately sized turning angle. In this particular instance, the LSTM model outperformed the others, with an absolute error of 0.0034 rad, and the CD model underperformed with an error of 0.1392 rad. The difference

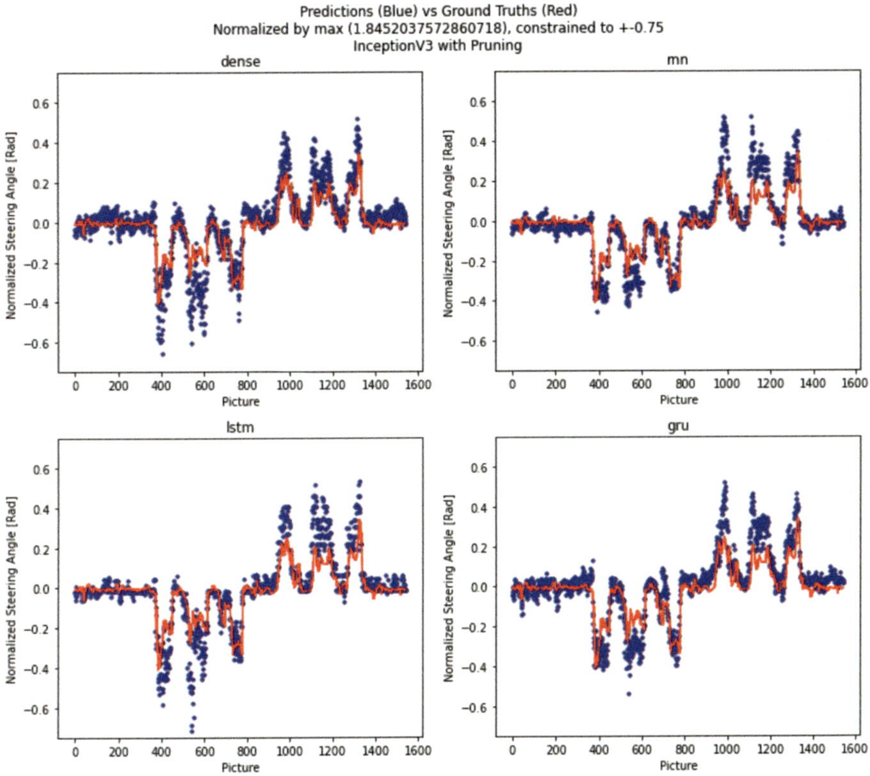

Figure 4.18 Comparison of model predictions (blue) with the true output values (red) for the test course **after pruning**. It is clear that there is better agreement than with the model trained on the non-pruned dataset. However, some bias is still present [17].

between the LSTM and the RNN layers, however, were 0.0394 rad, a very small difference. Figure 4.19b shows the RNN layer's predictions on four randomly chosen input images. Two images have high turning angles (0.4014 and 0.3613 rad), and two images have low turning angles (0.0908 and 0.0297 rad). The top-left image shows an uncommon road pattern, perhaps leading to the confusion. The bottom-left image does not show any strange road patterns. Although the dirt patch on the right of the image might contribute to the model's error, it is more likely that there is simply not enough input for high-angle training.

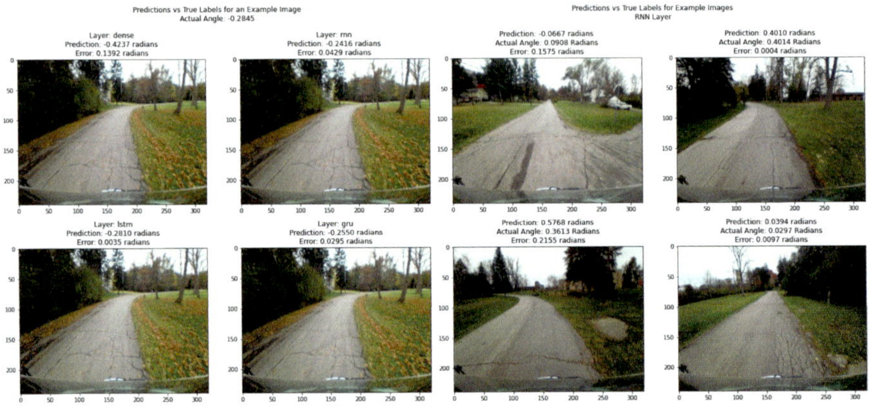

(a) (b)

Figure 4.19 Comparison of structures on sample images. (**a**) Comparison of recurrent structure predictions on a single image. (**b**) Comparison of RNN structure predictions on multiple images.

4.5 Discussion and Future Work

In this work, we have shown that deep learning is a viable method for autonomous lane navigation on unmarked roads. The use of pre-trained convolutional neural networks, fine-tuning, recurrent neural networks, and pruning to train models on pictures of roads collected using a vehicle-mounted webcam can produce predictions of steering wheel angles based on unseen data show that the InceptionV3 pre-trained network performs best out of eight pretrained networks (a mean absolute error of 0.2799 radians) and separately, it is shown that CDR (convolutional, dense, and recurrent neural networks) outperform CD+LSTM, CD+GRU, and plain CD networks after pruning with a mean absolute error of 0.1799 radians. It is thus demonstrated that CNN+DNN+RNN networks with data pruning produce the best result when applied on testing data, are accurate to less than 0.2 radians on average. Testing different structures, components, and algorithms have allowed us to develop a concrete method for data collection, processing, and analysis.

We believe that the CDR was more effective than the CD+LSTM and CD+GRU models because it is less useful to remember how the road looked at many timesteps prior; only the most recent timesteps, corresponding to the section of road the vehicle is presently traversing. With LSTM and GRU layers, "memories" of previous stretches of road began to confuse the model.

There are numerous applications for this work. First, deep learning-based steering is an alternative to traditional vision methods in autonomous vehicles. Methods that rely on lane markings or other visual color/contrast-based techniques can fall short in certain situations, such as if the road is unmarked or even unpaved. A deep learning-based strategy can serve to fill this gap. Moreover, deep learning-based steering can be applied in other areas. For example, it can be used on a factory floor to help robots navigate. It would require additional training, but it could be applied in situations where traditional path planning is unable to function.

There are numerous paths for the future study of deep learning-based steering. For a start, our system can be improved upon through the collection of additional data. This can include additional images with high turning angles for better balance, new routes to expand versatility, and different environmental conditions, as changes in weather, lighting, and traffic will need to be accounted for. More real-world testing is also necessary for further analysis of our work: despite theoretical feasibility based on test data, we have very few demonstrations of how the model responds in real time.

References

[1] Martín Abadi, Ashish Agarwal, Paul Barham, Eugene Brevdo, Zhifeng Chen, Craig Citro, Greg S. Corrado, Andy Davis, Jeffrey Dean, Matthieu Devin, Sanjay Ghemawat, Ian Goodfellow, Andrew Harp, Geoffrey Irving, Michael Isard, Yangqing Jia, Rafal Jozefowicz, Lukasz Kaiser, Manjunath Kudlur, Josh Levenberg, Dandelion Mané, Rajat Monga, Sherry Moore, Derek Murray, Chris Olah, Mike Schuster, Jonathon Shlens, Benoit Steiner, Ilya Sutskever, Kunal Talwar, Paul Tucker, Vincent Vanhoucke, Vijay Vasudevan, Fernanda Viégas, Oriol Vinyals, Pete Warden, Martin Wattenberg, Martin Wicke, Yuan Yu, and Xiaoqiang Zheng. TensorFlow: Large-scale machine learning on heterogeneous systems. https://www.tensorflow.org/, 2015. Software available from tensorflow.org.

[2] Aharon Bar Hillel, Ronen Lerner, Dan Levi, and Guy Raz. Recent progress in road and lane detection: a survey. *Machine vision and applications*, 25(3):727–745, 2014.

[3] Aylin Caliskan, Joanna J Bryson, and Arvind Narayanan. Semantics derived automatically from language corpora contain human-like biases. *Science*, 356(6334):183–186, 2017.

[4] Zhilu Chen and Xinming Huang. End-to-end learning for lane keeping of self-driving cars. In *2017 IEEE Intelligent Vehicles Symposium (IV)*, pages 1856–1860. IEEE, 2017.

[5] Kyunghyun Cho, Bart Van Merriënboer, Dzmitry Bahdanau, and Yoshua Bengio. On the properties of neural machine translation: Encoder-decoder approaches. *arXiv preprint arXiv:1409.1259*, 2014.

[6] François Chollet et al. Keras. https://keras.io, 2015.

[7] François Chollet. Xception: Deep learning with depthwise separable convolutions. In *Proceedings of the IEEE conference on computer vision and pattern recognition*, pages 1251–1258, 2017.

[8] Francois Chollet. *Deep learning with Python*. Simon and Schuster, 2018.

[9] Jia Deng, Wei Dong, Richard Socher, Li-Jia Li, Kai Li, and Li Fei-Fei. Imagenet: A large-scale hierarchical image database. In *2009 IEEE conference on computer vision and pattern recognition*, pages 248–255. Ieee, 2009.

[10] Giuseppe DeRose. Deep Steer Project Presentation, LTU MCS 5713, Fall 2021. https://youtu.be/PD9orAlHP1c, 2021. Accessed on July 19, 2022, Powerpoint, Video.

[11] Hesham M Eraqi, Mohamed N Moustafa, and Jens Honer. End-to-end deep learning for steering autonomous vehicles considering temporal dependencies. *arXiv preprint arXiv:1710.03804*, 2017.

[12] Hege Haavaldsen, Max Aasboe, and Frank Lindseth. Autonomous vehicle control: End-to-end learning in simulated urban environments. In *Symposium of the Norwegian AI Society*, pages 40–51. Springer, 2019.

[13] Kaiming He, Xiangyu Zhang, Shaoqing Ren, and Jian Sun. Deep residual learning for image recognition. In *Proceedings of the IEEE conference on computer vision and pattern recognition*, pages 770–778, 2016.

[14] Andrew G Howard, Menglong Zhu, Bo Chen, Dmitry Kalenichenko, Weijun Wang, Tobias Weyand, Marco Andreetto, and Hartwig Adam. Mobilenets: Efficient convolutional neural networks for mobile vision applications. *arXiv preprint arXiv:1704.04861*, 2017.

[15] Gao Huang, Zhuang Liu, Laurens Van Der Maaten, and Kilian Q Weinberger. Densely connected convolutional networks. In *Proceedings of the IEEE conference on computer vision and pattern recognition*, pages 4700–4708, 2017.

[16] Mark Kocherovsky. Comparison of Pre-Trained Models for DeepSteer, LTU MCS 5713, Fall 2021. http://qbx6.ltu.edu/chung/stuproj/2021/DeepSteer_ComparisonModels_MK.pdf, 2021. Accessed on July 21, 2022, Powerpoint.

[17] Mark Kocherovsky, Giuseppe DeRose, Nick Paul, and CJ Chung. DeepSteer: Autonomous Driving through Convolutional and Recurrent Learning. http://qbx6.ltu.edu/chung/stuproj/2021/DeepSteer_ComparisonModels_MK.pdf, 2022. Poster.

[18] Ana I Maqueda, Antonio Loquercio, Guillermo Gallego, Narciso García, and Davide Scaramuzza. Event-based vision meets deep learning on steering prediction for self-driving cars. In *Proceedings of the IEEE conference on computer vision and pattern recognition*, pages 5419–5427, 2018.

[19] Amr Mohamed, Jing Ren, Moustafa El-Gindy, Haoxiang Lang, and AN Ouda. Literature survey for autonomous vehicles: sensor fusion, computer vision, system identification and fault tolerance. *International Journal of Automation and Control*, 12(4):555–581, 2018.

[20] Nicholas Paul. Using end-to-end deep recurrent neural network architectures to steer a vehicle, Capstone project report, Fall 2018, LTU. http://qbx6.ltu.edu/chung/stuproj/f18/RNN_NickPaul_2018.pdf, 2022. Accessed on July 19, 2022.

[21] Nicholas Paul, Mitchell Pleune, ChanJin Chung, Ben Warrick, Sean Bleicher, and Charles Faulkner. Actor: A practical, modular, and adaptable autonomous vehicle research platform. In *2018 IEEE International Conference on Electro/Information Technology (EIT)*, pages 0411–0414. IEEE, 2018.

[22] Nicholas Paul, Ian Timmis, and Chan-Jin Chung. Using an end-to-end deep cnn to steer a vehicle in real time. https://www.youtube.com/watch?v=Okjwd1dms9A, 2018. Accessed on July 28, 2022.

[23] Dean A Pomerleau. Alvinn: An autonomous land vehicle in a neural network. *Advances in neural information processing systems*, 1, 1988.

[24] Hajira Saleem, Faisal Riaz, Leonardo Mostarda, Muaz A Niazi, Ammar Rafiq, and Saqib Saeed. Steering angle prediction techniques for autonomous ground vehicles: a review. *IEEE Access*, 9:78567–78585, 2021.

[25] Joseph Schulte, Mark Kocherovsky, Nicholas Paul, Mitchell Pleune, and Chan-Jin Chung. Autonomous human-vehicle leader-follower control using deep-learning-driven gesture recognition. *Vehicles*, 4(1):243–258, 2022.

[26] Karen Simonyan and Andrew Zisserman. Very deep convolutional networks for large-scale image recognition. *arXiv preprint arXiv:1409.1556*, 2014.

[27] Stanford Artificial Intelligence Laboratory et al. Robot Operating System. https://www.ros.org.

[28] Christian Szegedy, Vincent Vanhoucke, Sergey Ioffe, Jon Shlens, and Zbigniew Wojna. Rethinking the inception architecture for computer vision. In *Proceedings of the IEEE conference on computer vision and pattern recognition*, pages 2818–2826, 2016.

[29] Mingxing Tan and Quoc Le. Efficientnet: Rethinking model scaling for convolutional neural networks. In *International conference on machine learning*, pages 6105–6114. PMLR, 2019.

[30] Ian Timmis, Nicholas Paul, and Chan-Jin Chung. Teaching vehicles to steer themselves with deep learning. In *2021 IEEE International Conference on Electro Information Technology (EIT)*, pages 419–421. IEEE, 2021.

[31] Jun Wang, Li Zhang, Yanjun Huang, and Jian Zhao. Safety of autonomous vehicles. *Journal of advanced transportation*, 2020, 2020.

[32] Yue Wang, Dinggang Shen, and Eam Khwang Teoh. Lane detection using spline model. *Pattern Recognition Letters*, 21(8):677–689, 2000.

[33] Yue Wang, Eam Khwang Teoh, and Dinggang Shen. Lane detection and tracking using b-snake. *Image and Vision computing*, 22(4):269–280, 2004.

Authors' Biography

Mark Kocherovsky graduated from Lawrence Technological University with a BS degree in Computer Science with minors in Physics and Mathematics in 2021, and graduated from LTU with an MSCS in 2022. He is currently a PhD student at Michigan State University working on evolutionary computation.

Guiseppe DeRose Jr. is a Product Development supervisor in ADAS Virtual Verification and Validation at Ford Motor Company and is an adjunct faculty member at Lawrence Technological University. He graduated from Michigan State University with a BS in Computer Science and B.S., M.S. and Ph.D. in Mechanical Engineering. His main areas of expertise are in numerical methods and simulation of linear/non-linear mechanics, multibody dynamics, vehicle dynamics and assisted driving.

Nicholas Paul is an adjunct faculty member teaching robotics classes at Lawrence Technological University and an Artificial Intelligence Engineer at SoarTech. He graduated from LTU with a Masters in Computer Science. He has worked with autonomous software development with the LTU vehicles since 2017 as a member as well as a co-adviser of the LTU IGVC self-drive team. He co-authored two journal papers and seven conference papers on autonomous vehicle research using the LTU vehicles. He has also been involved in the US Army RTK project.

Ian Timmis Ian Timmis is currently a Senior Machine Learning Engineer at Bardavon Health Innovations working on AI for musculoskeletal health. His main areas of expertise are in Machine Learning and Computer Vision on the edge as well as in Machine Learning platform design. He received his Master's in Computer Science from Lawrence Technological University.

Chan-Jin "CJ" Chung is a Professor of Computer Science at Lawrence Technological University, Southfield, Michigan. He graduated from Wayne State University with a PhD degree in Computer Science. He is the founder of Robofest competition, faculty advisor of Intelligent Ground Vehicles Competition (IGVC) teams, and the director of the computer science autonomous robotics lab with two drive-by-wire electric vehicles for self-drive research projects.

Appendix

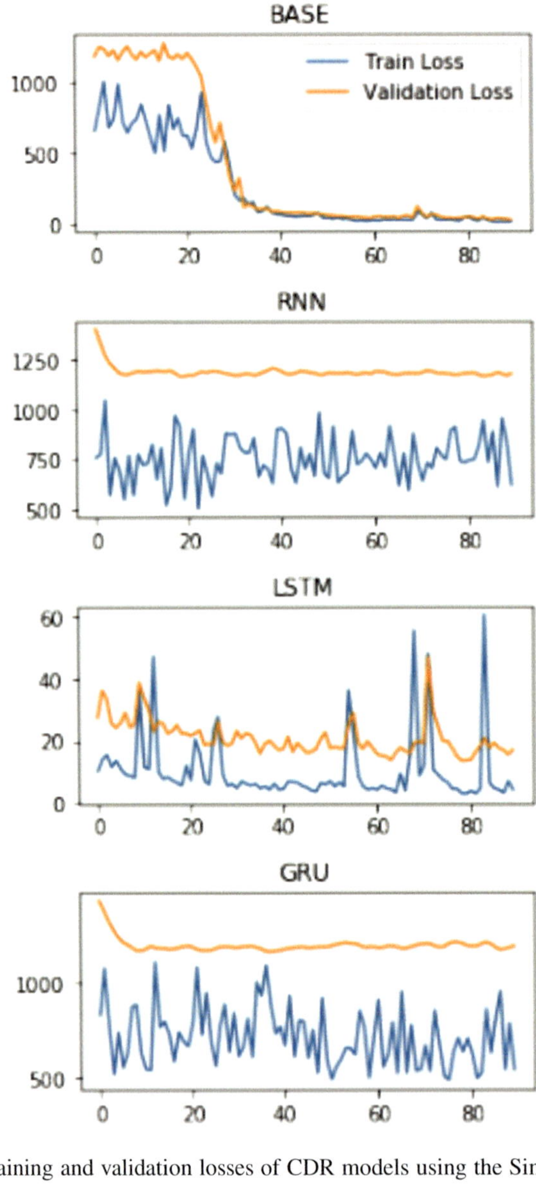

Figure A.1 Training and validation losses of CDR models using the SimpleRNN architecture [20].

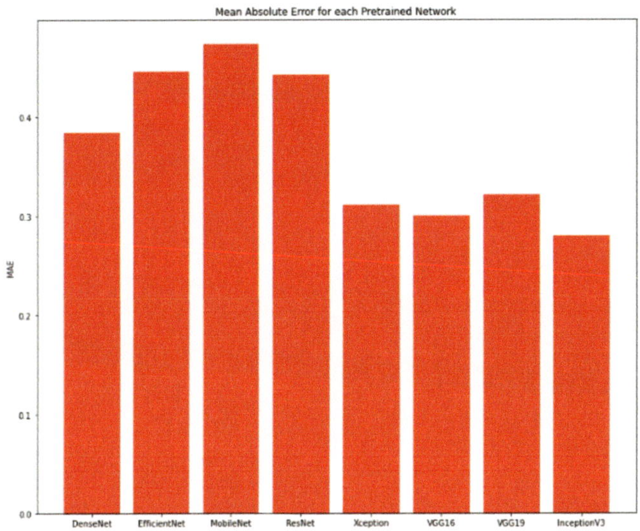

Figure A.2 Error of CD models with interchangeable pretrained networks.

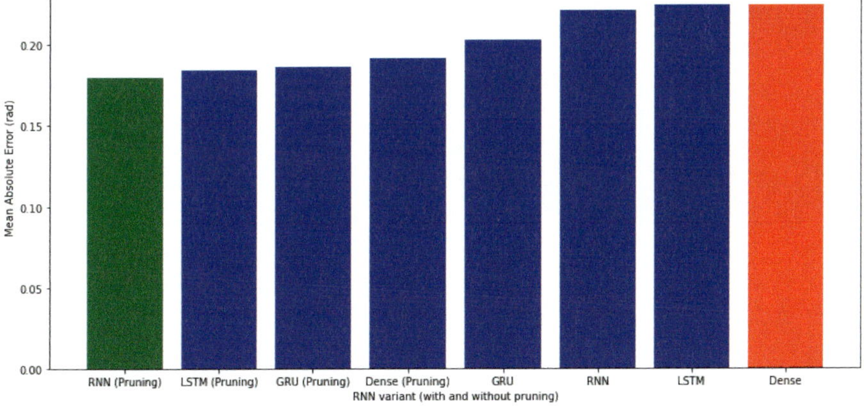

Figure A.3 Comparison of mean absolute errors of each model when applied to the test data, measured in radians. The DNN with pruning, on the far-left in green, has the best performance (lowest MAE) while the CD network without pruning, shown on the far-right in red, has the worst performance (highest MAE) [17].

5

2D and 3D Pose Estimation for Gesture Recognition in Deep-learning-driven Human–vehicle Leader–follower Systems

Joseph Schulte, Mark Kocherovsky, Justin Dombecki, Nicholas Paul, Mitchell Pleune, and Chan-Jin Chung

Department of Math and Computer Science,
Lawrence Technological University, USA

Abstract

The three decades of research into leader–follower autonomy (LFA) – in which the movement of one autonomous "leader" is followed by one or more autonomous "followers" – has mostly focused on systems in which vehicles follow other vehicles. In 2022, We demonstrated a human-led LFA system, in which a vehicle autonomously followed a human operator. The human operator is able to control the vehicle using body language, which is recognized by a gesture-recognition system. We had developed a modular pipeline using a combination of pre-built and custom-built components with high performance and high maintainability. The pipeline translates camera frames into pose data, which is then interpreted as a vehicle message. The message is sent to the vehicle, which performs the associated actions. In this work, we present improvements to our pipeline. First, we implemented a bipartite mapping algorithm for more consistent global target persistence to make the system more robust against overlapping detections, such as if a person walks in front of or behind the current operator. This improves operational safety. Second, we migrated the object detection module from YOLOv3 to YOLOv5 Nano, affording us substantially higher performance with much lower hardware requirements. Third, we exchanged the two-dimensional pose estimation network for a three-dimensional pose estimation network. This enables better precision during gesture detection, enabling increased detection accuracy, thus improving operation. Our results show that these new components improve the overall performance of the human–vehicle leader–follower system.

Keywords: Leader–follower, autonomous vehicles, self-driving car, machine learning, neural networks, deep learning, YOLO, posenet, pose estimation, gesture recognition, bipartite mapping.

5.1 Introduction

5.1.1 Leader–follower background

Leader–follower autonomy (LFA) systems, by which a single (or several) autonomous vehicle(s) follow other vehicles in the absence of a human driver, is a discipline having continued growth for the last few decades. Research of LFA systems include the development of mathematical models [9, 6, 26, 21, 16], testing in simulations [21, 26, 6, 9, 16, 32, 18, 33], and live experiments [16, 17, 33, 34, 32, 21] using configurations with two or more robots. There is evidence of usable triphibian applications of LFA systems, as well [32, 34]. Furthermore, the emergence of practical neural networks has replaced much of the traditional mathematical pathfinding solutions [17].

Much of the academic focus, regarding full-scale vehicles, has been on vehicle–vehicle (VV) systems. However, academia has seemingly disregarded LFA development using human–vehicle (HV) configurations on full-size autonomous vehicles. Albeit research has been conducted on human-robot (HR) following [31, 19] and with commercial applications [1], these focus on small robots configurations rather than full-scale vehicles.

5.1.2 Gesture recognition background

To solve gesture recognition (GR) problems, a classifying apparatus is required for various hand signals and body positions. Similar to LFA, GR has several decades of research. In 1991, Murakami and Taguchi developed such an apparatus using a recurrent neural network (RNN), which classified "signs" in Japanese Sign Language (JSL) by mapping a time-series context with the positional values of a signer's finger configuration and their hand orientation, using a glove with sensors [20]. More recent solutions use convolutional neural networks (CNN), which is another type of machine learning (ML) commonly used for image recognition. Contemporary research suggests implementing an array of neural network (NN) formations, along with CNNs, to accomplish GR tasks [12, 14, 15, 35]. Some studies use biological sensor data [14], and others use pose estimation (PE), whereby an image is decomposed into data representing the position of a body, or more specifically a hand [15].

5.1.3 Previous work

We demonstrated a LFA-HV system is possible [28] with our full-scale research vehicle, ACTor 1 (autonomous campus transport 1, Figure 5.1), which is a Polaris Gem 2 equipped with a research computer and sensory capabilities – including a LiDAR and camera. The computer, enabled with a DataSpeed drive-by-wire (DBW) system, is able to control the vehicle's

Figure 5.1 An image of ACTor 1 [10].

navigation, directly. Using the Robot Operating System (ROS), we program nodes to detect high-level user commands with the sensors, which are translated into low-level commands to the vehicle [24, 10]. This allows members of the research team to develop at higher abstraction levels without needing to work with the lower level messages to the vehicle's propulsion and navigation [25].

In the 2021 paper, we processed the video stream of a camera to identify "leaders" wearing ArUco markers, an optimized fiducial marker (similar to QR codes) [28, 2]. Detection of the markers and the humans wearing them was achieved using YOLOv3 object detection [27]. The result was the ability of a human leader to guide ACTor1, by simply wearing an ArUco marker and walking to a destination.

In contrast, this study is an exhibition of a LFA-HV system without the use of a traditional fiducial marker, by modeling a deep-learning gesture recognition classifier (GRC), using neural networks (NN). We also demonstrate this GRC can translate to vehicle controls in real-time and for real-life use cases. We believe this study sets research precedent in showcasing a leader–follower autonomy human–vehicle (LFA-HV) system implemented on a full-size vehicle. We envision applications for LFA-HV systems beyond just the automobile sector – such as transporting materials in loading bays, on construction sites and on factory floors.

Additionally, two improvements to the 2021 system will presented, which uses the 2D posenet, published in early 2022 [30]. First, bipartite mapping relations between human detections were implemented to improve target persistence [29]. The system could become confused if two potential targets were close enough that their detections overlapped. In bipartite mapping, each element in one set is linked to no more than one element in a second set. By linking detections of individuals in this manner, the system is now more robust to cases when a second person steps into frame, steps between the target and vehicle, or where the target is demonstrating before an in-frame audience.

Second, the upgrade to a 3D PE model in place of the previous two-dimensional (2D) model for pose estimation (PE) [23, 30] improved robustness to reliably distinguish gestures apart. The introduction of a 3D pose estimation model makes the LFA-HV system more reliable when presented with gestures, which are similar when projected in two dimensions, but distinct when projected into three dimensions, as proved by the increased accuracy of the gesture classifier.

Ahead, the concepts of these improvements are discussed in further detail: Section 5.2 will explain our use of Bipartite Mapping, and Section 5.3 will overview NNs, the chosen classification gestures for the GRC, and the GR system itself. Further ahead, the ACTor 1 platform is further outlined, by discussing ROS, our node architecture and its implementation in Section 5.4. In Section 5.5, we showcase the results of our demonstrations. Section 5.6 concludes, by summarizing the results and outlining avenues of future research.

5.2 Bipartite Mapping

The modification to our system solves the problem of target persistence. If a bystander was close enough in the frame to the target that their detections merged, the targeting system would tend to become confused. This would result in manual program interruption for safety. We thus chose to use a bipartite mapping algorithm, which constructs a relationship between two sets where each element in one set is linked to at most one element from another set.

The algorithm works by finding a minimal matching of a weighted bipartite graph representing the sets, hence the word "**bipartite**." When a new frame is captured, our algorithm uses data from previous frames to predict where each of the detected objects should be in the new frame. It then builds a list of links between the known objects. The detections in the new frame are sorted by how close they are to the prediction. The

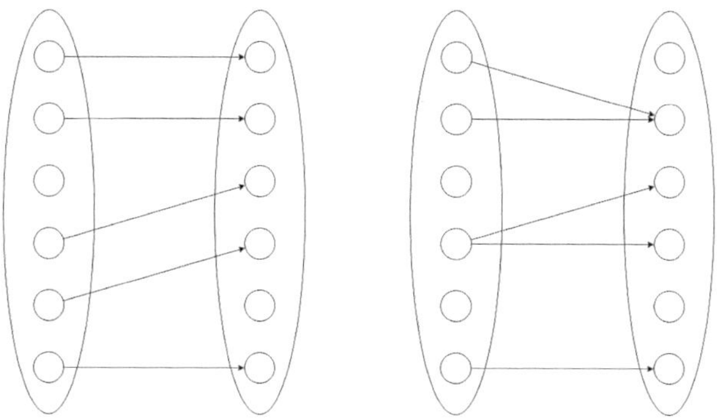

(a) (b)

Figure 5.2 (**a**) Example of a bipartite mapping. Notice that each element from one side is mapped to at most one element from the other side. (**b**) Example of non-bipartite mapping.

algorithm then goes through each link from closest to farthest and either keeps a link if neither object has been used or discards it if either node has already been linked. The remaining links form a minimal matching, which represents a mapping between existing and new detections. This is used to assign persistent object IDs and perform motion smoothing on all detections, improving the effectiveness and safety of the LFA system [29, 13].

5.3 Gesture Recognition

5.3.1 How to teach a machine

Artificial neural networks (NNs) are algorithmic structures designed to perform machine learning, whereby a computer program can be "trained" on a dataset in order to "learn" to make useful predictions. Neural networks are made up of layers of logically connected nodes called **neurons**. A densely-connected neural network (DNN) has an initial layer of **input neurons**, which represent the input data, a final layer of **output neurons**, which represent the result of processing, and layers of **hidden neurons** in between, where processing occurs. The final result of processing is called a **prediction**.

Processing translates to **weight values**. A connection between two neurons has a weight value, which is multiplied by the previous neuron's output when the data is fed forward. The output of each neuron is usually regulated by **activation functions** meant to ensure that the weights do not "explode" in value, leading to unbalanced results. We use rectified linear units for our hidden neurons and a softmax function for the output.

While training, the network iterates over a training dataset which contains sample data and the corresponding output. During each "epoch", the network makes predictions based on the training data and measures the prediction error (also called **loss**). These error values are then used to tweak the weights to produce more accurate predictions. We also use a subset of the training data as validation, which tests the model in-between epochs to better measure training performance. This is done to prevent **overfitting**, in which an NN model becomes so good at predicting the training data that it loses flexibility to new data [8, 11].

A common application of DNNs for image recognition is the **convolutional neural network** (CNN), which uses an iterative "matching" algorithm to identify features common to the images in the dataset. That is, the neural network learns abstract features (on its own), which can aid classification of one image to the next. For instance, a CNN classifier for facial recognition might abstract parts of a face to sieve, narrowing in on a prediction, instead of comparing one entire image to another [8].

5.3.2 Gesture classification

Gestures were trained with a neural network classifier, which controls the ACTor using corresponding commands. The mapping of gesture to command is in Table 5.1.

The begin following command is designated for when a potential leader gestures by laying their right-hand upon their chest over their heart. As this pose is recognized, the command is invoked, which begins tracing that person (now called **target**) within the camera frame. When the target moves, the ACTor vehicle follows with leeway, adducing the LIDAR to define a safe clearance between them (see Section 5.5.1 for more about the ACTor's systems).

The stop following command is designated for when the target places their right-arm into a right angle on their right side with an open palm. As this pose is recognized, this command is invoked, which forgets the target and no longer traces them in the camera frame.

If neither gesture is recognized, the effect of the previous command will continue. Figure 5.3 shows each of the three pose cases.

Publicly hosted, our dataset present here: https://www.kaggle.com/dashlambda/ltu-actor-gesture-training-images (accessed on August 7, 2022). Please, cite this work for any intention using the dataset images for any purpose. Included are 1959 follow images, 1789 stop images, and 1795 none images. To normalize the dataset for training, our GRC randomly samples the smallest image count (1789 images) from each of these subsets.

5.3.3 Trials with a CNN

Testing the concept of a real-time LFA-HV with a conventional CNN using Keras [7, 3], proved inadequate. Another paper [30] describes the CNN construction, training, and results, which showed a failure.

Table 5.1 Mapping of gestures and commands.

Gesture	Command
Right-hand over heart	Begin following
Open right-hand with arm in right angle	Stop following
Neither of these	–

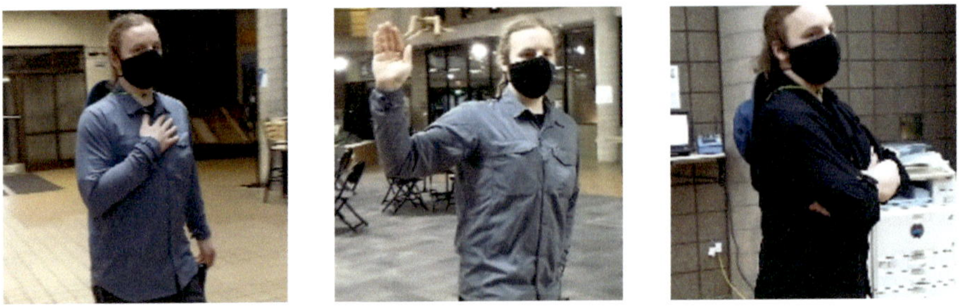

(a) (b) (c)

Figure 5.3 Author J. Schulte poses to display the three gestures. (**a**) Pose to target the user and initiate begin follow actions. (**b**) Pose for stop follow actions which forgets the target. (**c**) Not a pose: continue executing actions from last recognized pose [30].. Table 5.1

Using only a CNN for GRC would hinder our system in the long-term: firstly, a determination of the gesture's position in the image would be required, which is technically feasible but demanded too much development effort, at our scale. Secondly, the dataset limitation introduces **bias**, as the GRC would be trained solely on images of one of the authors.

Data engineers (and statisticians, alike) must minimize bias in their calculations. ML systems - especially a NN models - can learn to replicate prejudices shown by humans, even when it was not intended by design. For example, facial recognition software trained with images of people with one skin tone will likely misclassify people with a different skin tone, or automated hiring software might learn to prefer one gender over another [5]. Since our dataset consists of a single person, using only a CNN for GR introduces an extreme bias in gesture predictions. In the next section, an alternate method is described.

5.3.4 Modular data stream pipeline design

Iterating on our LFA-HV solution, a modular pipeline allows the reduction of complexity in a pure-CNN solution for GR. By instituting, rather complex, prefabricated components early in a data stream, the GRC derives its predictions from relatively simpler data structures (i.e., PoseNet coordinates vs. entire images). Other benefits of a modular pipeline include: fast test-and-learn cycles to discern optimal prebuilt components for specific task, focused development on custom nodes required for translation within the data stream and the ability to extend the solution for other research purposes in the future.

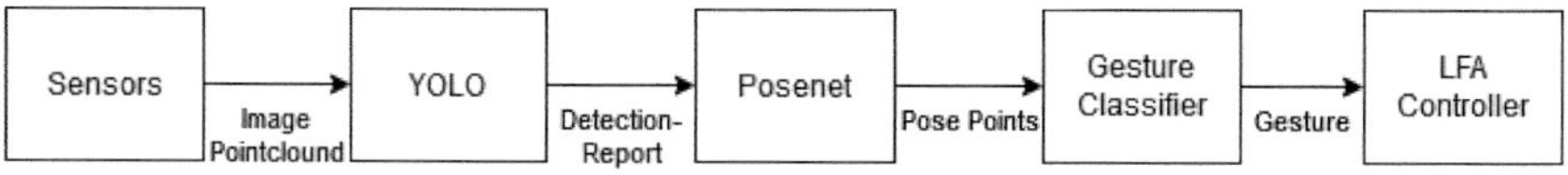

Figure 5.4 Modular data stream pipeline for LFA-HV – *(left-to-right)* The **sensors** pass a *data frame* (image and LiDAR) to the **YOLO** object detection (OD) CNN, which sends a *detection report* (a list of ODIDs isolated using our bipartite mapping algorithm) to the **PoseNet** CNN, which, in turn, sends lists of *pose coordinates* to the **gesture classifier** (GRC), which sends the *detected gesture* to the **LFA controller**, which translates the gesture into a *vehicle command.*

Figure 5.4 exhibits the pipeline. In practice, our design has elements of older data glove and modern "skeletonization" methods. Like other ML apparati, ours mingles a CNN with other additional components [20, 12, 14, 15, 35].

During target discovery, each OD is filtered for human detections (HD), which are processed through the posenet and GRC to determine a HD posing with the begin following gesture. Once a target is determined, the vehicle begins following that HD. At this point, the OD function adjusts to only feed the target's poses into the posenet and GRC stages, with a narrower scope of recognizing the stop following pose – when detected, resets back to target discovery.

In practice, our pipeline performs well in both lab and live environments (the latter is discussed in Section 5.5.1). To draw a comparison to our pure-CNN solution, we fed-in the dataset from that experiment into the pipeline, which yielded a test accuracy of 0.8500 and a test loss of 0.4010. The results are compiled in Table 5.2, showing the modular design outperforming the purely-convolutional design.

Figure 5.5 superimposes the classification of the begin following pose in a real-world scenario. The sensor dataframe containing the author is passed to YOLO, which perceives a HD and ascertains its position in the frame. The position translates to bounding box coordinates (BBC), visualized as a pink box around him. A cropped image, from the BBC, is sent to PoseNet, which calculates key features on the author's body, displayed as red squares. Finally,

Table 5.2 Juxtaposition of pure-CNN and modular designs evaluated against the same dataset. Higher accuracy and lower loss denote better models [30].

Model	Test loss	Test accuracy
CNN	5.2013	0.2684
Modular	0.4010	0.8500

Figure 5.5 Modular pipeline targets author M. Kocherovsky. [30]

the GR identifies the pose marking the author as the target, shown in green text.

The advantage of this pipeline is the separation of tasks. OD and PE are offloaded to robust pretrained networks, allowing the use of limited gesture datasets without introducing visual bias. Studies of the TensorFlow posenet or YOLOv3's biases in image recognition are outside the scope of this work.

Referring back to the data stream pipeline in Figure 5.4 – next, we will discuss two experiments regarding posenet in Sections 5.3.5 and 5.3.6. Later, more details about the Sensors and the LFA Controller will be covered in Section 5.4.

5.3.5 2D pose estimation

A TensorFlow 2D posenet model (available here: https://tfhub.dev/google/movenet/singlepose/lightning/4 (accessed on August 8, 2022)
was trialed as a single-pose estimation model (SPE) for a single HR.

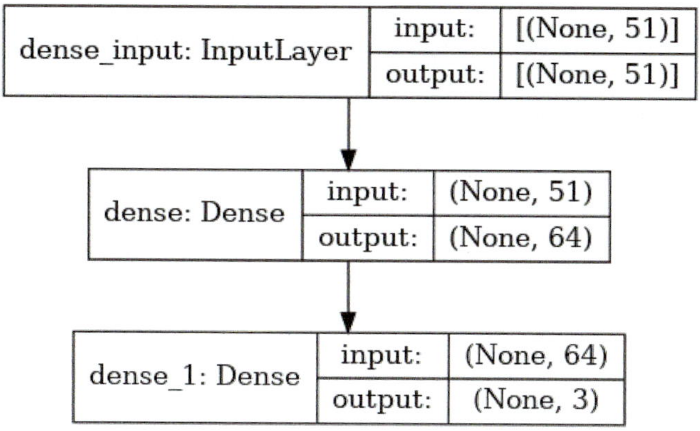

Figure 5.6 Pose classifier (SPEC) structure: a 51 neuron input layer (SPE output is 51 points), one 64 neuron hidden layer and an output layer with three neurons (for each command) [30].

The SPE accepts an image with 192×192 pixels, returning data for 17 parts of the human body [3, 4] (i.e., *x,y* positions and confidence scores), requiring further cropping of the BBC outputs from YOLO [27] to square frames at the required resolution.

The classifier (SPEC) structure is described in Figure 5.6. We used the categorical crossentropy function to measure the loss. Figure 5.7 displays the training progress, which concludes after five consecutive epochs without

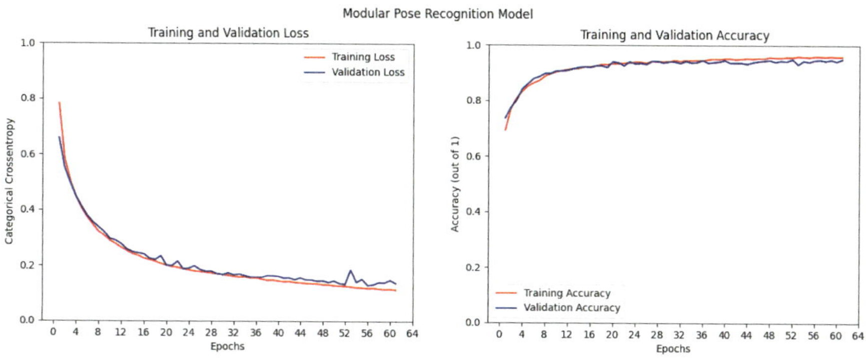

Figure 5.7 Graph illustrations of the SPEC training process. [30] (*left*) Each horizontal epoch is plotted against the categorical crossentropy value. (*right*) The vertical accuracy plots for each epoch. Training and validation sets are superimposed.

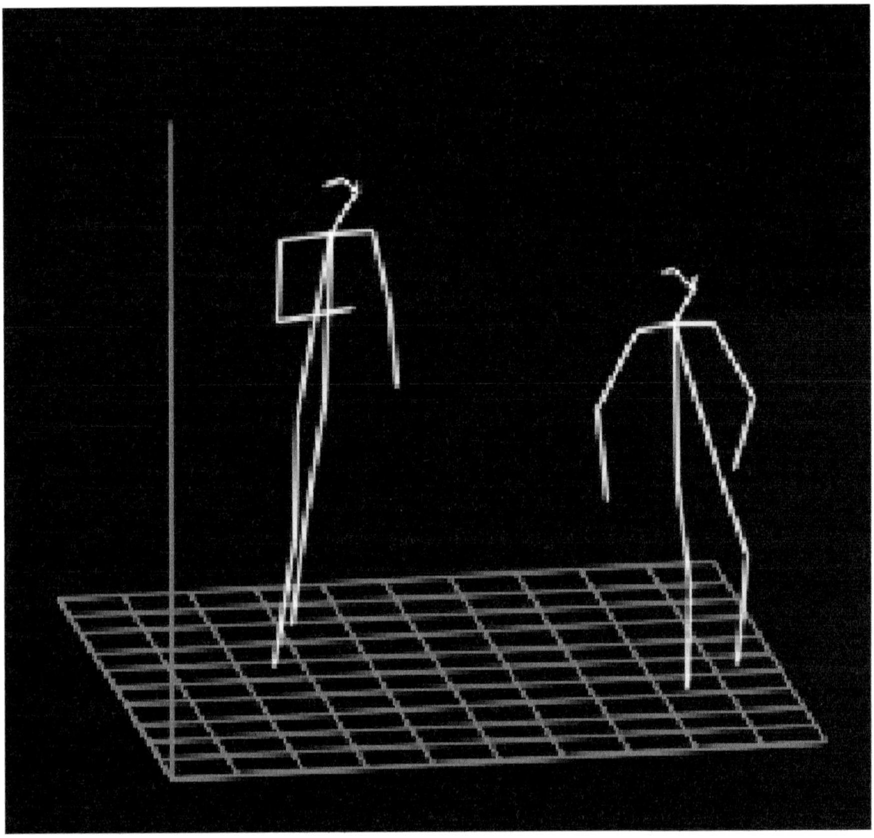

Figure 5.8 Wire representation of 3D poses from the new network.

validation loss improvement. Only the model with the least validation loss was saved during training.

5.3.6 3D pose estimation

2D pose estimation works very well as described in Section 5.3.5, but reducing the pose to two dimensions means some relevant information from the image is lost, particularly body orientation and separation between body parts in the direction of the camera. This was significant in the start the following gesture (hand-over-chest), where the 2D posenet has difficulty differentiating between a hand touching the chest and a hand-held away from the chest when facing the camera.

To solve this problem, we took advantage of our modular pipeline and migrated to a 3D posenet. This new posenet uses a monocular image to estimate *X, Y,* and Z coordinates for each joint in space (see Figure 5.8) under the assumption that they are 1.8 m tall [23]. This 3D pose estimation gives us the ability to recover body orientation and limb separation information, which could improve the potential accuracy of our classifier.

5.4 Integration of Models with the ACTor 1 Vehicle

5.4.1 ACTor 1 overview

The ACTor 1 (Figure 5.1) is a Polaris Gem e2 modified with a suite of sensors and computer control. ACTor 1 can be controlled using any one of three computing units: An Intel NUC, a small off-the-shelf desktop computer equipped with an NVIDIA 1070Ti GPU, or a Raspberry PI 3B. These computers can interface with the drive-by-wire system developed by DataSpeed. The vehicle's sensors include a Velodyne VLP-16 360° 16 line 3D LIDAR, Allied Vision Mako G-319C camera with a 6 mm 1stVision LE-MV2-0618-1 lens, and a Piksi Multi GNSS Module. The ACTor's systems are connected

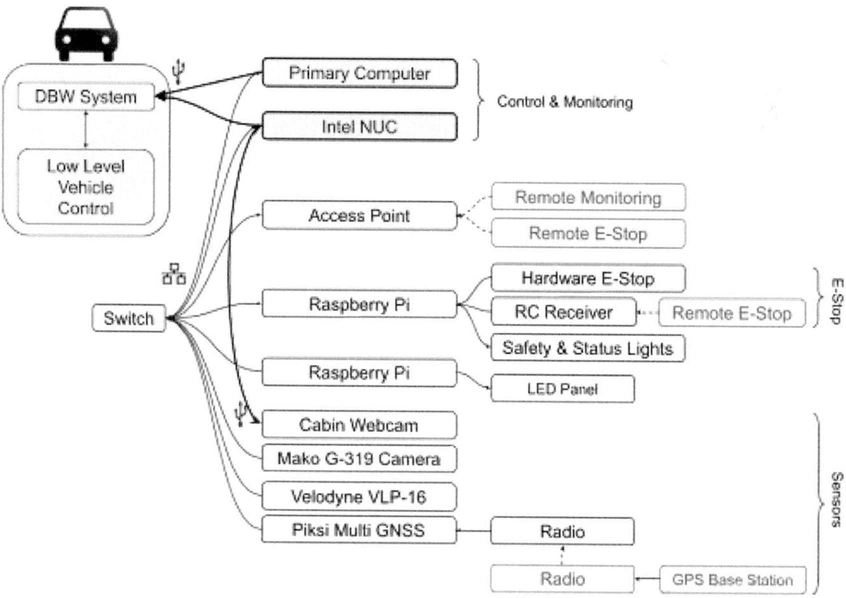

Figure 5.9 An image of the ACTor 1 hardware setup [10, 30]..

by ethernet and CAN buses. A diagram of our hardware modules can be seen in Figure 5.9 [10].

5.4.2 ROS fundamentals

The robot operating system (ROS) is used to interface with the ACTor's hardware [24]. ROS running under LINUX OS is a middleware platform that provides an abstraction of all hardware with ROS support and internal software that allows algorithm developers to write modular control programs.

ROS systems are structured as a set of implicitly connected **nodes**. Each component program has its own set of nodes through which it can indirectly interact with other nodes. Nodes interact using logical channels called **topics**. The act of broadcasting the topic is called **publishing**. Nodes listen for messages on topics, which is called **subscribing**. Neither publishers nor subscribers check to see which other nodes are publishing or subscribing. Nodes can publish or subscribe to as many topics as necessary. Topics have specific **message types** which generally correspond to established data types,

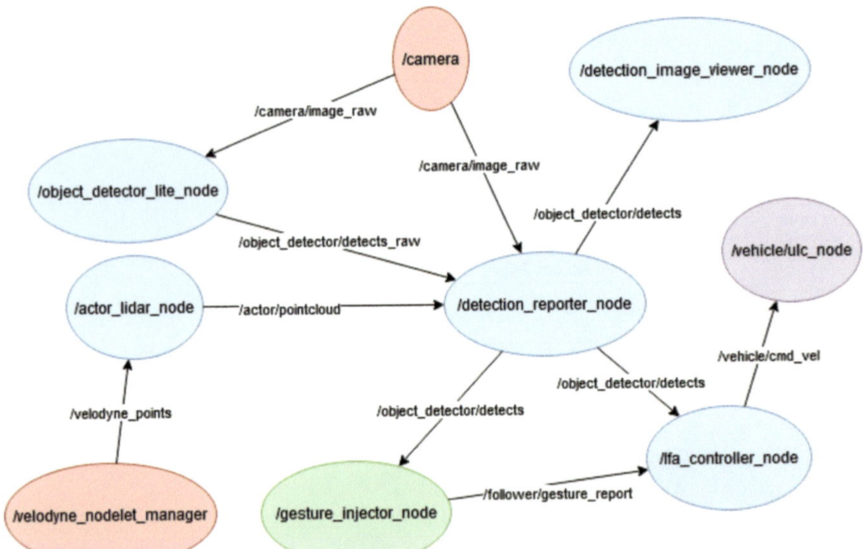

Figure 5.10 A diagram of the main ROS nodes (ellipses) and topics (arrows) specific to our leader follower system. Nodes marked in red indicate hardware nodes that publish sensor data. Blue nodes are nodes we created for the LFA systems, the green node implements our gesture recognition pipeline, and the purple node accepts commands to operate the vehicle's drive-by-wire.

Figure 5.11 Author M. Kocherovsky gives ACTor 1 the following command. Author J. Schulte sits in the driver's seat as a safety precaution.

such as integers or timestamps. If the default types are not sufficient, then developers may easily create their own message types [25, 22].

5.4.3 ROS node design and implementation

We have nine nodes running concurrently; a full diagram of our ROS nodes can be seen in Figure 5.10. Our code is also available here https://github.com/jschulte-ltu/ACTor_Person_Following (accessed on February 26, 2023). We ask that you cite this paper if you intend to use the code for any purpose. Each node is written in ROS for C++, except for Gesture Injection, which is written in ROS for Python.

1. Velodyne Nodelet Manager: This node provides an interface from our control unit to the LIDAR. It publishes the LIDAR sensor data for each point on the Velodyne Points topic;
2. Mono Camera: This node provides an interface from our control unit to the camera. It publishes the camera frames on the Image Raw topic;
3. LIDAR Reporter: This node receives raw input from the Velodyne Points topic, packages it into a convenient format, and publishes the reformatted data on the LIDAR Points topic;

4. Object Detection: This node subscribes to the Camera topic and runs object detection on camera frames using YOLO. It then publishes the location of each detected object in the image on the Bounding Boxes topic;
5. Detection Reporter: This node subscribes to the Bounding Boxes, LIDAR Points, and Image Raw topics and integrates their data to produce a coherent stream of information. It identifies the human detections reported by YOLO, superimposes their location in the image onto the 3D LIDAR point cloud to find their true location in three dimensions, identifies targets based on the given criteria, and attempts to keep track of the target from frame to frame. It publishes the consolidated information to the Detects Firstpass topic;
6. Detection Image Viewer: This node subscribes to the Detects topic to produce a visualization of the system's state. For each detection in the image, it draws the bounding box given by YOLO, draws the 17 pose points, and writes their distance from the vehicle's LIDAR system, the gesture they are performing, and whether or not they are a pose target. It can also superimpose the LIDAR point cloud onto the image and report the current action being performed by the vehicle. This node is purely for monitoring and visualization;
7. Gesture Injection: This node subscribes to the Detects topic, implements our gesture recognition pipeline as described in Section 5.3.4 to identify each target's gesture and the corresponding commands, then republishes the detection information with these new identifications to the Gesture Report topic. This node serves as a convenient and effective way to splice in the gesture detection pipeline with minimal alterations to our existing code;
8. LFA (Leader Follower Autonomy) Controller: This node subscribes to the Detects, Gesture Report, Follower Start and Follower Stop topics and publishes to the Display Message, Display RGB, Enable Vehicle, and ULC command topics. This is the last node in our LFA pipeline, which takes the detection and gesture information generated by the prior nodes and determines the actual commands sent to the vehicle. Those commands are published on the Command Velocity topic;
9. ULC (Universal Lat-Long Controller): This node provides an interface between our control unit and the drive-by-wire system. It takes the command from the Command Velocity topic and translates them into action by the vehicle.

Figure 5.12 ACTor 1 autonomously follows author M. Kocherovsky in the atrium of Lawrence Technological University. Author J. Schulte sits in the driver's seat as a safety precaution.

5.5 Testing Results

5.5.1 Results using the pipeline with 2D posenet

We ran our experiments on ACTor 1 using its main computing unit, equipped with an AMD Ryzen 72700 Processor and a ZOTAC GeForce GTX 1070 Ti GPU [10]. A video demonstration of our initial testing is available on YouTube at the following link: https://www.youtube.com/watch?v=A_CExLAEiB0 (accessed on August 7, 2022).

Later, we performed 10 trials to find out how reliable our LFA system is. Figures 5.11 and 5.12 demonstrate our system in operation.

Our testing area was the atrium at Lawrence Technological University. The atrium is a large open indoor area where we park ACTor 1 and ACTor 2. The atrium is the location of the LTU cafeteria, a coffee shop, and a convenience store. It is often the site of university and student-run events. There are numerous seating arrangements, vegetation, noticeboards, and other obstacles. More pertinent to the experiment is that there are usually humans sitting in or walking through the atrium, allowing us to test the system's robustness to multiple potential sources of input. Another important

Figure 5.13 Author M. Kocherovsky poses with ACTor 1 before a trial. The blue star denotes the approximate target location

factor is that due to the large skylight in the building, the atrium is subject to different natural lighting conditions depending on the weather and time of day, which provides us chances to test the robustness of YOLO and our posenet in a variety of lighting conditions.

On the day of the experiment, the atrium had an event, thus providing us with a slightly more complex environment. This can be seen in Figure 5.13. There were several booths in the atrium and slightly more students than on an average day. Tests were performed between 15:00 and 16:30, during which the outside weather was cloudy. The lighting conditions saw an average luminosity of 402 lx, with a maximum of 625 lx and a minimum of 238 lx.

Our experiments of 10 trials [30] were relatively simple; with the ACTor starting at one end of the atrium, the LFA system would be initialized with a user standing in front of ACTor 1. Once the system is ready, the user would give a stop command to reset any lingering variables, and then give a start command. They would then walk to the other end of the atrium and give a stop command once they have reached the target location, marked with a blue star in Figure 5.13. To be a successful trial, the LFA system had to recognize the start gesture, accurately follow the user, and recognize a stop gesture. A video example of our experimental trails is available on YouTube at the following link: https://www.youtube.com/watch?v=6ToP_Kuebj0 (accessed on August 7, 2022). Drive-by-wire system failures were not counted, as the DBW system is outside the scope of this work. There was also a single case where the human operator stopped the vehicle manually as a premature safety decision. This case was also not counted.

We report 100% success out of 10 trials, not counting the aforementioned cases. Detailed test results are described in [30]. In all trials that were not ended due to unrelated system failures and safety precautions, the gesture-driven LFA system was able to recognize the Start and Stop gestures as well as follow accurately, even when other people were present. However, we have also found two main avenues for future improvement: reducing false-positive gesture detections and improving target persistence.

Figure 5.14 ROS bag recording before overlap with detection IDs in blue.

Figure 5.15 ROS bag recording during overlap.

False-positive gesture detections occur when the gesture recognition classifier predicts that a gesture command is given to it when no actual gesture is being performed. This is caused by information loss from using a two-dimensional pose estimation model. This can make poses difficult to tell apart depending on the user's orientation, environmental factors, and how the user chooses to enact the pose. We intend to mitigate this issue using a three-dimensional pose estimation model, which we can integrate into the system more easily due to our modular design.

Losses of target persistence occurs when a bystander is close enough to the user so their bounding boxes overlap, such as if the bystander walks in front of the user or if there is a crowd behind the user. Our current architecture uses a simple closest-match algorithm to decide which detection in a new frame corresponds to the target from the previous frame. Thus, if the boxes overlap or the object detection module misses the target for more than one frame, the system can become confused. We intend to overcome this issue using a bipartite mapping algorithm to maintain detection identities between frames. Bipartite mapping is a relationship between two sets in which elements in one set are linked to elements in the other such that no two connections share an element. By finding the best mapping of detections

Figure 5.16 ROS bag recording after overlap. Notice detection IDs are consistent with the IDs before overlap.

from one frame to the next, taking position, size, and speed into account, it should be able to predictably identify all objects at once rather than only the target.

5.5.2 Bipartite mapping results

To test the bipartite mapping algorithm in a controlled setting, we recorded a ROS bag of two people walking in a circle with significant visual overlap in the center. Without bipartite mapping, this scenario resulted in unpredictable behavior as the system could not differentiate the two detections during the overlap. With the bipartite mapping algorithm, the system consistently tracked both people through the overlap without issue as shown in Figures 5.14 - 5.16. We observed similar performance in real-world scenarios tracking a single target with multiple people in frame and frequent overlaps.

5.5.3 3D pose estimation results

To compare the performance of 2D and 3D pose estimation, we built up a larger dataset, then trained them both under identical conditions. We trained

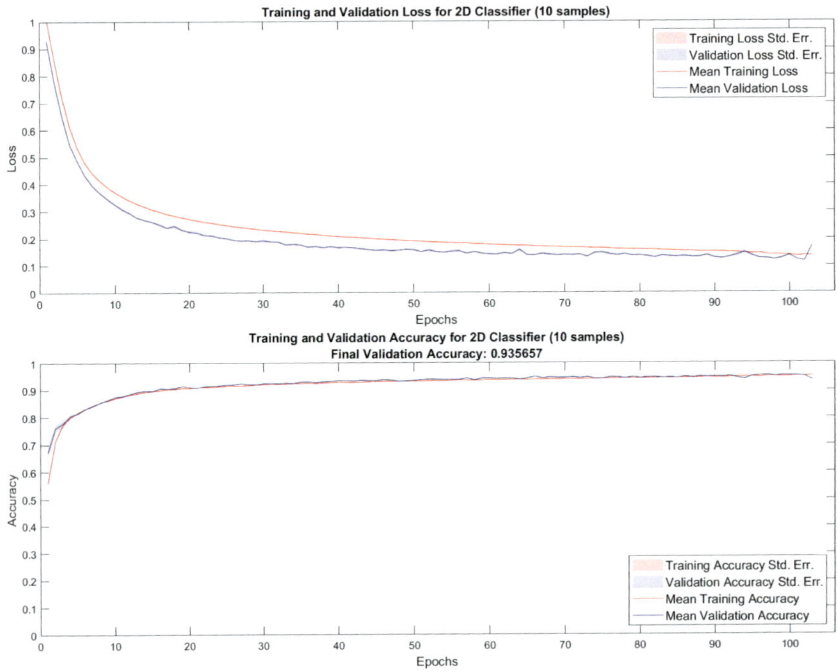

Figure 5.17 Loss and accuracy results from 2D network with standard error confidence bounds.

the classifiers ten times with different random samplings of the dataset and collected the accuracy and loss history as shown in Figure 5.17 and 5.18. Comparing the training history for the 2D and 3D networks, we see that the 3D network converges significantly earlier than the 2D network, reaching approximately 5% higher accuracy nearly 20 epochs sooner (93.6% vs. 98.4%).

We see a similar improvement in the confusion matrices (Figure 5.19 and 5.20 generated by the validation data. The 3D classifier shows a roughly 5% improvement in F1 score (0.946 vs. 0.984) and shows noticeably stronger (in fact perfect on the validation data) performance differentiating the "stop" and "start" gestures, with only a very small tendency to mix up "start" and "none."

These results show that 3D human pose estimation can recover significantly more information from an image than 2D pose estimation for the

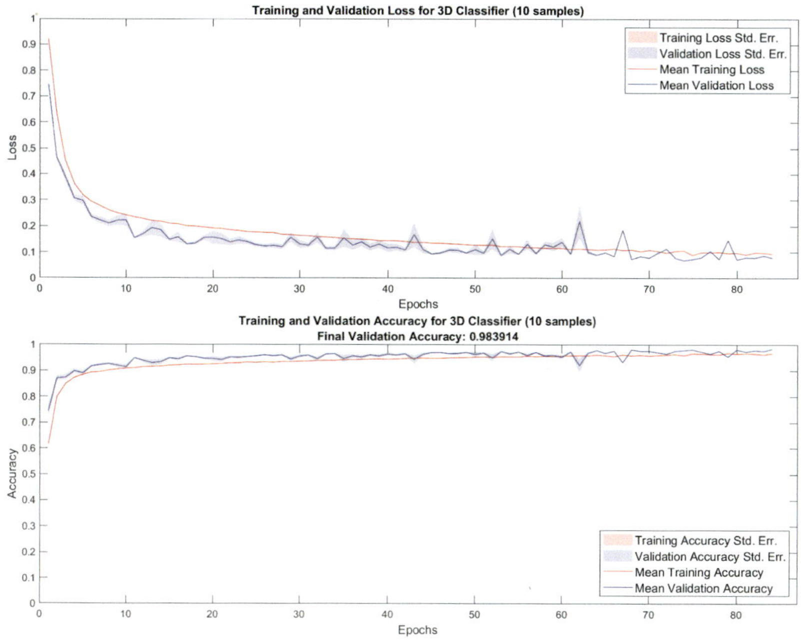

Figure 5.18 Loss and accuracy results from 3D network with standard error confidence bounds.

purpose of gesture recognition, and further that this increase in information can enable noticeable improvements in classifier performance, increasing stability and decreasing false positives.

5.6 Discussion

5.6.1 Summary

We have demonstrated that practical and reliable autonomous human-vehicle leader–follower systems are possible and feasible with current technology. We have demonstrated that it is possible to use pre-trained neural network models, namely YOLO, and newly trained DNN classifiers to recognize gesture commands in dynamic environments and translate those commands into actions on an integrated self-driving car computer platform through a pipeline architecture. We use modular software design and software reuse principles to accelerate the development, add new features or functions, improve quality, and limit technical difficulties.

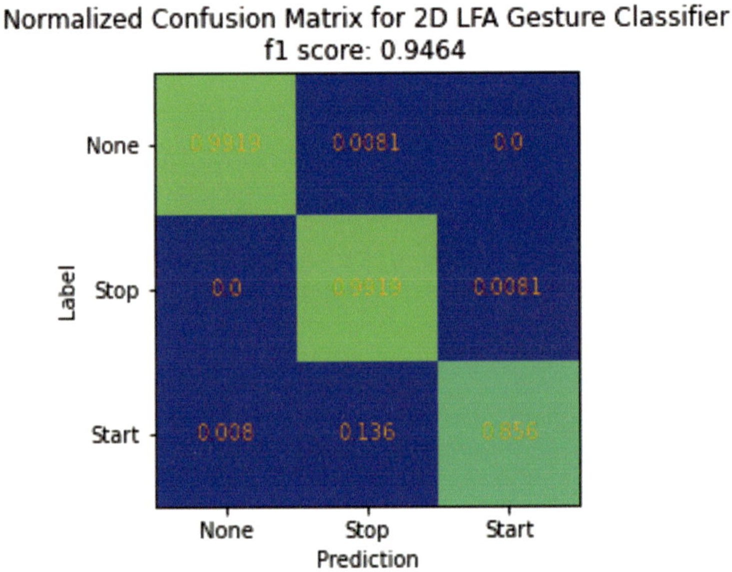

Figure 5.19 2D confusion matrix.

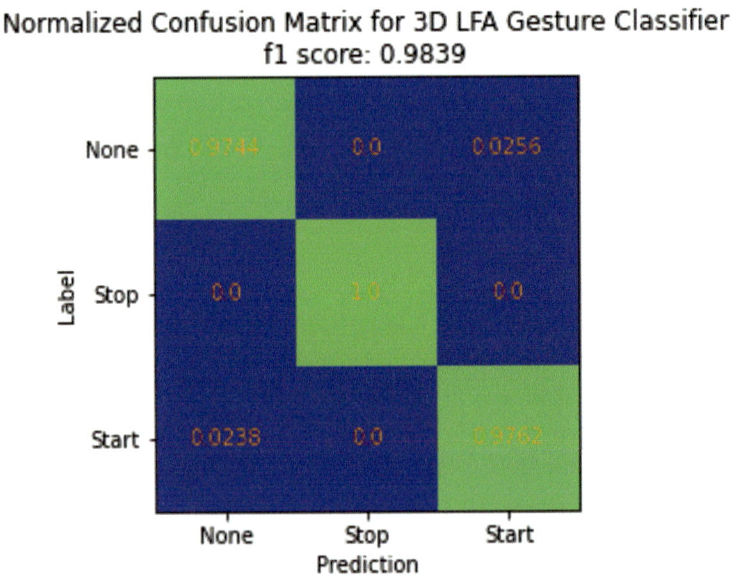

Figure 5.20 3D confusion matrix.

The core of this project is our human–vehicle leader–follower architecture based on YOLOv5 object detection. For the purpose of gesture-based control, in this system, we designed a modular gesture recognition pipeline incorporating a 2D pose estimation model and a simple DNN classifier. While 2D pose estimation allowed us to create multiple control gestures, it posed serious limitations for gesture variety and detection. The lack of depth information makes it difficult to impossible to differentiate between certain gestures, such as an outstretched palm and a hand on the chest, which both limits the potential number of usable gestures and increases the occurrence of false positives. To solve this problem, we exploit the modular nature of our pipeline to introduce 3D pose estimation. This improved detection accuracy by approximately 5% across the board. We also implemented bipartite mapping to improve target persistence, making tracking individual humans more reliable when multiple humans are in close proximity.

5.6.2 Future work

We envision several areas of future development and study. Improvements include further optimizations of our modules to achieve faster processing speeds, thus making the system more responsive. We also recommend adding obstacle detection and path planning modules to make the system safer and more versatile. In addition, we seek to add more gestures to add additional functionality to the system, namely an emergency stop (E-Stop) that can be activated by anyone in the frame.

We mainly envision applications in industrial contexts. The primary use of such a system would be to transport materials, equipment, and parts across factory floors, loading bays, and construction sites with minimal human operation and, in some cases, far less physical exertion from the human workers. If the proper safety features are present, the use of our system would decrease the chances for human error and improve the health of the workers in these occupations, thus increasing productivity. In non industrial settings, the human–vehicle leader follower system could be used in valet parking services. Expanding the gestures to include an automated parking system would allow valet parking-and parking in general-to be more efficient and free from human error.

References

[1] More info about Travelmate. https://travelmaterobotics.com/more-info-about-travelmate/, 2021. Accessed: 2021-09-26.

[2] OpenCV: Detection of ArUco Markers. https://docs.opencv.org/master/d5/dae/tutorial_aruco_detection.html, 2021.

[3] Martín Abadi, Ashish Agarwal, Paul Barham, Eugene Brevdo, Zhifeng Chen, Craig Citro, Greg S. Corrado, Andy Davis, Jeffrey Dean, Matthieu Devin, Sanjay Ghemawat, Ian Goodfellow, Andrew Harp, Geoffrey Irving, Michael Isard, Yangqing Jia, Rafal Jozefowicz, Lukasz Kaiser, Manjunath Kudlur, Josh Levenberg, Dandelion Mané, Rajat Monga, Sherry Moore, Derek Murray, Chris Olah, Mike Schuster, Jonathon Shlens, Benoit Steiner, Ilya Sutskever, Kunal Talwar, Paul Tucker, Vincent Vanhoucke, Vijay Vasudevan, Fernanda Viégas, Oriol Vinyals, Pete Warden, Martin Wattenberg, Martin Wicke, Yuan Yu, and Xiaoqiang Zheng. TensorFlow: Large-scale machine learning on heterogeneous systems. https://www.tensorflow.org/, 2015. Software available from tensorflow.org.

[4] Alphabet Inc. movenet/singlepose/lightning. https://tfhub.dev/google/movenet/singlepose/lightning/4, 2021.

[5] Aylin Caliskan, Joanna J Bryson, and Arvind Narayanan. Semantics derived automatically from language corpora contain human-like biases. *Science*, 356(6334):183–186, April 2017.

[6] Ka Cheok, Kiran Iyengar, and Sami Oweis. Smooth trajectory planning for autonomous leader-follower robots. In Gordon Lee and Ying Jin, editors, *Proceedings of 34th International Conference on Computers and Their Applications*, volume 58 of *EPiC Series in Computing*, pages 301–309. EasyChair, 2019.

[7] François Chollet et al. Keras. https://keras.io, 2015.

[8] Francois Chollet. *Deep learning with Python*. Manning, 1. edition, 2017.

[9] Luca Consolini, Fabio Morbidi, Domenico Prattichizzo, and Mario Tosques. Leader–follower formation control of nonholonomic mobile robots with input constraints. *Automatica*, 44(5):1343–1349, 2008.

[10] Giuseppe DeRose, Thomas Brefeld, Justin Dombecki, Mitchell Pleune, Joseph Schulte, Nichola Paul, and Chan-Jin Chung. ACTor: IGVC Self-Drive Design Report, 2021.

[11] AD Dongare, RR Kharde, Amit D Kachare, et al. Introduction to artificial neural network. *International Journal of Engineering and Innovative Technology (IJEIT)*, 2(1):189–194, 2012.

[12] Wei Fang, Yewen Ding, Feihong Zhang, and Jack Sheng. Gesture recognition based on cnn and dcgan for calculation and text output. *IEEE access*, 7:28230–28237, 2019.

[13] AMH Gerards. Matching. *Handbooks in operations research and management science*, 7:135–224, 1995.

[14] Yu Hu, Yongkang Wong, Wentao Wei, Yu Du, Mohan Kankanhalli, and Weidong Geng. A novel attention-based hybrid cnn-rnn architecture for semg-based gesture recognition. *PloS one*, 13(10):e0206049, 2018.

[15] Du Jiang, Gongfa Li, Ying Sun, Jianyi Kong, and Bo Tao. Gesture recognition based on skeletonization algorithm and cnn with asl database. *Multimedia Tools and Applications*, 78(21):29953–29970, 2019.

[16] N. Kehtarnavaz, N.C. Griswold, and J.S. Lee. Visual control of an autonomous vehicle (BART)-the vehicle-following problem. *IEEE Transactions on Vehicular Technology*, 40(3):654–662, 1991.

[17] N. Kehtarnavaz, N. Groswold, K. Miller, and P. Lascoe. A transportable neural-network approach to autonomous vehicle following. *IEEE Transactions on Vehicular Technology*, 47(2):694–702, 1998.

[18] Sridhar Lakshmanan, Yuedong Yan, Stan Baek, and Hesham Alghodhaifi. Modeling and simulation of leader-follower autonomous vehicles: environment effects. In Charles M. Shoemaker, Hoa G. Nguyen, and Paul L. Muench, editors, *Unmanned Systems Technology XXI*, volume 11021, pages 116 – 123. International Society for Optics and Photonics, SPIE, 2019.

[19] K. Morioka, Joo-Ho Lee, and H. Hashimoto. Human-following mobile robot in a distributed intelligent sensor network. *IEEE Transactions on Industrial Electronics*, 51(1):229–237, 2004.

[20] Kouichi Murakami and Hitomi Taguchi. Gesture recognition using recurrent neural networks. In *Proceedings of the SIGCHI conference on Human factors in computing systems*, pages 237–242, 1991.

[21] Teck Chew Ng, J.I. Guzman, and M.D. Adams. Autonomous vehicle-following systems : a virtual trailer link model. In *2005 IEEE/RSJ International Conference on Intelligent Robots and Systems*, pages 3057–3062, 2005.

[22] Jason M O'Kane. *A gentle introduction to ROS*. Jason M. O'Kane, 2014.

[23] Daniil Osokin. Real-time 3D Multi-person Pose Estimation Demo. Online, 2021.

[24] Nicholas Paul, Mitchell Pleune, ChanJin Chung, Ben Warrick, Sean Bleicher, and Charles Faulkner. Actor: A practical, modular, and adaptable autonomous vehicle research platform. In *2018 IEEE International Conference on Electro/Information Technology (EIT)*, pages 0411–0414, 2018.

[25] Morgan Quigley, Ken Conley, Brian Gerkey, Josh Faust, Tully Foote, Jeremy Leibs, Rob Wheeler, and Andrew Ng. Ros: an open-source robot operating system. volume 3, 01 2009.

[26] Amr Ramadan, Ahmed Hussein, Omar Shehata, Fernando Garcia, Hadj Tadjine, and Elmar Matthes. Dynamics platooning model and protocols for self-driving vehicles. In *2019 IEEE Intelligent Vehicles Symposium*, pages 1974–1980, 06 2019.

[27] Joseph Redmon and Ali Farhadi. YOLOv3: An Incremental Improvement. *arXiv*, 2018.

[28] Joseph Schulte, Justin Dombecki, Mitchell Pleune, Giuseppe DeRose, Nicholas Paul, and Chan-Jin Chung. Developing leader follower autonomy for self-driving vehicles using computer vision and deep learning. In *Lawrence Technological University Eighth Annual Research Day*, 2021. Poster, Online.

[29] Joseph Schulte, Mark Kocherovsky, and Chan-Jin Chung. Gesture-Based Human-Vehicle Leader-Follower Autonomy Systems with a Bipartite Mapping Algorithm, 2022. Poster.

[30] Joseph Schulte, Mark Kocherovsky, Justin Dombecki, Nicholas Paul, Mitchell Pleune, and Chan-Jin Chung. Autonomous human-vehicle leader-follower control using deep-learning-driven gesture recognition. *Vehicles*, 4(1):243–258, 2022.

[31] S.A. Setiawan, J. Yamaguchi, Sang Ho Hyon, and A. Takanishi. Physical interaction between human and a bipedal humanoid robot-realization of human-follow walking. In *Proceedings 1999 IEEE International Conference on Robotics and Automation (Cat. No.99CH36288C)*, volume 1, pages 361–367 vol.1, 1999.

[32] Aleksander S. Simonsen and Else-Line M. Ruud. The application of a flexible leader-follower control algorithm to different mobile autonomous robots. In *2020 IEEE/RSJ International Conference on Intelligent Robots and Systems (IROS)*, pages 11561–11566, 2020.

[33] Alexandru Solot and Andrea Ferlini. Leader-follower formations on real terrestrial robots. In *Proceedings of the ACM SIGCOMM 2019 Workshop on Mobile AirGround Edge Computing, Systems, Networks,*

and Applications, MAGESys'19, page 15–21, New York, NY, USA, 2019. Association for Computing Machinery.

[34] Qing Tang, Yanqiu Cheng, Xianbiao Hu, Chenxi Chen, Yang Song, and Ruwen Qin. Evaluation methodology of leader-follower autonomous vehicle system for work zone maintenance. *Transportation Research Record*, 2675(5):107–119, 2021.

[35] Xiao Yan Wu. A hand gesture recognition algorithm based on dc-cnn. *Multimedia Tools and Applications*, 79(13):9193–9205, 2020.

Authors' Biography

Joseph Schulte graduated from Lawrence Technological University with a BS in computer science and a BS in mathematics in 2021, then with an MS in computer science in 2022. He is working as a software engineer for The MathWorks, Inc.

Mark Kocherovsky graduated from Lawrence Technological University with a BS degree in Computer Science with minors in Physics and Mathematics in 2021, and graduated from LTU with an MSCS in 2022. He is currently a PhD student at Michigan State University working on evolutionary computation.

Justin Dombecki graduated from Lawrence Technological University with a BS in Mathematics and Computer Science in 2008. He applied his degree to the emerging mobile application development industry by working on contract with more than 20 published iOS applications. In 2017, he joined General Motors to focus on vehicle integration with their suite of myBrand applications. He returned to LTU in 2018 working toward a MS in Computer Science with a focus on Intelligent Systems – AI/ML and autonomous vehicles. He is planning to graduate in Spring 2023. His contributions to academic papers include topics: Vehicle Integrated Deep Learning Gesture Recognition and integrating the US Army RTK project in an autonomous vehicle.

Nicholas Paul is an adjunct faculty member teaching robotics classes at Lawrence Technological University and an Artificial Intelligence Engineer at SoarTech. He graduated from LTU with a Masters in Computer Science. He has worked with autonomous software development with the LTU vehicles since 2017 as a member as well as a co-adviser of the LTU IGVC self-drive team. He co-authored two journal papers and seven conference papers on autonomous vehicle research using the LTU vehicles. He has also been involved in the US Army RTK project.

Mitchell Pleune Mitchell Pleune graduated from Lawrence Technological University with a BS in computer science. He has four years of experience working on autonomous software development with LTU vehicles. He is currently a machine learning engineer at Veoneer, a worldwide leader in automotive technology, producing software, sensors, and control units for restraint control systems, advanced driving assistance systems (ADAS), and other technologies that improve driver safety.

Chan-Jin "CJ" Chung is a Professor of Computer Science at Lawrence Technological University, Southfield, Michigan. He graduated from Wayne State University with a PhD degree in Computer Science. He is the founder of Robofest competition, faculty advisor of Intelligent Ground Vehicles Competition (IGVC) teams, and the director of the computer science autonomous robotics lab with two drive-by-wire electric vehicles for self-drive research projects.

6

Adversarial Attacks and Defenses on Autonomous Vehicle Systems

Sunny Raj

Department of Computer Science and Engineering Oakland University,USA
Email: raj@oakland.edu

Abstract

Advances in autonomous vehicle systems (AVS) have been enabled by the incredible real-world performance achieved by deep neural networks (DNNs), and DNNs are a major component behind the decision-making process of AVS. Despite this success, DNNs have also been shown to be brittle and are known to misbehave in the presence of artificial and natural adversarial noise. Fog, rain, and other natural phenomenon have been shown to affect DNN systems adversarially. Artificial adversarial attacks on DNNs range from imperceptible changes to the input that can produce vastly different output from the DNN to physical patch attacks, where an attacker can put a small patch on a traffic sign to make the AVS treat it as an entirely different signal. Adversarial attacks on DNN are pervasive, and all architectures are affected by it. This lack of consistency in safety-critical applications such as AVS is concerning as minor deviation from the expected behavior can lead to massive implications. Various empirical and certified defense methods have been developed to guard against these adversarial attacks. This chapter will survey various strategies to generate natural and artificial adversarial attacks and techniques to mitigate these attacks.

Keywords: Adversarial attacks, adversarial training, natural adversarial attacks, sampling defence, certified defence.

6.1 Introduction

The promise of fully autonomous vehicles is closely tied to machine learning and artificial intelligence. Machine learning, specifically deep neural networks (DNN), is deeply integrated into existing autonomous vehicle systems. The occasional erroneous output of a machine learning system might not be much of a problem in nonsafety critical applications, like, for example, when trying to classify different varieties of flowers. But it becomes an issue in safety-critical applications such as autonomous driving, where a wrong output can be catastrophic, leading to loss of life and property. A head-up display showing the speed limit and various other traffic signals detected on the road is a common feature in high-end cars. Imagine a scenario when an attacker wants to crash a car with this feature and applies an adversarial patch on a speed limit 60 sign that makes the DNN in the car classify it as a stop sign. The car will read the adversarial sign and will try to signal the driver to stop the car; if the driver is not paying attention to traffic signals, he might think that the car is correct and stop, leading to a potentially dangerous crash. This scenario is even more concerning in the case of fully autonomous cars, where the car would stop on its own without warning the driver. These are all very real scenarios that can happen to current state-of-the-art DNNs. An example of this attack is shown in Figure 6.1.

(a)

(b)

Figure 6.1 Adversarial patch attack: A small patch applied to the traffic sign can produce a classification of stop sign from the DNN and can lead to severe and unintended consequences. A human observer would have no problem identifying both signs as speed limit 60. (a) speed limit 60 and (b) stop sign.

Deep neural networks were not designed with correctness and safety as the primary goals, and this oversight manifests in various ways that make a straightforward and vanilla implementation of these systems unsuitable for safety-critical applications. For a long time, the primary goal of neural network research has been to improve the accuracy of classification on various datasets. Initial efforts were spent on improving the accuracy of the MNIST dataset [18], which contains 28×28 images of digits from 0 to 9. When sufficient accuracy on MNIST was achieved, the focus shifted towards more complex CIFAR-10 and ImageNet datasets [9, 16]. ImageNet consists of 14 million images and consists of around 20,000 categories ranging from different breeds of dogs to various birds and common objects. A significant portion of the current advancement in machine learning can be attributed to the ILSVRC competition [31], where the primary goal is to improve the accuracy of classification of 1000 classes in the ImageNet dataset. The rate of progress of DNNs has been impressive, and most ILSVRC submissions outperform humans and get top 5 accuracy greater than 95%.

Given that the SOTA DNNs outperform humans on the ILSVRC challenge, there is an expectation that they will also be as or more robust than humans. This expectation has not come to fruition as the mechanism for achieving this superhuman performance differs from how humans process images. Humans have a remarkable ability to understand scenes and context, and achieving this understanding has been challenging for DNNs. Furthermore, around 2013–2014, it was shown that DNNs could break down under imperceptible noise and that they are not robust against these so-called "adversarial examples," which are pockets of misclassified instances near correctly classified input [36]. The existence of adversarial examples is not unique to certain architecture but pervasive, and all DNNs are susceptible to it. Therefore, designing high-accuracy DNNs that do not suffer from it has proved unsuccessful. The future of AVS depends on the advancements of DNNs, so these adversarial examples are concerning, and there is an express need to manage them for the safety of these systems.

Adversarial attacks can be broadly categorized into white-box and black-box attacks. In the white-box setting, an attacker is aware of the architecture and the weights of the DNN. This knowledge allows the use of gradients and enables unlimited queries to the DNN to generate powerful adversarial examples. These adversarial examples can be images that look similar to a natural image but are incorrectly classified with a very high probability, or they can be images that have a small patch applied to them which confuses the DNN into making an incorrect classification. These patches would not normally confuse a human observer but break DNN classifiers. A family of natural adversarial examples allows carefully crafted fog, rain, and hail to be added to natural images and have also been shown to fool DNNs and can be used to attack AVS while avoiding detection from human observers.

The availability of neural network weights and architecture for white-box attacks might not always be feasible. The family of black-box adversarial attacks works only by observing the output of the DNN and does not require knowledge of weights and the neural network's architecture. It has been shown that white-box attacks made on DNNs trained on the same dataset transfer well between different architectures and can be used for black-box attacks. Various strategies to optimize these transfer-based attacks have been proposed in the literature. Another strategy for black-box attacks relies on minimizing the number of queries being made to the DNN under attack. STOA black-box attacks are almost as powerful as white-box attacks and can successfully attack and fool any DNN model.

Various strategies to defend against adversarial attacks have been developed though non of them are satisfactory and work in all scenarios. The most popular adversarial defense consists of training the DNN models on adversarial examples; this training makes DNNs robust against both white-box and black-box attacks. Various types of input prepossessing that remove adversarial perturbations before giving it as an input to the DNN have also been shown to be effective in the black-box setting. Recently, there has been an effort to move away from empirical defenses that do not provide guarantees and toward certified defenses that can provide certificates stating that no adversarial examples exist in the neighborhood of certain input images. SOTA certificate generation methods only generate certificates of a small radius around the input, and adversarial examples outside the radius can still be visually indistinguishable from human observers. Nevertheless, these efforts are a key step toward robust and safe DNNs.

This chapter will explore the different types of adversarial attacks. We will start with optimization and gradient-based attacks, then look at black-box attacks, patch attacks, and natural adversarial attacks. Toward the end of the chapter, we will look at some defenses, including transformation defenses, adversarial training, and certified defenses. The illustrated examples in the chapter mainly focus on the German traffic sign recognition benchmark dataset (GTSRB) [34], but these attacks are equally applicable to other machine learning objectives in the AVS, including pedestrian detection and lane-keeping assistance.

6.2 Attacks

Confounding properties of the neural network were first discovered while investigating the interpretability and semantic meaning of individual neurons in the DNN [36]. It was realized that the input-output mapping was discontinuous, and pockets of misclassified images existed around correct classification and looked in-distinguishable from the original image. The initial attacks assumed the availability of the model architecture and weights for finding adversarial examples and thus were mostly focused on white-box attacks. Later work focused on black-box, patch, and natural attacks. Understanding the attack methodology is critical for designing systems that are robust against adversarial attacks, and in this section, we will look at the various popular ways of generating adversarial attacks against DNNs.

6.2.1 White-box attacks

The initial attack discovered by Szegedy et al. [36] was primarily concerned with the existence of adversarial examples, and the optimization problem reflects this approach. Given a neural network f, a class set $K = 1 \cdots k$, the loss function L, and a target label $y \in K$, these pockets of adversarial examples in the neighborhood of the input can be found using the following box-constrained optimization:

$$\begin{aligned} &\text{Minimize } ||r||_2 \\ &\text{subject to } f(x+r) = y \\ &\qquad\qquad x + r \in [0,1]^m \end{aligned} \tag{6.1}$$

This optimization is designed to find the closest adversarial example to the original image; in practice, as long as the adversarial example is indistinguishable to the human observer and within a certain distance from the input, the attack is considered successful. The attack used the Limited-memory Broyden-Fletcher-Goldfarb-Shanno (L-BFGS) optimization as an approximation of the optimization in Eqn (6.1) to find adversarial examples in the neighborhood of the input x:

$$\begin{aligned} &\text{Minimize } c|r| + L(f(x+r), y) \\ &\text{subject to } x + r \in [0,1]^m \end{aligned} \tag{6.2}$$

This initial paper coined the term “adversarial examples” and led to subsequent efforts at creating faster and stronger adversarial examples. The

L-BFGS optimization, although successful at finding adversarial examples, is slow in practice; later white-box attacks have focused on using the gradient of backpropagation to generate adversarial examples quickly.

Fast gradient sign method: In the course of training, gradients are computed for all of the neural network weights using backpropagation. The input, which is not a trainable weight, does not require gradients; a major insight of the fast gradient sign method (FGSM) method is to backpropagate the gradient to the input layer and move the input data toward the negative direction of the gradient [11]. It was found that instead of using the exact gradient direction, using the sign of the gradient of each feature is sufficient for obtaining a misclassification. The adversarial example using FGSM can be generated using the following equation:

$$x^* = x + \epsilon \text{ sign}(\nabla_x L(f(x + r), y)), \tag{6.3}$$

where the ∇_x function returns the gradient at the input layer, and the sign function returns 1 when the input is greater than 0 and -1 otherwise.

Iterative methods: The FGSM method described previously generates adversarial examples using just one computation of gradient; in some instances, this computation alone does not guarantee a successful adversarial attack. Various variations have been proposed that apply eqn (6.3) multiple times to generate adversarial examples with more consistency. The basic iterative method (BIM) [17] applies the FGSM methodology shown in eqn, (6.3) iteratively to generate attacks with higher success probability:

$$\begin{aligned} x^0 &= x \\ x^{t+1} &= \Pi_{x+\mathcal{S}}(x^t + \alpha \text{ sign}(\nabla_{x^t} L(f(x^t), y))), \end{aligned} \tag{6.4}$$

where the Π function projects its input to be within the set of allowed perturbations S. A minor variation of BIM, projected gradient descent (PGD) [21], chooses the starting x^0 randomly from within $x+S$; this random initialization significantly improves the performance of the adversarial attack. An example of the PGD attack is presented in Figure 6.2. Using the assumption that the classification boundaries of the neural network are linear can allow for searching for the closest adversarial example and is used by the DeepFool [23] method; since the assumption that the classification boundary is linear is not true, an iterative approach is used. Another iterative attack method is the

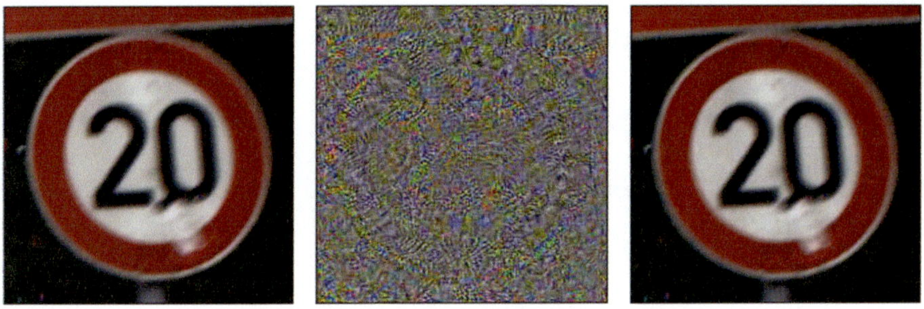

(a) (b) (c)

Figure 6.2 White Box Attack: (a) The original image is correctly classified as speed limit 20. (b) Perturbations are generated using the PGD method. The perturbations have been scaled for better visibility. (c) PGD attacked image classified as a stop sign. The original and the attacked images are visually similar but produce different classifications from the DNN.

moment iterative fast gradient sign method (MI-FGSM) [10], which uses momentum in each iteration to produce robust adversarial examples that transfer well to other network architectures. MI-FGSM adversarial examples can be generated using the following equations:

$$\begin{aligned} x^0 &= x \\ g^0 &= 0 \\ g^{t+1} &= \mu g^t \frac{\nabla_{x^t} L(f(x^t), y))}{|\nabla_{x^t} L(f(x^t), y))|} \\ x^{t+1} &= x^t + \alpha \ \mathrm{sign}(g^{t+1}) \end{aligned} \tag{6.5}$$

Carlini Wagner attack: A key requirement in the FGSM and the iterative methods described previously is the availability of gradients. Methods to obfuscate the gradient or to make the gradients noncontinuous after training can break these attacks. Defensive distillation is one such method that trains a DNN on the original dataset to train another more robust network by using the data and the output probabilities from the previously trained network; this distillation decreases the size of the gradient by three orders of magnitude and can be problematic for gradient dependent methods to attack [26]. The Carlini Wagner attack (C&W) [5] uses optimization to find adversarial examples and does not require gradients. It is an improvement over the L-BFGS method and is able to use more efficient optimizers such as Adam and gradient descent that do not require supporting box constraints. The original perturbation r in eqn (6.2) is defined using a change of variable on w:

$$r = \frac{1}{2}(\tanh(w) + 1) - x, \tag{6.6}$$

and is used to solve the following optimization to find adversarial examples classified as y:

$$\text{Minimize } ||\frac{1}{2}(\tanh(w) + 1) - x||_2^2 + cZ(\tanh(w) + 1), \tag{6.7}$$

where c is the model parameter and Z is

$$Z(x') = \max_{i \neq y}(f(x')_i - f(x')_y). \tag{6.8}$$

6.2.2 Black-box attacks

The powerful white-box attacks we have discussed, like FGSM, PGD, and MI-FGSM, utilize gradients to generate attacks. However, using gradients is only possible if the architecture and the weights of the model under attack are available to the attacker. The other set of attacks, such as L-BFGS and C&W, do not rely on the gradient but are optimization-based attacks that assume that the attacker can make an unlimited number of queries to the model. These assumptions might not be valid in all scenarios; in several instances, the weight of the model under attack might be private and not accessible to an attacker. In another example, a public model might have a subsystem to monitor the number of queries being made to it and only allow a certain fixed number of queries, thereby thwarting optimization-based attacks.

A family of black-box attacks that do not utilize the gradients of the network under attack and only make a limited number of queries to the model under attack has been proposed to overcome the limitations of white-box attacks in the scenarios mentioned earlier. A simple and effective black-box attack method uses transfer-based attacks, where an ensemble of networks is used to generate white-box adversarial examples [20]. These adversarial examples transfer effectively to a black-box model whose weight is unavailable. Another strategy is to train a substitute model having the same decision boundaries as the model under attack. Adversarial examples generated using white-box attacks on the substitute model then transfer effectively to the intended model. A successful training methodology involves querying the original model with intelligent samples generated using Jacobian-based heuristics [25]. This heuristic generates queries based on directions in which

(a) (b) (c)

Figure 6.3 Black box attack: (a) The original image is correctly classified as speed limit 20. (b) Perturbations are generated using the HSJA method. The perturbations have been scaled for better visibility. (c) HSJA attacked image classified as Speed Limit 70 sign. Black box attacks do not require knowledge of the weight and gradients of the model under attack but still produce adversarial images that look indistinguishable from the original.

the original model output is varying and outperforms generating random Gaussian queries. Training a separate network or family of networks for generating transferable attacks is still expensive, and other more efficient methods have been proposed.

Boundary attack (BA) [1] and its variants are strong black-box attacks that have been shown to utilize fewer samples and still produce a strong attack. This attack is initialized with a sample x^0 having a different label $f(x^0) \neq f(x)$, to the image x under attack and moves towards x, by sampling from a proposal distribution $\mathcal{P}$, while ensuring that the attacked samples x^t at each step t, never have the label $f(x)$. Finding the initial sample is trivial x^0; it can be generated randomly, or the attacker can use a known image with a different label to start the search. An overview of the different steps used for decision boundary attack is shown in Figure 6.4.

In the BA attack, the next adversarial sample x^{t+1} is chosen randomly from the distribution; HopSkipJump [6] presents an improvement over the randomly chosen sample by estimating the gradient of the decision boundary around the sample x^t and finding a new sample along the decision boundary that decreases the distance from the original sample x. A binary search finds the boundary x^t separating the correct class and adversarial class. Then a Monte Carlo estimate of the direction of the separating boundary is done by sampling around the decision boundary. A new sample is generated in the direction of the gradient $\hat{x}^{t+1}$ and the binary search is again conducted to find the decision boundary x^{t+1}; the process continues until a sufficiently similar adversarial sample is found. Estimating the gradient instead of using random

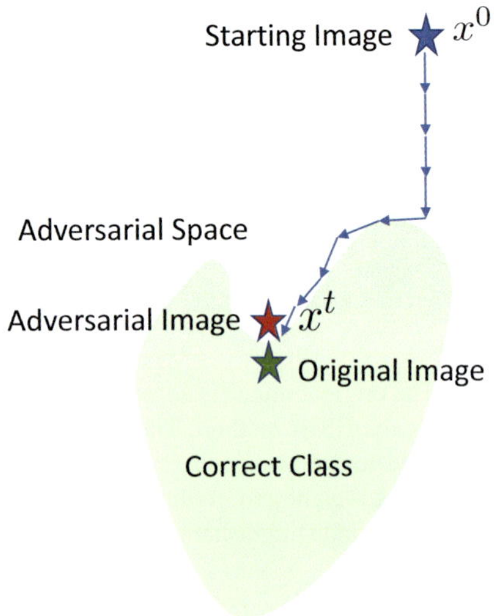

Figure 6.4 Decision boundary attacks find adversarial images by performing a random walk using rejection sampling to stay in the adversarial region and get sufficiently to the original image so that they look identical but have different classifications.

sampling allows this method to reduce the number of queries being made to the model. An overview of the binary search process used for HSJA is shown in Figure 6.5, and an example of an HSJA attack on a GTSRB image showing the indistinguishable attacked image is presented in Figure 6.3.

6.2.3 Patch attacks

The attacks that we have discussed so far cover the whole image, and the main objective of these attacks is to be identical to the original image and for the perturbations to be invisible to a human observer. Another class of attacks called patch attacks does not follow this approach; rather, it only changes a small portion of the input and is visible to a human observer but still manages to confuse the classifier into producing an incorrect classification. These adversarial patches are a convenient form of attack on an autonomous vehicle system, where a printed patch can be applied to the traffic signal to produce incorrect predictions. The adversarial patch method proposed by Brown et al. [2] generates patches that are invariant to transformations and can be printed and then applied to photographs and objects in the physical world. A patch application operator $A(p, x, l, t)$ is proposed that can apply a patch p after transforming it using transformation t to image x at location l.

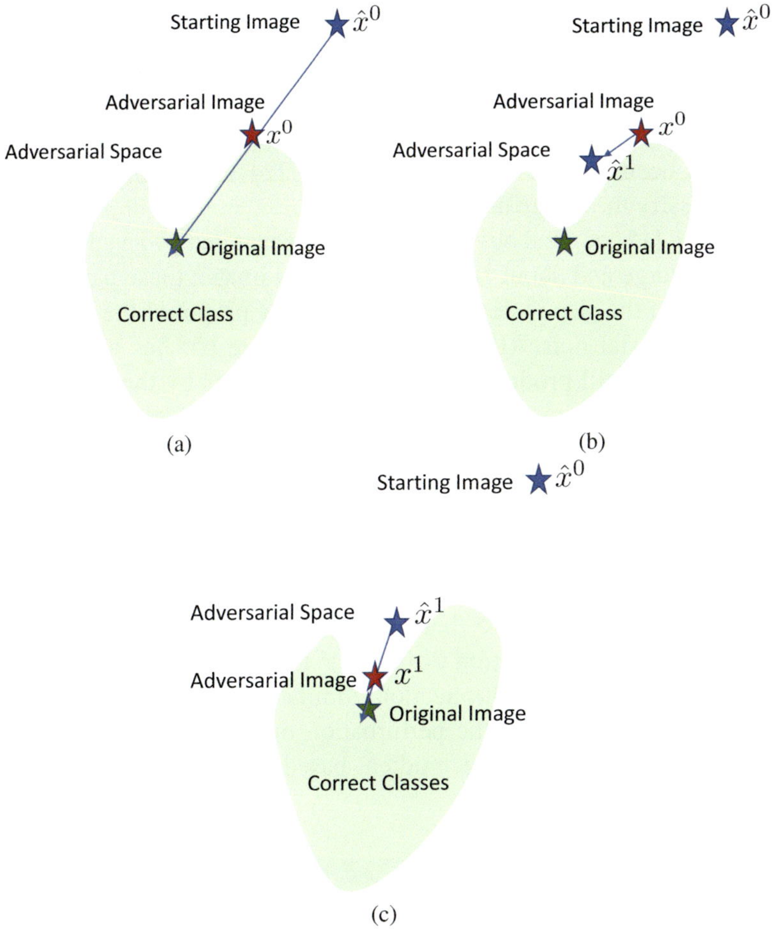

Figure 6.5 HopSkipJump Attack overview. Using binary search to obtain decision boundaries and then estimating gradients using Monte Carlo search allows for an efficient black-box attack with fewer queries to the model. (a) Decision boundary x^0 is obtained using a binary search between the original image and $\hat{x}^0$. (b) The decision boundary gradient is obtained using Monte Carlo estimation, and a new sample $\hat{x}^1$ is generated in the direction of the gradient. (c) Decision boundary x^1 is again obtained using a binary search.

The adversarial patch $\hat{p}$ is obtained by solving the following optimization:

$$\hat{p} = \arg\max_{p} \mathbb{E}_{x \sim X, t \sim T, l \sim L}[\log \mathbb{P}(y|A(p, x, l, t))], \tag{6.9}$$

where X is the set of training images, T is the set of transformations, and L is the set of positions where the patch can be applied. The expectations over images, transformations, and position provide a patch $\hat{p}$ that can be applied universally to any image at any position and is robust against common transformations such as rotation and scaling. An example of an adversarial patch attack is shown in Figure 6.6. In the figure, a small adversarial patch of size 30×30 has been applied to an image of size 224×224, which confuses the DNN to classify the image as a stop sign.

Karmon et al. [15] pointed out that the patches generated by an adversarial patch attack are large and about 10% of the original image; these patches are not effective when the size is decreased, and so they proposed the localized and visible adversarial noise (LaVAN) method, where patches as small as 2% of the input can still produce incorrect classifications by the DNN. The attack strategy uses the same formulation as gradient-based attacks but only perturbs a small area of the input image repeatedly until a wrong classification is obtained. In this method, the penultimate layer of the DNN, denoted by M, is used for training and optimization instead of the usual softmax layer. A mask $m \in [0,1]^n$, where n is the size of the input image, is introduced, which determines the location of the mask. Positions having a 1 in m have the mask applied to them, while positions having a 0 are left unchanged. The unchanged portion of image x is denoted by $(1-m) \odot x$, where $\odot$ is the elementwise multiplication. Similarly, the changed portion is denoted by $m \odot \delta$, where δ is the perturbation over the whole image and can be obtained using any of the gradient-based methods. The following

(a) (b) (c)

Figure 6.6 Adversarial Patch Attack: (a) The original image is correctly classified as speed limit 20 with a probability of 1. (b) 30×30 Adversarial patch. (c) Adversarial patch attacked image classified as a stop sign with a probability of 0.9. Only the patch portion of the image has changed; this patch can be applied to any image and will lead to a stop sign classification from the DNN.

optimization function generates a compact patch $m \odot \hat{\delta}$ that can fool the classifier:

$$\hat{\delta} = \arg\max_{\delta}[M(y|(1-m) \odot x + m \odot \delta) \\ -M(f(x)|(1-m) \odot x + m \odot \delta)]. \quad (6.10)$$

This optimization moves the patch towards the target label y and simultaneously moves it away from the original classification $f(x)$. Eqn 6.10 generates patch attacks that are not robust against transformation and are not universal in the sense that they can not be applied to any image. A similar strategy to eqn 6.9 of taking an expectation over various images, transformations, and locations generates universal and transferable attacks.

6.2.4 Natural adversarial attacks

Neural networks have also been shown to be vulnerable to natural-looking noise such as hail, fog, and rain [29, 22]. Carefully crafted natural noise has been shown to fool neural networks. Ozdag et al. [24] used stereo images to generate realistic-looking foggy images that can fool neural networks into making wrong classifications. An example of a foggy image used to fool the network is shown in Figure 6.7. Natural perturbation-based attacks have the potential to be applied to physical objects and disrupt autonomous vehicle systems while remaining undetected by humans.

Figure 6.7 Natural adversarial attack. Carefully crafted foggy images can fool state-of-the-art DNNs and lead to incorrect classification.

6.3 Defenses

Defending against adversarial attacks has proved to be a challenging endeavor and is still a subject of ongoing research. Multiple strategies that proved to be initially successful against existing attacks have then later been shown to be not robust against new attacks. Some of the early defenses relied on gradient masking, where the gradient around the prediction is obfuscated by making them zero or noncontinuous [26]. In these methods, the decision boundaries of the neural network are kept the same to achieve the same accuracy while making the gradient unsuitable for attacks. Most powerful white-box attacks rely on the gradient of the network to find adversarial examples, so masking the gradient should have allowed a powerful defense against attacks. Unfortunately, this line of thinking had several issues. First, defenses that make the gradient very small but still point in the same direction can be broken by increasing the step size of perturbations. Second, defenses that truly obfuscated the gradient are still vulnerable to transfer attacks, as the decision boundaries remain the same, and an attack on a different DNN transfers well to the defended DNN.

As can be observed from the above argument, the attack model is continuously evolving and thwarting existing defense strategies. To date, no satisfactory defense exists; nevertheless, the defenses that we will discuss are useful and provide some benefits in limited scenarios.

6.3.1 Preprocessing defenses

Preprocessing-based defenses rely on the idea that the adversarial perturbations produced by the gradient-based attacks with a fixed perturbation budget can be removed by preprocessing and denoising the input and then giving this modified image as input to the DNN. Multiple variants of preprocessing have been proposed, and these defenses work well against black-box attacks. Spatial smoothing, including median filtering, rotation, blurring, and taking the output of multiple models and averaging it, has been shown to provide some robustness against adversarial attacks [8, 41]. These techniques are designed to change the adversarial perturbation to make them ineffective while preserving the overall nature of the input for correct predictions. A naive strategy would be to use multiple weak preprocessing defenses to have a strong defense, but this has been shown to not work in practice [13].

A successful strategy is to use randomness during the preprocessing stage. Randomly resizing the input image and adding random padding around the image have been shown to provide protection against adversarial attacks [40]. Random changes in the input provide protection as the attacker might not be able to guess the preprocessing and tailor the attack specific to it. Determining the amount of deviation from the original image and the amount of padding to be added to the image has to be done carefully so that it does not lead to decreased accuracy on processed images. However, it was found that a small change in the input size does not adversely affect the original accuracy of the model but still provides robustness against attacks. It is hypothesized that the transformations destroy the adversarial perturbations of the input, and the randomness confounds the attacker, thus allowing for good robustness against black-box and some white-box attacks. The preprocessing steps used by Xie et al. [40] are shown in Figure 6.8.

Neural network-based denoisers have also been used to preprocess the input to generate clean images. Denoising autoencoder (DAE) [38] and denoising U-net (DUNET) [30] have been shown to be successful in defending against adversarial attacks. Lio et al. [19] used DUNET in their high-level representation guided denoiser (HGD) method to get the first prize in the NIPS 2017 adversarial defense competition [17]. To train the DAE and DUNET denoisers, a subset of input from the training dataset is first selected from the training dataset and then attacked using adversarial attacks; then, the attacked images x^* are given as the input to the DAE and DUNET networks, the output of the DAE network is the clean image $\hat{x}$, whereas the output of the DUNET network is the adversarial perturbation $d\hat{x}$. For the DAE, the loss is the L1 distance between the original image x and the predicted image $\hat{x}$. For DUNET, the loss is obtained by first getting the clean image $\hat{x}$ by subtracting the output of DUNET from the input and then finding the L1 distance from

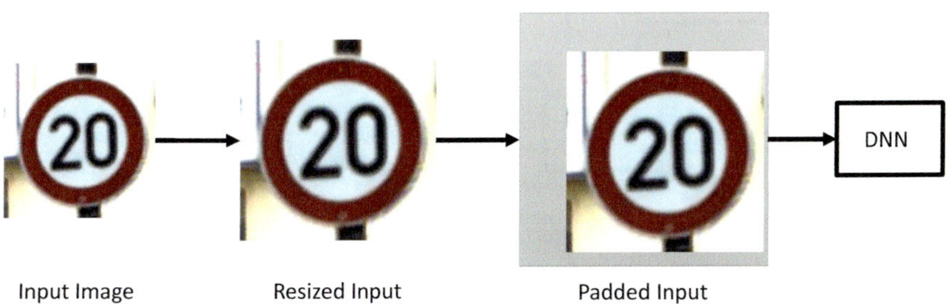

Figure 6.8 Randomization as a defense against adversarial attacks. The input image is scaled to random sizes and then randomly padded with zeros around the border before classification using the DNN.

the original image. An overview of the architecture used for training DAEs and DUNETs for denoising the input is shown in Figure 6.9.

Preprocessing-based defenses are suitable for use when the attackers do not know the preprocessing steps and work best in the black-box setting. However, these defenses fail in the white-box setting and do not provide satisfactory robustness; though using randomness in the preprocessing has been hypothesized to be effective, the idea has not been fully validated.

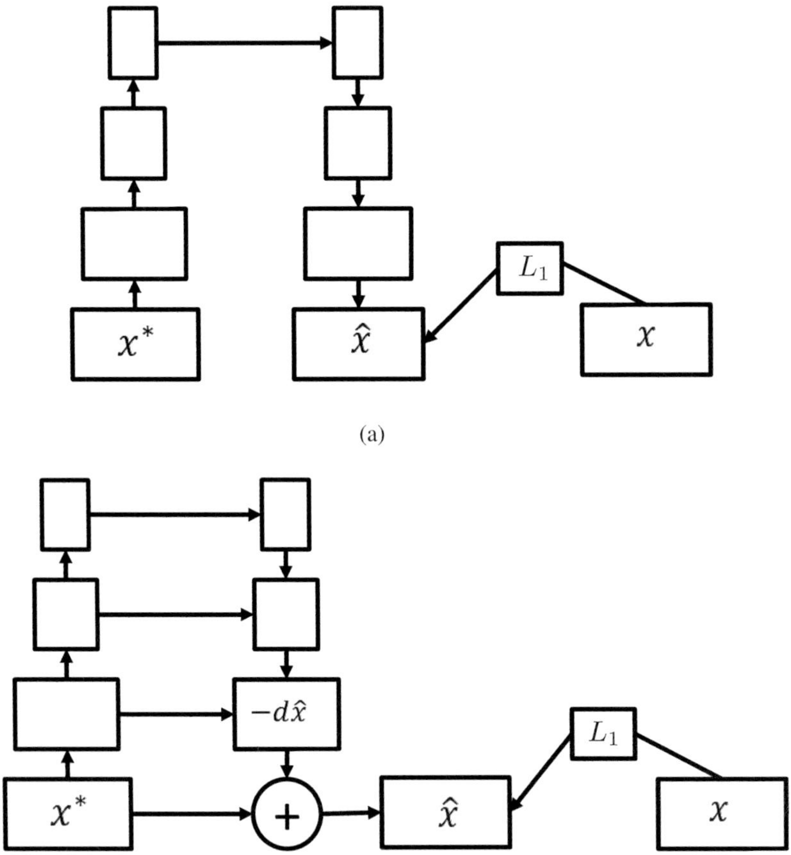

(b)

Figure 6.9 Network architectures for training DAE (a) and DUNET (b) for denoising images to defend against adversarial attacks. x is the original image, x^* is the attacked image, $\hat{x}$ is the clean image, $d\hat{x}$ is the predicted adversarial noise, and L_1 is the loss function.

6.3.2 Input-space sampling

Various input-space sampling strategies have been proposed as a defense against adversarial attacks. Intuitively, as adversarial examples are small pockets of misclassified data near the original image, sampling in the area around the adversarial image should return more samples belonging to the correct class. This sampling strategy does not always work in practice, as the correct distance to be used for sampling is not always apparent. A simple sampling strategy used by the SATYA method that does not need to search for a suitable sampling distance is to sample on the decision boundary between the adversarial and the correct image [28]. This sampling strategy returns samples classified as belonging to the correct class with a higher probability and can be used to detect and recover the original class of adversarial examples. An overview of the SATYA sampling strategy on the decision boundary is presented in Figure 6.10.

The SATYA method requires that the correct classification exists in the vicinity of the adversarial image and works well against attacks that generate adversarial examples by finding the classification boundary and staying close to it. Targeted attacks that push the adversarial example deeper into the adversarial territory are not suitable for defense using this method. Another method called the attribution-based confidence (ABC) [14] generates samples in the input space guided by the attribution of the input and then measures the

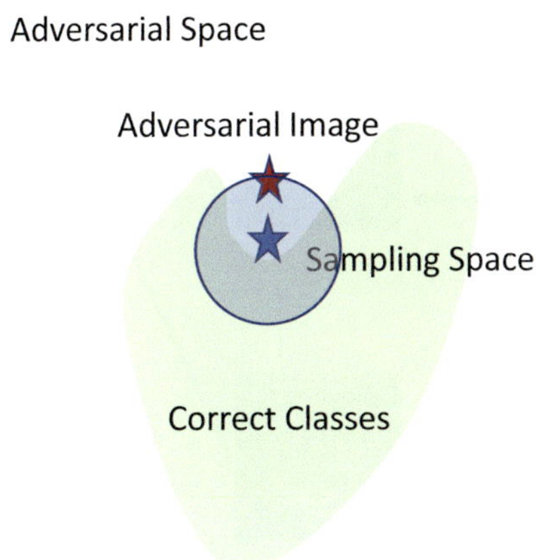

Figure 6.10 Input sampling based Satya defense. The decision boundary nearest to the input image is obtained, and then samples are generated and tested in a small hypersphere around the decision boundary. The majority of the samples belong to the correct class. This defense works when the adversarial examples are near the space of the correct class.

agreement among the classification labels of the sample. A higher agreement between the samples and the original image indicates an unmodified natural image, while a lower agreement between the sample classification and the original image indicates an attacked image. In ABC, important pixels of the input leading to the current classification are obtained using the Integrate Gradients attribution method [35]. Input samples are then generated by clamping the most important pixels of the input probabilistically, with the most important pixels being more likely to be clamped. The clamping, according to attributions, removes most of the problematic perturbations and makes the clamped image more likely to be assigned to the correct label. An overview of the method is presented in Figure 6.11, and an example of the ABC defense against an adversarial patch attacked image is shown in Figure 6.12. In the case of the adversarial patch attack, most of the pixels responsible for the attack are situated on the patch, which has a higher probability of being clamped, and so there is a change in the class of the clamped image and lower conformance of the samples. The natural image has important pixels distributed throughout the image and is less likely to change classification upon clamping.

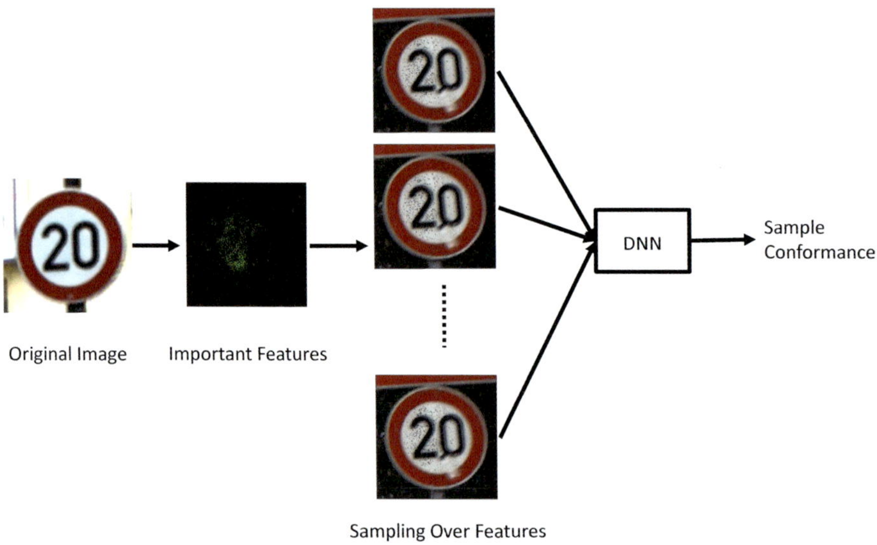

Figure 6.11 ABC metric. Important input features are found using Integrated Gradients and are then used for generating samples by probabilistically clamping the most important pixels. Natural images have a high agreement between the label of the samples and the original label, while perturbed images have a lower agreement.

6.3.3 Adversarial training

The previous section introduced techniques that mainly provided robustness in the black-box setting. This section presents a family of techniques that provide robustness in the white-box setting, where the weights, the gradients, and any preprocessing steps of the model are all available to the attacker.

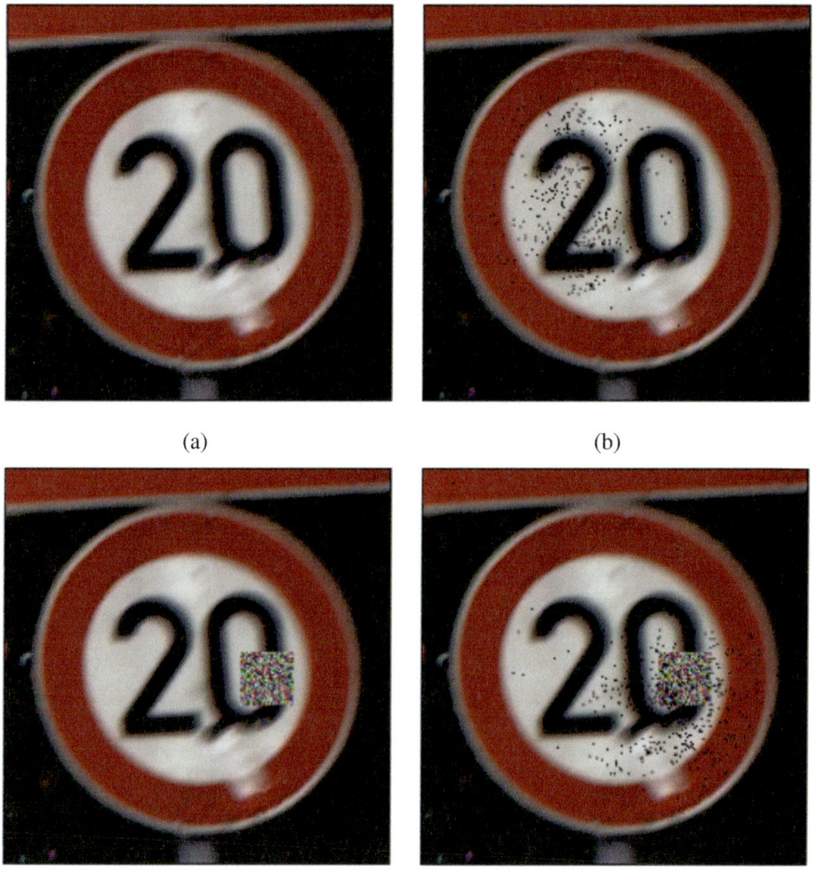

(a) (b)

(c) (d)

Figure 6.12 ABC defenses against adversarial patch attack: (a) The original image is correctly classified as speed limit 20 with a probability of 1. (b) Samples generated for checking conformance. ABC confidence on 100 samples is 0.6. (c) Adversarial patch attacked image classified as a stop sign with a probability of 0.9. (d) Samples generated to check conformance. ABC confidence on 100 samples is 0.01. The low ABC confidence value on the attacked image indicates potential tampering with it.

Defending against white-box attacks is more challenging compared to black-box attacks, as any number of the several efficient and powerful gradient-based or optimization-based attacks can be used by the attacker. The most successful practical defense in the white-box setting is adversarial training. The simple idea is to use both adversarial and clean data during the training phase to obtain robustness against future attacks on the DNN.

Several variations of adversarial training exist, but almost all solve the min-max problem, where the "min" problem is to minimize the loss of the network and the "max" problem is to find the worst adversarial example given the perturbation budget. An ideal solution to the min-max problem would have the lowest loss or the highest accuracy, even for the worst possible attacks. Minimizing the neural network loss can be achieved using standard training procedures. However, finding the worst-case perturbation is challenging, and there is no clear definition of what constitutes the max problem, though different approximations have been tried and shown to work. The min-max problem for adversarial training solves the following optimization:

$$\min_{\theta} \sum_{i=1}^{N} (\max_{\delta \in \mathcal{S}} L(\theta, x_i + d, y_i)), \tag{6.11}$$

where θ is the DNN parameter, N is the size of the dataset, $\mathcal{S}$ is the set of allowed perturbations, L is the loss function, x is the input data, and y is the target. PGD-based adversarial training (PGD-AT) [21], where the max problem is solved by obtaining adversarial data using the PGD attack, is among the most popular adversarial training-based defense and is usually the baseline for comparison with other methods. The $\max_{\delta \in \mathcal{S}} L(\theta, x_i + d, y_i))$ component of the above equation is obtained by solving the following equation iteratively:

$$x^{t+1} = \Pi_{x+\mathcal{S}}(x^t + \alpha \text{ sign}(\nabla_{x^t} L(\theta, x^t, y))). \tag{6.12}$$

The Π function projects the input to be within the allowed perturbation set S, and the ∇_x function returns the gradient of the input layer after backpropagation. WRN-34-10 network trained on CIFAR-10 with PGD-AT helps achieve adversarial accuracy of around 47.04% and normal accuracy of 87.30%. Without PGD-AT, the adversarial and normal accuracies are 0% and 95.29%, respectively. The PGD-AT provides good accuracy on attacked examples but significantly decreases accuracy on natural, benign examples, leading to a tradeoff between robust and natural accuracy.

The theoretically principled tradeoff between robustness and accuracy (TRADES) is an adversarial training methodology that decomposes the error on adversarial examples as the sum of natural error and boundary error. Natural error is the error on natural data that has not been perturbed. Boundary errors are observed in DNNs having natural images close to the classification boundary in the input space. It is hypothesized that this proximity is the main reason for the existence of adversarial examples as input data points closer to the classification boundary are more easily attacked as the change required to cross the boundary is low. In TRADES, the optimization is designed as a regularization problem where the primary objective is to decrease the natural error while a regularization term decreases the boundary error. The optimization problem for TRADES is as follows:

$$\min_{f} \sum_{i=1}^{N} \{L(f(x^i), y^i) + \max_{\delta \in S} L(f(x^i), f(x^i + \delta))/\lambda\}, \tag{6.13}$$

where f is the DNN, and λ is the regularization coefficient. The first term in the optimization minimizes the natural loss, and the second term minimizes the distance between the predictions of natural and adversarial data. The second term has the effect of smoothing the prediction around the input, thus increasing the distance of the input data from the boundary by removing abrupt decision boundaries around the natural input, leading to lower boundary errors and higher robust or adversarial accuracy. In addition, the regularization coefficient λ balances the importance of natural and boundary error and allows for finetuning the relative importance of the two errors during optimization.

WRN-34-10 network trained on CIFAR-10 with TRADES helps achieve adversarial accuracy of around 56.61% and normal accuracy of 84.92%. This is a gain of around 9% over PGD-AT on adversarial examples, though some decrease in natural accuracy accompanies this gain.

The PGD-AT algorithm only uses adversarial images during the training process, and this can lead to overfitting on adversarial examples. Furthermore, it was found that worst-case adversarial samples that are furthest from the original natural input might not be suitable for initial training. A curriculum-based approach was developed (CAT) [3] that trains the DNN on milder adversarial samples initially and introduces stronger and stronger adversarial samples as the training progresses. The iterations of PGD attacks determine the strength of the attack, with fewer iterations creating weaker adversarial examples and higher iterations creating stronger adversarial examples. In the CAT method, the initial training starts with adversarial images created using lower iterations, and it is increased as the training progresses. Another strategy to create weaker attacks is to stop the iteration as soon as a misclassification is achieved; this strategy is used by the friendly adversarial training (FAT) [43] method. These strategies have the additional benefit of being faster than the PGD-AT method as they require less time to generate adversarial samples. Ensemble adversarial training (EAT) [37], where adversarial samples are generated from a family of models instead of the model under training, has also been shown to be effective.

Besides improving accuracy on adversarial data, there has also been a focus on improving the training time. Adversarial training can take anywhere between 3 and 30 times longer than normal training. Free-AT [33] has been proposed as a technique to decrease the training time by reusing the gradient during the training phase; in the PGD-AT method, a training phase is followed by an attack phase to obtain adversarial data. Free-AT merges the two phases to reuse the gradient for both updating the model parameter and for obtaining the adversarial data leading to significant speedup while maintaining the same adversarial and natural accuracy as PGD-AT.

Adversarial training has proved to be a popular defense against adversarial attacks as they are among the only defenses that work consistently in the white-box setting. In spite of this popularity, there are a number of shortcomings that need to be addressed in order for it to be suitable for safety-critical systems. A major issue with AT is the lack of safety guarantees. Since autonomous vehicle systems should be provably safe and have expected behavior, and ATs do not provide these guarantees, they are not a suitable defense for use in AVS, where safety guarantees are required. Other issues with AT include a large generalization gap, with high accuracy on adversarial training samples and low accuracy on adversarial test samples. This large generalization gap limits the accuracy that can be achieved on adversarial test samples. AT-trained models also tend to be more robust against the type of attacks used during training and might not be as resilient against new attacks, though using an ensemble of attacks during training has shown to make DNNs more robust against new attacks.

6.3.4 Certified defense

The defenses mentioned earlier provide some protection against adversarial attacks on the DNN, but they do not provide any guarantees about the prediction. A failed attack in those cases does not imply that an adversarial sample does not exist; rather, it implies that a sample was not found using the existing attack methodologies. This is underscored by the fact that as defenses have evolved, so have the attack methodologies, with newer and better attacks generally thwarting previous defenses. Certified defenses have been proposed as a solution to break this cat and mouse game of evolving attacks and defenses.

A certified defense provides a guarantee or certificate that there is no adversarial example within a fixed neighborhood around the input sample. These defenses can generally be classified into two categories a) exact defenses and b) conservative defenses. Given an input image x, a neighborhood $\mathcal{S}$, and a network f, an exact defenses method generates a certificate with states if an adversarial sample is present in the $\mathcal{S}$ neighborhood or not. On the other hand, conservative defense methods generate certificates if no adversarial examples exist in the $\mathcal{S}$ neighborhood or produce no such certificate. Exact methods utilize mixed integer linear programming and satisfiability theory to create these certificates and are computationally intractable for large SOTA neural network architectures used in AVS systems.

Conservative certificate defenses, on the other hand, are more scalable, and current methods can generate certificates for large models, including state-of-the-art ImageNet models. Early efforts at conservative defense were based on bounding the Lipschitz constant of the neural network [27, 39, 12, 42]. This bound ensures that a slight change in the input can only produce a bounded change in the output and allows making assumptions about the neighborhood of the input to create certificates. Obtaining the Lipschitz constant of a neural network is intractable and usually requires some local approximations. Some methods estimate this bound by using optimization [27, 39], and other methods propagate a perturbation through the layers of the neural network to find the maximum change in the last layer to estimate the Lipschitz constant [12, 42].

A growing field of work has moved away from estimating the Lipschitz constant as it is still not effective for large architectures and has pivoted towards randomized Gaussian sampling for generating certificates. Randomized smoothing [7] is a popular technique that provides certificates for input in the L2 norm using Gaussian noise. In the randomized smoothing technique, a new smooth and robust classifier g can be constructed from a pre-existing classifier f. The new robust classifier g which predicts the most probable output for an input x perturbed by Gaussian noise with a standard deviation of σ and can be formulated as follows:

$$g(x) = \arg\max_{c \in C} \mathbb{P}(f(x+\epsilon) = c) \tag{6.14}$$
$$\text{where } \epsilon \sim \mathcal{N}(0, \sigma^2 I)$$

The standard deviation σ controls the noise of the perturbation and is a hyperparameter affecting the robustness and accuracy tradeoff, with smaller σ allowing for higher accuracy and less robustness and larger σ allowing for lower accuracy but more robustness. In the above formulation, to generate a certificate for input x using the robust classifier g, samples are generated from $\mathcal{N}(x, \sigma^2 I)$ around the input. If C_A is the most probable class with a probability P_A and C_B is the second most probable class with probability P_B. Then for $P_A > P_B$, it can be shown that the classification C_A of input x does not change in the radius R obtained using the following equation:

$$R = \frac{\sigma}{2}(\Phi^{-1}(P_A) - \Phi^{-1}(P_B)), \tag{6.15}$$

where Φ is the Gaussian CDF. Obtaining the exact probabilities P_A and P_B using sampling is not possible, so random smoothing methods obtain a lower bound on P_A and an upper bound on P_B by sampling sufficient data from $\mathcal{N}(x, \sigma^2 I)$. It has been shown that eqn 6.15 is valid when P_A and P_B are replaced with their lower and upper bounds, respectively. The above formulation has the advantage that it can work with any underlying classifier f irrespective of its complexity and type. Furthermore, the certified radius is large if the tested σ is large or if there is a larger difference between the most (P_A) and the second most probable classes (P_B). Estimating bounds of P_A and P_B is done by random sampling and rank verification hypothesis testing. As can be expected of sampling in a high dimensional input space typical of SOTA DNNs, certifying a bigger radius requires a large number of samples. For example, certifying a radius of 4 with 99.9% confidence requires 100,000 samples; even then, the radius is too small for practical applications.

The assumption that the classifier f will work well with images perturbed with Gaussian noise is not always effective and provides loose certificate bounds. Pretrained off-the-shelf DNNs are not trained on noisy images and have lower natural accuracy on images with Gaussian noise. A workaround is to use Gaussian noise during the training phase, but it can make the network less accurate compared to a network trained only on natural data. Another technique, called Denoised Smoothing [4, 32], allows the use of off-the-shelf trained networks and can provide better robustness certificates with a larger L2 radius. In this technique, sample images are generated in a similar manner to the random smoothing but are then denoised to remove the Gaussian perturbations that had been added previously. These denoised image samples can then be given as input to the off-the-shelf network to find the lower and upper bounds of P_A and P_B, respectively. The denoised smoothing classifier can be obtained by changing eqn 6.15 to include a denoising function $\mathcal{D}$ as follows:

$$g(x) = \arg\max_{c \in C} \mathbb{P}(f(\mathcal{D}(x + \epsilon)) = c) \quad (6.16)$$
$$\text{where } \epsilon \sim \mathcal{N}(0, \sigma^2 I).$$

The function $g(x)$, which is made of the composition $f \circ D$ is provably robust and can generate larger L2 certificates when compared to randomized smoothing classier. An overview of the method is shown in Figure 6.13. Again, the important distinction to be made here is that the $g(x)$, which produces a prediction after sampling from a Gaussian distribution, is the

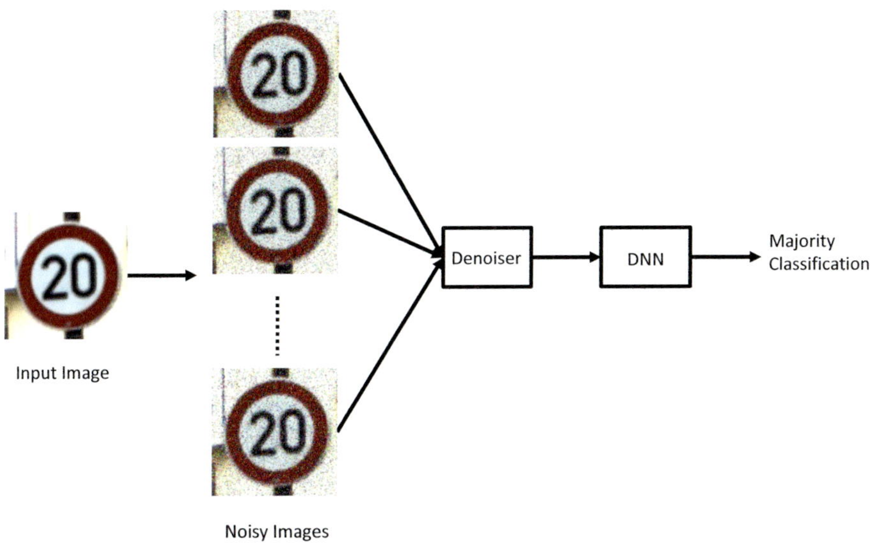

Figure 6.13 Gaussian noise is added to the input image and subsequently removed using a Denoiser. The denoised images are classified using the DNN, and the input image is assigned to the class having the highest number of samples. The composition of the denoiser and the DNN is provably robust against adversarial perturbations.

robust classifier and not the underlying function f. In practice, the effectiveness of the new denoising smoothing depends on the effectiveness of the denoiser. Earlier attempts used a DnCNN and MemNet-based denoiser, and it helped achieve a certified accuracy of 33% on ImageNet with an L2 radius of 0.5 [32]. Improvements in denoising and the introduction of various denoising diffusion probabilistic models have helped improve the adversarial accuracy to 71.1% on ImageNet with an L2 radius of 0.5 [4].

These advancements in certified defenses are impressive but still very limited as they have a small L2 radius and do not provide any certificates in the Linf norm. This deficiency can be seen in a simple example, where a slight change in hue can have a very small change in Linf but a very large change in the L2 norm. These small changes should not affect the performance of a DNN-based classifier, but since there are no practical certificate methods for the Linf norm, no guarantees can be provided for such inputs. Even though current defense methods do not have strong practical applicability, they are still an important step toward a safe and secure neural network future.

6.4 Conclusion

Throughout this chapter, we looked at an aspect of DNNs that is critical for its applicability in safety-critical applications such as autonomous vehicle systems. There are various types of adversarial attacks that can fool AVS systems; these attacks include powerful white-box attacks that require the model weight and architecture to compute the gradient or optimization-based attacks that assume unlimited access to the model. Even when white-box access is not available powerful black-box attacks can generate adversarial examples that only require a limited number of queries to the model. We also looked at attacks that transfer well to the physical world by using patches and natural phenomena such as fog. Finally, the existence of adversarial examples is universal, and all DNNs are affected by it.

Given the troubling nature of these adversarial attacks, numerous efforts have been made to protect against these attacks. Input preprocessing-based defenses provide some protection in the black-box setting. Randomizing the preprocessing steps provides even more protection and can also be used in the white-box setting. The most popular defense being used right now is adversarial training, where adversarial data is generated during training time to train a robust DNN model. AT DNNs are robust against adversarial attacks, but there is still a significant gap between the test and train accuracy. AT DNNs perform well in practice, but a major downside of the methods is they do not provide any safety guarantees. Certificate-based methods have been developed that provide a certificate stating that no adversarial attack exists in the vicinity of the input. Initial attempts to generate certificates did not scale to large DNNs, but the existing randomized smoothing-based method can scale to large networks and provide L2 certificates around the input. A major downside of these methods is that they do not provide safety guarantees in the Linf norm, and whatever guarantees are provided in the L2 norm are still very limited. Nevertheless, certificate-based methods have the promise to put an end to the constant warfare between sophisticated empirical defenses and ever-evolving adversarial attacks.

The key takeaway of this chapter is that DNNs in the current state is not suitable for safety-critical applications, but the advent of improved certified defense methods in conjugation with adversarial training will allow for more robust and safe DNNs in the future.

References

[1] Wieland Brendel, Jonas Rauber, and Matthias Bethge. Decision-based adversarial attacks: Reliable attacks against black-box machine learning models. In *International Conference on Learning Representations*, 2018.

[2] Tom B Brown, Dandelion Mané, Aurko Roy, Martín Abadi, and Justin Gilmer. Adversarial patch. *arXiv preprint arXiv:1712.09665*, 2017.

[3] Qi-Zhi Cai, Chang Liu, and Dawn Song. Curriculum adversarial training. In *Proceedings of the Twenty-Seventh International Joint Conference on Artificial Intelligence, IJCAI-18*, pages 3740–3747. International Joint Conferences on Artificial Intelligence Organization, 7 2018.

[4] Nicholas Carlini, Florian Tramer, J Zico Kolter, et al. (certified!!) adversarial robustness for free! *arXiv preprint arXiv:2206.10550*, 2022.

[5] Nicholas Carlini and David Wagner. Towards evaluating the robustness of neural networks. In *ISSP*, pages 39–57. IEEE, 2017.

[6] Jianbo Chen, Michael I Jordan, and Martin J Wainwright. A query-efficient decision-based attack. In *2020 IEEE Symposium on Security and Privacy*, pages 1277–1294. IEEE, 2020.

[7] Jeremy Cohen, Elan Rosenfeld, and Zico Kolter. Certified adversarial robustness via randomized smoothing. In *International Conference on Machine Learning*, pages 1310–1320. PMLR, 2019.

[8] Nilaksh Das, Madhuri Shanbhogue, Shang-Tse Chen, Fred Hohman, Li Chen, Michael E Kounavis, and Duen Horng Chau. Keeping the bad guys out: Protecting and vaccinating deep learning with jpeg compression. *arXiv preprint arXiv:1705.02900*, 2017.

[9] Jia Deng, Wei Dong, Richard Socher, Li-Jia Li, Kai Li, and Li Fei-Fei. Imagenet: A large-scale hierarchical image database. In *2009 IEEE Conference on Computer Vision and Pattern Recognition*, pages 248–255, 2009.

[10] Yinpeng Dong, Fangzhou Liao, Tianyu Pang, Hang Su, Jun Zhu, Xiaolin Hu, and Jianguo Li. Boosting adversarial attacks with momentum. In *Proceedings of the IEEE Conference on Computer Vision and Pattern Recognition*, pages 9185–9193, 2018.

[11] Ian Goodfellow, Jonathon Shlens, and Christian Szegedy. Explaining and harnessing adversarial examples. In *International Conference on Learning Representations*, 2015.

[12] Sven Gowal, Krishnamurthy Dvijotham, Robert Stanforth, Rudy Bunel, Chongli Qin, Jonathan Uesato, Relja Arandjelovic, Timothy Mann, and Pushmeet Kohli. On the effectiveness of interval bound propagation for training verifiably robust models. *arXiv preprint arXiv:1810.12715*, 2018.

[13] Warren He, James Wei, Xinyun Chen, Nicholas Carlini, and Dawn Song. Adversarial example defense: Ensembles of weak defenses are not strong. In *11th USENIX workshop on offensive technologies (WOOT 17)*, 2017.

[14] Susmit Jha, Sunny Raj, Steven Fernandes, Sumit K Jha, Somesh Jha, Brian Jalaian, Gunjan Verma, and Ananthram Swami. Attribution-based confidence metric for deep neural networks. In *Advances in Neural Information Processing Systems*, pages 11826–11837, 2019.

[15] Danny Karmon, Daniel Zoran, and Yoav Goldberg. Lavan: Localized and visible adversarial noise. *arXiv preprint arXiv:1801.02608*, 80:2507–2515, 10–15 Jul 2018.

[16] Alex Krizhevsky, Geoffrey Hinton, et al. Learning multiple layers of features from tiny images. 2009.

[17] Alexey Kurakin, Ian J Goodfellow, and Samy Bengio. Adversarial examples in the physical world. In *Artificial intelligence safety and security*, pages 99–112. Chapman and Hall/CRC, 2018.

[18] Yann LeCun, Léon Bottou, Yoshua Bengio, and Patrick Haffner. Gradient based learning applied to document recognition. *Proceedings of the IEEE*, 86(11):2278–2324, 1998.

[19] Fangzhou Liao, Ming Liang, Yinpeng Dong, Tianyu Pang, Xiaolin Hu, and Jun Zhu. Defense against adversarial attacks using high-level representation guided denoiser. In *Proceedings of the IEEE conference on computer vision and pattern recognition*, pages 1778–1787, 2018.

[20] Yanpei Liu, Xinyun Chen, Chang Liu, and Dawn Song. Delving into transferable adversarial examples and black-box attacks. *arXiv preprint arXiv:1611.02770*, 2016.

[21] Aleksander Madry, Aleksandar Makelov, Ludwig Schmidt, Dimitris Tsipras, and Adrian Vladu. Towards deep learning models resistant to adversarial attacks. *arXiv preprint arXiv:1706.06083*, 2017.

[22] Alberto Marchisio, Giovanni Caramia, Maurizio Martina, and Muhammad Shafique. fakeweather: Adversarial attacks for deep neural networks emulating weather conditions on the camera lens of autonomous systems. *arXiv preprint arXiv:2205.13807*, 2022.

[23] Seyed-Mohsen Moosavi-Dezfooli, Alhussein Fawzi, and Pascal Frossard. Deepfool: a simple and accurate method to fool deep neural networks. In *CVPR*, pages 2574–2582, 2016.

[24] Mesut Ozdag, Sunny Raj, Steven Fernandes, Laura L Pullum, and Sumit Kumar Jha. On the susceptibility of deep neural networks to natural perturbations. Technical report, Oak Ridge National Lab.(ORNL), Oak Ridge, TN (United States), 2019.

[25] Nicolas Papernot, Patrick McDaniel, Ian Goodfellow, Somesh Jha, Z Berkay Celik, and Ananthram Swami. Practical black-box attacks against machine learning. In *Proceedings of the 2017 ACM on Asia conference on computer and communications security*, pages 506–519, 2017.

[26] Nicolas Papernot, Patrick McDaniel, Xi Wu, Somesh Jha, and Ananthram Swami. Distillation as a defense to adversarial perturbations against deep neural networks. In *2016 IEEE Symposium on Security and Privacy (SP)*, pages 582–597. IEEE, 2016.

[27] Aditi Raghunathan, Jacob Steinhardt, and Percy S Liang. Semidefinite relaxations for certifying robustness to adversarial examples. *Advances in Neural Information Processing Systems*, 31, 2018.

[28] Sunny Raj, Laura Pullum, Arvind Ramanathan, and Sumit Kumar Jha. Satya: Defending against adversarial attacks using statistical hypothesis testing. In *International Symposium on Foundations and Practice of Security*, pages 277–292. Springer, Cham, 2017.

[29] Arvind Ramanathan, Laura Pullum, Zubir Husein, Sunny Raj, Neslisah Torosdagli, Sumanta Pattanaik, and Sumit K Jha. Adversarial attacks on

computer vision algorithms using natural perturbations. In *2017 Tenth International Conference on Contemporary Computing (IC3)*, pages 1–6. IEEE, 2017.

[30] Olaf Ronneberger, Philipp Fischer, and Thomas Brox. U-net: Convolutional networks for biomedical image segmentation. In *International Conference on Medical image computing and computer-assisted intervention*, pages 234–241. Springer, 2015.

[31] Olga Russakovsky, Jia Deng, Hao Su, Jonathan Krause, Sanjeev Satheesh, Sean Ma, Zhiheng Huang, Andrej Karpathy, Aditya Khosla, Michael Bernstein, Alexander C. Berg, and Li Fei-Fei. ImageNet Large Scale Visual Recognition Challenge. *International Journal of Computer Vision (IJCV)*, 115(3):211–252, 2015.

[32] Hadi Salman, Mingjie Sun, Greg Yang, Ashish Kapoor, and J Zico Kolter. Denoised smoothing: A provable defense for pretrained classifiers. *Advances in Neural Information Processing Systems*, 33:21945–21957, 2020.

[33] Ali Shafahi, Mahyar Najibi, Mohammad Amin Ghiasi, Zheng Xu, John Dickerson, Christoph Studer, Larry S Davis, Gavin Taylor, and Tom Goldstein. Adversarial training for free! *Advances in Neural Information Processing Systems*, 32, 2019.

[34] Johannes Stallkamp, Marc Schlipsing, Jan Salmen, and Christian Igel. Man vs. computer: Benchmarking machine learning algorithms for traffic sign recognition. *Neural networks*, 32:323–332, 2012.

[35] Mukund Sundararajan, Ankur Taly, and Qiqi Yan. Axiomatic attribution for deep networks. In *ICML*, pages 3319–3328. JMLR. org, 2017.

[36] Christian Szegedy, Wojciech Zaremba, Ilya Sutskever, Joan Bruna, Dumitru Erhan, Ian Goodfellow, and Rob Fergus. Intriguing properties of neural networks. In *2nd International Conference on Learning Representations, ICLR 2014*, 2014.

[37] Florian Tramèr, Alexey Kurakin, Nicolas Papernot, Ian Goodfellow, Dan Boneh, and Patrick McDaniel. Ensemble adversarial training: Attacks and defenses. In *International Conference on Learning Representations*, 2018.

[38] Pascal Vincent, Hugo Larochelle, Yoshua Bengio, and Pierre-Antoine Manzagol. Extracting and composing robust features with denoising autoencoders. In *Proceedings of the 25th international conference on Machine learning*, pages 1096–1103, 2008.

[39] Eric Wong and Zico Kolter. Provable defenses against adversarial examples via the convex outer adversarial polytope. In *International Conference on Machine Learning*, pages 5286–5295. PMLR, 2018.

[40] Cihang Xie, Jianyu Wang, Zhishuai Zhang, Zhou Ren, and Alan Yuille. Mitigating adversarial effects through randomization. In *International Conference on Learning Representations*, 2018.

[41] Weilin Xu, David Evans, and Yanjun Qi. Feature squeezing: Detecting adversarial examples in deep neural networks. *arXiv preprint arXiv:1704.01155*, 2017.

[42] Huan Zhang, Tsui-Wei Weng, Pin-Yu Chen, Cho-Jui Hsieh, and Luca Daniel. Efficient neural network robustness certification with general activation functions. *Advances in neural information processing systems*, 31, 2018.

[43] Jingfeng Zhang, Xilie Xu, Bo Han, Gang Niu, Lizhen Cui, Masashi Sugiyama, and Mohan Kankanhalli. Attacks which do not kill training make adversarial learning stronger. In Hal DaumÃĪ III and Aarti Singh, editors, *Proceedings of the 37th International Conference on Machine Learning*, volume 119 of *Proceedings of Machine Learning Research*, pages 11278–11287. PMLR, 13–18 Jul 2020.

Authors' Biography

Sunny Raj is an Assistant Professor in the Computer Science and Engineering department at Oakland University. Dr. Raj received his Ph.D. in Computer Science from the University of Central Florida. Part of his research effort is devoted to increasing trust in machine learning algorithms by improving explainability and by calibrating the output of these systems. Another focus area is to design ultra-energy-efficient ML systems to allow these systems' broader use and applicability. Dr. Raj has papers in top machine learning and design automation venues like NeurIPS and ICCAD. His work has led to two best paper awards.

7

Multi-sensor Fusion for Multi-target Detection and Tracking

Li Liu, Jiang Tian, Zhongchao Shi, Jianping Fan, and Yong Rui

Lenovo Research, China
liuli16, tianjiang1, shizc2, jfan1, yongrui@lenovo.com

Abstract

Sensors (LiDAR, camera, radar, etc.) not only intelligentize vehicles in autonomous driving systems but also can be set up at the intersection, on both sides of the road, to monitor all traffic participants on the road in real time. Different range sensors have different advantages and disadvantages. Multi-target detection and tracking, a higher level task which provides decision-making data for intelligent transportation, can be performed by a fusion system combining multiple sensors. Such fusion constructs a more secure and complete collaborative sensing environment. In this chapter, a survey on multi-sensor fusion for multi-target detection and tracking will be conducted, which includes homogeneous and heterogeneous sensors' information fusion. The former helps on long-distance tracking across multi-camera, and the latter reduces uncertainty through the complementary advantages of different sensors.

Keywords: Multi-sensor fusion, autonomous driving, intelligent transportation, 3D object detection, multi-target tracking.

7.1 Introduction

The level of vehicle autonomy has been greatly enhanced because of tremendous advancements in the autonomous driving system. Sensors, as the primary means of replacing people in perceiving the external environment, have sparked a lot of interests in the field of autonomous driving, from scientific research to industry.

To construct a trustworthy environment perception, the autonomous driving system employs a variety of sensors, which perform imaging, detection, tracking, localization, and other operations [1, 2]. It allows autonomous vehicles to recognize and predict obstacles in the driving environment, as well as to move without inflicting casualties. In addition to vehicles, sensors are also deployed on the roadside to monitor the status of the road, cars, pedestrians, and other traffic participants in real time, which can reduce road accidents and congestion, reallocate driving time, and achieve safe and smooth driving by communicating with vehicle controllers. Figure 7.1 depicts a typical roadside sensor distribution and an autonomous vehicle sensor configuration, including Light Detection and Ranging (LiDAR), radar, and cameras. However, a single sensor can only provide limited information and is influenced by its own properties [3, 4]. For example, a camera with rich texture information will be influenced by light. Millimeter-wave (MMW) radar can compensate for the camera's light sensitivity but cannot distinguish the precise shapes of obstacles. When sensors' data is handled separately and in isolation, the internal connection between the data is broken, and some essential environmental data is lost. In addition, a wide range of weather, road, and traffic circumstances must be considered to accomplish true autonomous driving. The sensing system must offer accurate results under a variety of situations, which is challenging to achieve with a single sensor. Multi-sensor fusion technology improves the dependability and resilience of the overall system by increasing information transmission and expanding the time and space coverage of different sensors. The core principle of multi-sensor fusion is similar to the human brain's technique of comprehensive information processing. Several sensors are processed in multi-level and multi-space, and a consistent interpretation of the environment is achieved.

In most autonomous driving systems, the combination of cameras, LiDAR, and MMW radar forms the core set of sensors that enables imaging, detection, tracking, and sensing of the driving environment [5]. In recent years, multi-sensor fusion technologies for autonomous driving have improved quickly. Particularly in the object perception module, multi-target detection and tracking, a higher level task that offers decision-making data for intelligent transportation can be improved by a fusion system combining multiple sensors [6]. As a result, multi-sensor fusion algorithms are now practically essential for all autonomous driving perception modules in order

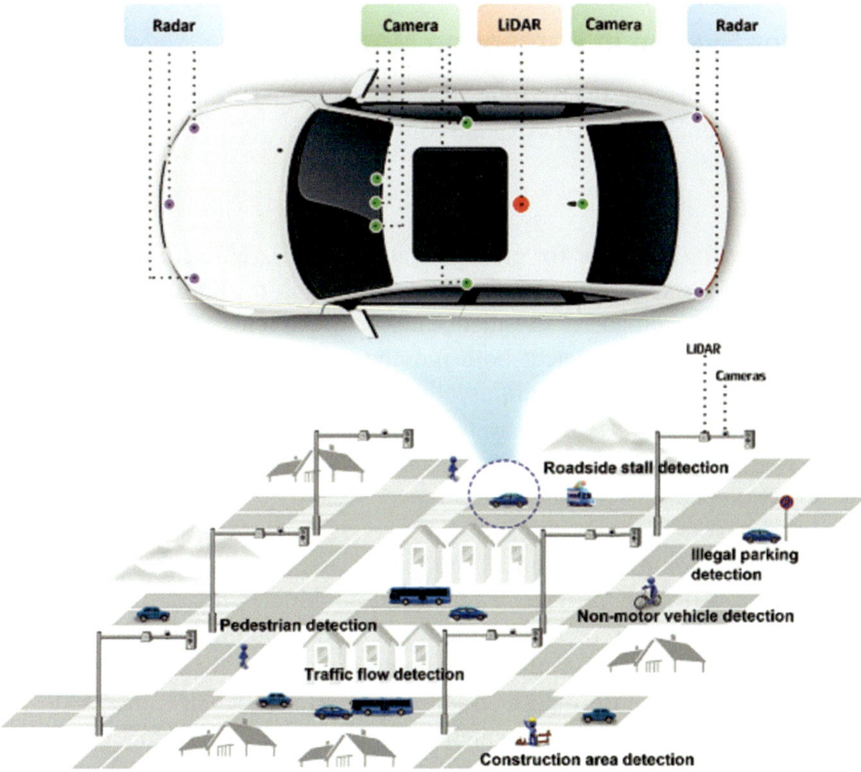

Figure 7.1 Vehicle side and roadside sensor distributions.

to overcome the limits of a single sensor and enhance overall efficiency and dependability.

This chapter mainly focuses on the multi-sensor fusion algorithms in the perception module, particularly in multi-target detection and tracking. The second section introduces the three main sensors in the vision module: camera, LiDAR, and MMW radar, as well as their application situations, benefits, and drawbacks. Datasets that are frequently utilized in the field of autonomous driving are introduced in Section 7.3. Section 7.4 discusses the main issues that multi-sensor fusion algorithms confront, evaluates and summarizes previous papers, and offers some current solutions. Section 7.5 goes over how to apply multi-sensor fusion algorithms for various visual tasks depending on the task's characteristics and suggests potential remedies. Section 7.6 outlines some conclusions with current multi-sensor fusion methods.

7.2 Sensor System in Perception Module

This section focuses on the introduction of three major sensors in the perception module: camera, LiDAR, and MMW radar.

7.2.1 Camera

Cameras are the finest sensor for giving an autonomous vehicle an accurate visual image of its surroundings. Cameras have high-resolution and rich texture information compared to other sensors. Meanwhile, they are relatively inexpensive and can detect both moving and static obstacles within their field of view. These capabilities allow the perception system of the vehicle to identify road signs, traffic lights, road lane markings, and barriers. Therefore, in the realm of autonomous driving, cameras are the most extensively equipped sensor. Pure vision is even used by some manufacturers to accomplish autonomous driving.

Cameras are often mounted on all four sides of self-driving cars – front, rear, right, and left – to provide a 360° vision. Monocular or binocular cameras, or a combination of both, are always used in autonomous driving. A monocular camera system, as the name implies, only uses a single camera. Because they lack depth information, traditional RGB monocular cameras are more limited than stereo cameras, which can estimate depth information via acquiring two cameras' disparity. In self-driving automobiles, two cameras are usually configured side by side to form a binocular camera system. Figure 7.2 shows eight cameras distribution of Tesla Model Y [7].

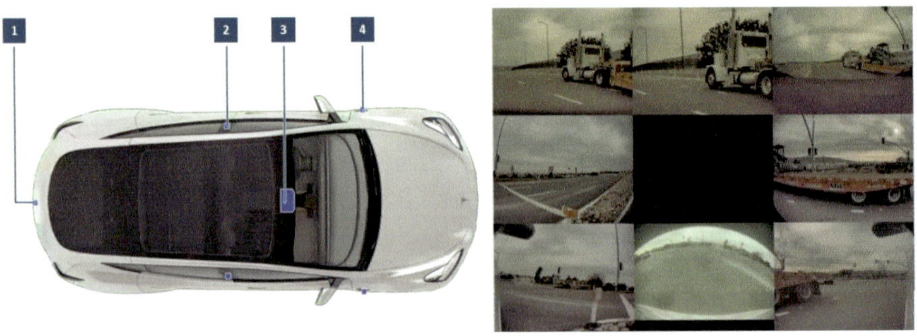

Figure 7.2 Tesla Model Y eight cameras' locations and corresponding captured images. 1 indicates a camera mounted above the rear license plate; 2 indicates two cameras mounted in door pillars; 3 indicates three cameras mounted to the windshield; 4 indicates two cameras mounted to fenders (adopted from Tesla's promotional material available at [7, 8]).

Camera vision, on the other hand, is susceptible to illumination and environmental conditions. Poor environmental circumstances, such as high contrast, glare, dim light, fog, snowfall, and heavy rain, can result in lower-quality images or videos, making object detection algorithms less accurate. Thus, image data are often fused with other sensor data such as radar and LiDAR data, to generate reliable and accurate environment perception.

7.2.2 LiDAR

LiDAR is a mapping technology that can emit laser pulses and detect reflections. The interval taken between emission and receiving of the light pulse enables the estimation of position [9]. As LiDAR scans its surroundings, it creates a point cloud, which is a 3D representation of the environment, as shown in Figure 7.3.

Mechanical LiDAR and solid-state LiDAR (SSL) are two types of LiDAR sensors. The angle of laser emission is controlled by spinning lenses in mechanical LiDAR. Rotating the lens enables a 360° horizontal FoV that covers the vehicle's surroundings. Instead of spinning lenses, SSL depends on electrical components to direct the laser beams, preventing mechanical failure. SSLs have gained popularity in recent years due to their durability, reliability, and generally lower cost than mechanical LiDAR. However, the

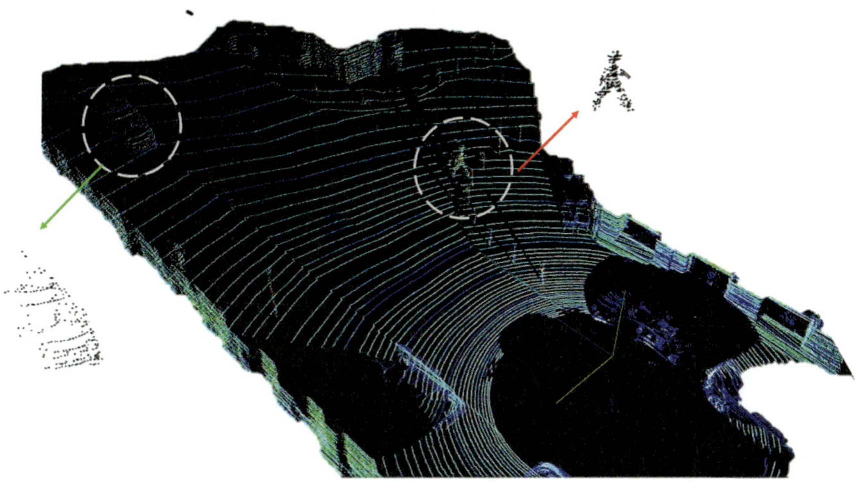

Figure 7.3 Point cloud captured by Velodyne HDL-64E LiDAR. The circled area in the image represents the point signal.

horizontal FoV is smaller and limited, usually 120° or less, and the accuracy is relatively lower. Mechanical LiDAR is currently more extensively utilized in the autonomous system because of its wider field of vision, longer detection range, and greater depth perception, which may provide dependable and precise day and night perceptions.

LiDAR is always mounted on top of autonomous vehicles or at crossroads to collect and measure millions of data points in real time, delivering relatively high-resolution 3D data. As LiDAR costs fall, it is becoming a more fundamental part of the perception system, assisting in multi-target detection and tracking. However, while LiDAR can measure the environment more accurately in 3D than camera systems, it cannot offer color information of the surrounding environment and has a limited visual resolution. LiDAR is fundamentally a distance measurement sensor, regardless of how accurate it is or how dense the resulting point cloud is. At the same time, air absorption, scattering, and refraction impact the detection beam of LiDAR. The detection range and accuracy of LiDAR are decreased in severe weather such as rain, fog, ice, and snow.

7.2.3 MMW radar

MMW radar is a type of radar that employs electromagnetic waves with short wavelengths. MMW radar measures the range, speed, and azimuth angle of many targets at the same time by transmitting electromagnetic waves via the antenna and receiving electromagnetic waves reflected from the target. MMW radar has a frequency range of 30–300 GHz and a wavelength range of 1–10 mm, which is between microwave and centimeter wave and has several advantages over microwave and photoelectric radar. In fact, the system components required to process MMW signals, such as antennas, are small. Short wavelengths also provide the advantage of high precision. Motion as small as a fraction of a millimeter can be detected by MMW systems operating at 77 GHz (corresponding wavelengths about 4 mm).

Radar in autonomous vehicles mainly operates at the frequencies of 24, 74, 77, and 79 GHz, corresponding to short-range Radars (SRR), medium-range radars (MRR), and long-range radars (LRR). They have slightly different functions: SRR technology enables blind-spot monitoring, lane change assistance, and parking assistance in autonomous vehicles. MRR sensors are used for cross-traffic alerts when obstacle detection is in the range of 100–150 m. The automatic distance control and brake assistance are supported by LRR sensors [10].

MMW radar has high-accuracy velocity estimates based on the Doppler effect, as well as reliable and accurate obstacle detection during the day and night because it is not affected by light or bad weather, and it is far less expensive to manufacture than LiDAR. Radar sensors, on the other hand, have a limited resolution, are unable to distinguish object outlines, and lack height information, making them unsuitable for object recognition, as shown in Figure 7.4.

7.2.4 4D MMW radar

4D MMW radar, also known as imaging radar, improves the detection and interpretation of target elevation compared with standard radar, which only provides target velocity, distance, and azimuth angle information [12]. 4D high-precision imaging MMW radar considerably improves detection resolution, the field of view, and detection range by adding virtual channels. It has higher resolution and enhanced signal processing capabilities allowing it to better detect and track objects that are moving, barely moving, or stopped. At the same time, because elevation information is included, 4D MMW radar can produce 3D point clouds similar to LiDAR, allowing for a wide range of perception enhancement applications. The emergence of 4D MMW radar has targeted the performance deficiencies of traditional MMW radar. It not

Figure 7.4 Radar point signal and corresponding image segmentation features in RadarScenes dataset [11, 61].

Figure 7.5 4D imaging radar signals and corresponding images (adopted from Arbe's promotional material available at [13]).

only adds 3D information to autonomous driving but also improves detection accuracy, sensitivity, resolution, and performance. As shown in Figure 7.5, the red box depicts the same object in the image as well as in the imaging radar [13]. It can be seen that the radar signal contains elevation information and the resolution is much higher than conventional radar.

In 2019, an Israeli company proposed the first 4D MMW radar system. In early 2020, Waymo announced the launch of 4D MMW radar in the fifth-generation autonomous driving perception kit. When working in tough weather circumstances such as rain or fog, 4D MMW radar, like regular radar, will not deviate much. Cost-effectiveness is another benefit of MMW radar. 4D MMW radar, on the other hand, is still in its early stages of development and testing, and it will take a long time to meet vehicle-level standards.

7.2.5 Conclusion

Cameras, LiDARs, and MMW Radars serve as the eyes of the perception module, which are the basis for ensuring the high-precision, high-determinism, and high-reliability operation of autonomous driving systems. However, as mentioned above, each sensor has its own set of strengths and shortcomings, which are displayed in Table 7.1. Cameras, for example, may gather complex environmental information at a high frame rate and high resolution, but they are passive sensors that do not emit light. The brightness of the environment has a significant impact on image quality. In tough environments, the difficulty of completing the perceptual task will be substantially increased. LiDAR is an active sensor that provides accurate depth accuracy, has good anti-interference capabilities, and is unaffected by light. However, LiDAR data is sparse and disorganized, and the monochromatic qualities of laser make color and texture information impossible to obtain. In theory, combining these two sensor types can make tasks like high-precision detection and recognition easier to accomplish in all-day and multi-environment settings. Although Tesla adopted the pure vision solution Autopilot to complete the automatic driving of the car, and Baidu proposed the pure vision L4 urban road solution Apollo Lite, most car manufacturers, including Audi, Mercedes-Benz, Waymo, Lixiang, NIO, and others, continue to use the multi-sensor combination mode to improve the accuracy and redundancy of the autonomous driving system. As a result, multi-sensor fusion technology-based perception has been a research focus in recent years.

7.3 Datasets

There are a lot of datasets linked to autonomous driving perception. A comparison of some datasets is shown in Table 7.2. KITTI [58], nuScenes [59], and Waymo [60] are the most popular three datasets.

Karlsruhe Institute of Technology in Germany and Toyota American Institute of Technology collaborated to create the KITTI dataset. It includes one LiDAR point cloud data, images from two gray cameras, and two color cameras. KITTI gathered 7481 training images and 7518 test images, as well as point clouds for each. Scenes captured include urban, rural, and highways. The original dataset is divided into five categories: road, city, residential,

Table 7.1 Camera, radar, and LiDAR pros and cons

Sensor	Advantages	Disadvantages	Distance	Cost (USD)
Camera	1. Color and texture information 2. High frame rate	1. No depth information 2. Susceptible to weather and light conditions	~250 m (rely on lens)	15
Radar	1. Applicable for all weather conditions 2. .Velocity and depth information	1. Low resolution 2. Generating false alarms	5m–200 m	150
LiDAR	1. Accurate depth 2. Less affected by external illumination	1. High price 2. Weather susceptible	~250 m	1000–80,000

Table 7.2 Datasets comparison: Waymo Open Datasets, NuScenes and KITTI for LiDAR; RadarScene and TJ4DRadSet for Radar

Dataset	**Environment**	**Road**	**Camera**	**LiDAR**	**Radar**	**Annotated frames**	**Annotated objects**
Waymo	Rain Night Dusk	Urban Suburban	$5\times$HD cameras	$5\times$LiDAR	-	230k(LiDAR)	12M(3D boxes)
nuScenes	Rain Night Dusk	Urban	$6\times$cameras	$1\times$Velodyne HDL-32E	$5\times$long range Radars	40k(LiDAR)	1.4M(3D boxes)
KITTI	-	Urban Highway Suburban	$2\times$gray cameras $2\times$RGB cameras	$1\times$Velodyne HDL-64E	-	15k(LiDAR)	200K(3D boxes)
RadarScene	Rain	Urban Suburban	$1\times$Camera	-	$4\times$Radar	-(Radar)	7.5k(point-wise)
TJ4DRadSet	Dim Dark	Urban	$1\times$Camera	$1\times$ LiDAR	$1\times$4D Radar	8k(Radar)	30k(3D boxes)

campus, and person. Labels are grouped into eight categories for 3D object detection.

nuScenes data collection takes place in Boston and Singapore, which delivers a complete set of sensor data for self-driving cars with one spinning 32-beam LiDAR, five long-range radar sensors, and six cameras. Its 3D boxes annotation number is seven times higher than that of KITTI dataset. Meanwhile, it has 1000 driving scenes, 700 for training, 150 for validation, and 150 for testing. nuScenes delivers a 360-degree camera field of view, allowing a full-scale perception of the surrounding area, unlike KITTI and Waymo, which only provide partial perspectives. In terms of data, nuScenes offers 2D and 3D object annotation, point cloud segmentation, high-precision maps, and more, with a total of 1.4 million image frames, 400,000 frames of LiDAR point cloud, 23 object categories, and 1.4 million 3D boxes. The class-imbalance problem is more serious in nuScenes where 23 types of objects have been labeled. In nuScenes, the largest object named imbalance is 1:10k; however, in KITTI it is only 1:36.

Waymo Open Dataset (WOD) is one of the largest open datasets frequently used for autonomous driving benchmark, which was made available by Google's self-driving car company Waymo in 2020. Five high-resolution cameras and five LiDAR sensors were used to acquire the Waymo Open Dataset. There are 150 scenarios for testing, 202 scenarios for validation, and 798 scenarios for training with each scene lasts for 20 seconds. There are about 12 million 2D box annotations and 12 million 3D box annotations in WOD. Meanwhile, there are around 250k camera image trajectories and about 113k LiDAR object trajectories.

MMW radar is reasonably priced and capable of measuring both distance and speed. In addition, it can operate in bad weather. As a result, MMW radar is considerably more common in self-driving automobiles than LiDAR [56]. However, the contour of the object cannot be identified since the resolution of a typical MMW radar signal is significantly lower than that of LiDAR, and it is challenging to increase the accuracy of the radar-based perception algorithm. Therefore, radar datasets [61]–63] are currently substantially smaller than LiDAR datasets. Among KITTI, nuScenes, and WOD, only nuScenes provides MMW radar signals, and the provided radar signals are quite sparse. This section will introduce the MMW radar datasets, including the high-resolution 4D imaging radar datasets, which may be useful in the later development of radar-based fusion algorithms.

RadarScenes [61] is equipped with four 77-GHZ radar sensors and one camera, including multiple real driving scenarios as towns, tunnels, and countryside. RadarScenes dataset contains around 7500 annotated objects and the corresponding track-ids. It is more than 4 hours and 100 km of driving. The annotated data has 11 categories. In addition, the author combined 11 categories into five groups (car, large vehicle, two-wheeler, pedestrian, and pedestrian group) based on the similarity between each category. Due to the resolution, it is challenging for radar to distinguish between individual pedestrians and pedestrian groups, and hence the dataset contains two classes (pedestrian and pedestrian group) that both represent pedestrian entities.

TJ4DRadSet [64], equipped with a 4D imaging radar, a camera, and a LiDAR, has 40K synchronized frames of data, in which 7757 frames are annotated with about 30K high-quality 3D bounding boxes and track-ids. Various road situations are covered by TJ4DRadSet, including elevated roads, complicated intersections, one-way roads, and urban roads. It also contains a variety of lighting situations, such as dim light, bright light, and regular light. This dataset's high-resolution 4D radar signals can be used to create 3D perception algorithms based on 4D radar.

7.4 Multi-sensor Fusion Challenges and Solutions

7.4.1 Challenges

Although multi-sensor fusion has several benefits, the fundamental problem for fusion algorithms is to properly fuse sensor data from multiple time periods, different types, and diverse perspectives. Challenges that multi-sensor fusion algorithms tackle are listed as follows.

Challenge 1. Different Viewing Angles of the Sensor

Sensors, based on different principles, are installed in various locations to collect data. As a result, the information collected by various sensors has a completely distinct perspective. Using cameras and LiDAR as examples, the information obtained by a camera is based on the principle of "pinhole imaging," where information is obtained from a viewing cone, whereas the information obtained by the LiDAR is obtained in the real 3D world, retaining the 3D information of objects, which can be projected to a variety of views. As a result, the representation of the same object by cameras and LiDAR is vastly different. For multi-sensor fusion, figuring out how to fuse data from many perspectives and how to achieve unified processing of data from several perspectives is a challenge.

Challenge 2. Different Data Representations

Point cloud and MMW radar data are irregular, sparse, and disordered, whereas camera data is regular, dense, and ordered. Meanwhile, the resolution of MMW radar data is much lower than cameras and LiDAR. It's difficult to match these three types of data precisely, whether in the original input space or the feature space.

Challenge 3. Difficulty of Information Fusion Implementation

LiDAR, camera, and MMW radar data are complementary in principle. However, there is no best strategy for multi-sensor information fusion in different applications. At this stage, fusion algorithm designs for various tasks are highly varied. In order to address this issue, section 4 will go over each task in depth.

Challenge 4. The Time of the Sensors Is Not Synchronized

It's difficult to ensure that multi-sensors collect data at the same time. Assume a car travels at 120km/h on highway. The frequency of camera collection is 30 frames per second (FPS), while the frequency of LiDAR collection is 10 FPS. After obtaining timestamps and constructing the closest match, the largest time gap between the laser and the camera's latest data is 16ms if the collection is totally independent, and the car can go forward about 0.5m, resulting in great changes in the collected surroundings. The scene gap created by the time difference will make data and feature alignment even more difficult.

Challenge 5. Lack of High-Quality Datasets

Multi-sensor fusion algorithms based on deep learning are dominant nowadays. However, the amount of high-quality, publicly available multi-modality datasets, particularly for MMW radar, is limited. Meanwhile, the existing datasets suffer from small size, class imbalance, and mislabeling problems. This makes multi-sensor fusion methods difficult to develop.

7.4.2 Solutions

The multi-sensor fusion method is still in its development stage. Most works focus on tackling problems 1, 2, and 3 using time-synchronized datasets like KITTI, nuScenes, and Waymo Open Datasets. Meanwhile, different tasks are developed quite differently for challenge 3, which will be introduced and discussed in detail in the next section. This section mostly covers the solutions to challenges 1 and 2. Solutions for challenge 4 are also provided in this section. Solutions for inadequacy of datasets, which still necessitates the ongoing efforts of professional collectors, will not be covered here.

7.4.2.1 Solutions for Challenges 1 and 2

Existing solutions can be grouped into three groups to address the challenges posed by diverse views and varied data types as follows.

1. Data augmentation. Because multi-sensors offer data from many perspectives and dimensions of the target, this method enhances the original data using multi-modality data fusion and then extracts features from the enhanced data.
2. Feature-level fusion. This method gathers target features from various sensor inputs and then fuses them to perform the classification and tracking tasks. Both global feature fusion and ROI feature fusion are possible when combining features.
3. Object-level fusion. This type of method is making preliminary position, categories, or other properties of objects using a single or multiple sensor data, then using a specific fusion strategy to comprehensively combine multiple object-level information.

These three types of fusions are then described, along with some typical examples. Many works employ more than one feature fusion strategies.

7.4.2.1.1 Data Augmentation

Data augmentation methods improve the final task performance by directly enhancing the images or point cloud data. This sort of method is separated into two types based on enhancing strategies. The first is a sensor calibration-based data augmentation approach. These algorithms calibrate the internal and exterior parameters of numerous sensors before performing spatial alignment, which means that the same target detected by several sensors corresponds to a single coordinate system. The data is then enhanced at the original data level from the same perspective to make the following tasks easier to complete. The second way is based on generating data enhancement, such as employing deep learning algorithms to directly generate data in the same perspective.

Data augmentation based on sensor calibration

In autonomous driving systems, sensor calibration is one of the most basic modules for sensor data processing. After the sensor detects the relative position of known features, calibration can be used to acquire the absolute coordinates of the relative position in real-world coordinates [14]. To achieve data augmentation, it is also possible to convert multi-sensor data to the same coordinate system and fuse multiple perspective data gathered by many sensors. Typical structure is shown in Figure 7.6. Because correct spatial alignment can only be obtained by accurate calibration, it is difficult to align data precisely just by utilizing calibration. As a result, works based solely on calibration usually add information to the original data, such as employing concatenation after spatial alignment, making it easier for further algorithms to extract multi-dimensional data.

PointPainting [15], a typical calibration-based data augmentation algorithm, projects LiDAR points into image plane using sensor internal and external parameters from public datasets, and then fuses the projected point cloud with image semantic segmentation results by concatenating the class score to each point. Any network that is based on LiDAR point can utilize the painted point cloud data. By combining LiDAR point clouds and camera images, a deep learning method for road detection is proposed in [16]. The LiDAR point cloud is converted into three 2D images with the same spatial resolution as the camera images. The input LiDAR and camera tensors are then simply concatenated in the depth dimension, acquiring a 6-channel tensor. Using the deep learning architecture, the augmented data can be used immediately for road detection. The authors [17] converts the LiDAR point cloud to the image plane, and converts the point cloud information into pixel-level depth information through upsampling, and the depth information is concatenated with the red, green, and blue data and fed into the deep CNN.

The data augmentation based on multi-sensor calibration has a clear physical meaning. It is convenient for the subsequent algorithm construction to enhance the amount of data information. Simultaneously, the calibration algorithm has been developed for a long time, and there are a variety of toolkits available. Fine calibration of multi-sensors, on the other hand, still needs a significant amount of efforts. The calibration effort for roadside cameras on city-scale road net or a highway with a length of thousands of kilometers is enormous. Even if the calibration is done, the expense of maintenance is prohibitively expensive. As a result, the majority of the work that relies on calibration is to complete the viewing angle alignment, followed by compensation for the deviation caused by calibration.

Data augmentation based on generation

In order to obtain a unified expression of surrounding features, generation-based data augmentation is to generate data directly using deep

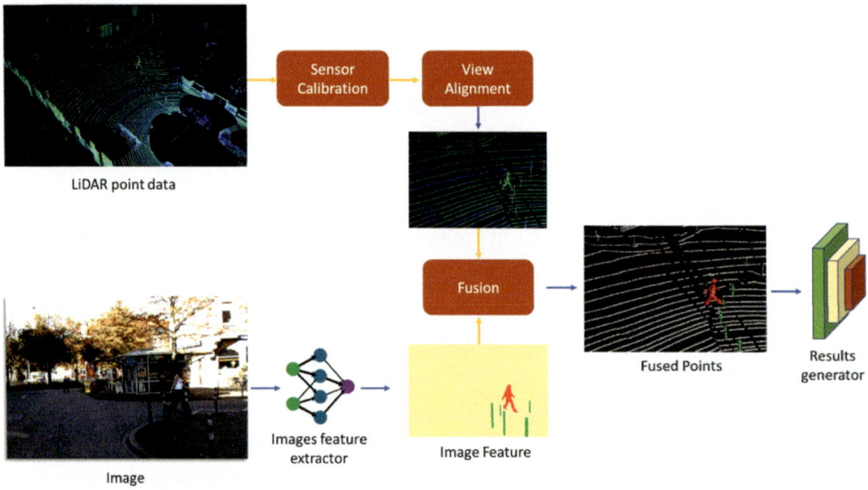

Figure 7.6 Typical sensor calibration-based data augmentation structure.

learning or other approaches. This type of algorithm is often more complex than that relying on calibration. The methods for expressing enhanced data are also diversified.

BEVFormer [18] is a typical generation-based data augmentation algorithm. It uses the transformer (spatial) and a temporal (time-domain) structures to fuse images from multi-cameras and create a bird's-eye view (BEV) feature that contains information from various fields of view. BEVFormer employs queries to identify geographical and temporal domains and generates a fused feature of the surrounding environment that can support many autonomous driving perception tasks. BEVFormer does not employ LiDAR or MMW radar data, but it does combine multi-camera images into a BEV. Both LiDAR and MMW radar point cloud data can be translated to a BEV, and the features can be mixed and matched to fuse multiple types of sensor data. PI-RCNN [19] fuses image features on LiDAR point clouds by proposing the Point-based Attentive Cont-conv Fusion (PACF) module. The fusion process is similar to PointPainting. The difference is that PointPainting converts the perspective using the public dataset's internal and external sensor settings before immediately fusing the data, while PI-RCNN not only uses the camera's calibration parameters to transform the point cloud to the image perspective, but also employs continuous convolution, Point-Pooling, and Attentive Aggregation operation to generate more expressive fused data.

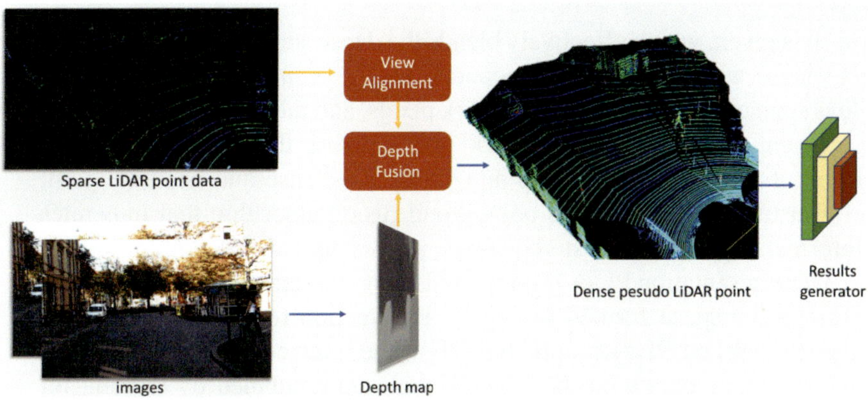

Figure 7.7 Typical generation-based data augmentation structure.

There's a subset of works that creates pseudo-point clouds from raw images and then fuses sparse LiDAR data to create dense point clouds, enhancing the performance of subsequent tasks, as shown in Figure 7.7. The development of a pseudo-point cloud [20–22] from raw images needs the depth information, which is also the primary source of mistake. The original pseudo-point cloud production relies solely on images, with depth maps calculated using stereo images. LiDAR is commonly used to guide and update the generated pseudo-point cloud by correcting the depth map, thereby making the dense point cloud more accurate. The authors [23] proposes a detection framework sparse fuse dense (SFD) that integrates the image and LiDAR point representation. One of the framework's branches transforms LiDAR points to the image plane to gain sparse depth information, and then utilizes a depth completion network to generate dense depth map, which transforms the 2D image into a 3D colorful pseudo-point cloud (pseudo-point painted with RGB). KDA3D [24] develops an image-based density augmentation approach to increase the sparsity and inhomogeneity of LiDAR point clouds. Some key points from the original point cloud are selected based on the LiDAR foreground scores and their distances, and then projected onto the image plane to obtain key pixels and extended pixels using k-means. Key and extended pixels are back-projected into 3D space to generate target pseudo-point cloud, resulting in dense point cloud data and improved subsequent detection accuracy. Pseudo-LiDAR++ (PL++) [25] uses inexpensive but sparse LiDAR signals (e.g., four beams) to reduce stereo visual depth estimation bias. This work starts by generating raw pseudo-point clouds from stereo image pairs. Following that, a graph-based depth correction (GDC) approach is given, which effectively blends the dense stereo depth that renders object shapes with the sparse but accurate LiDAR depth information. GDC employs sparse LiDAR point as landmark pixels, and then diffuses the precise measurement values of the landmark pixels across the whole depth map to correct the original depth map and produce a dense and accurate point cloud. Frustum fusion [26] is a point cloud fusion algorithm that integrates 3D information taken from stereo image pairs with LiDAR points. First, a dense 3D representation PL++ is computed using the stereo image pairs. Then frustums in 3D space formed by the RGB detection results are calculated, and the LiDAR and PL++ points that fall at the intersection are discovered. Finally, the PL++ data with the LiDAR point are combined by looking for points in the PL++ that are at the intersection of frustums but far from the LiDAR. As a result, the original LiDAR data is better retained to benefit following tasks.

Data augmentation based on generation can fuse multi-dimensional data to generate a coherent representation of the environment, making it a prominent topic in multi-sensor fusion research.

7.4.2.1.2 Feature-level fusion

Feature-level fusion fuses multi-target features taken from sensor data, such as target shapes, textures, time-frequency features, and color distributions from images or position features from MMW radar and LiDAR. Traditional machine learning approaches require extracting standard features, such as utilizing the filter like HAAR [27], and then classifying these features using SVM, Adaboost, and other methods [28–30]. Methods based on deep learning, on the other hand, have better recognition accuracy and require no manual feature extraction. Instead, the feature fusion mechanism is built into the neural network architecture [31–33]. As a result, the majority of work uses deep learning in feature-level fusion. Because of the huge differences in resolution and amount of distinct sensor data types, feature-level fusion based on deep learning usually uses multi-branch processing to extract features from different sensors before fusing them. The fused features are processed uniformly by the succeeding network when utilizing the methods like concatenation and element-wise operator to integrate numerous features, as shown in Figure 7.8.

Pointfusion [34] is a standard feature-level fusion approach, which fuses images with LiDAR point cloud. A CNN and PointNet extract the features individually for the two types of data. The different features are then integrated using concatenation. The integrated features can directly regress box corner positions, or generate spatial anchors to predict multiple 3D box hypotheses and their confidence levels. MMF [35] leverages multiple related tasks (2D detection, ground estimation, and depth completion) for accurate multi-sensor 3D object detection. The framework consists of two backbone branches that extract image and LiDAR BEV feature maps, respectively. Under different scales, by using a continuous fusion layer to project the image feature map into the BEV space, the LiDAR BEV feature and the image feature are combined via element-wise summation. EPNet [36] proposes a two-stream RPN. Two-stream RPN consists of LiDAR stream and image stream, which produce point features and semantic image features, respectively. At various scales, the LI-Fusion module is proposed to improve point features with related semantic image features, resulting in more discriminative feature representations. The authors [37] provides a new perception-aware multi-sensor fusion scheme (PMF), which can achieve the ideal semantic segmentation under unfavorable lighting (such as a dark night) with extremely sparse point clouds. The PMF approach uses a two-stream network to extract features from multi-modality input and then fuses them using multiple residual-based fusion (RF) blocks. RF block first concatenates image and LiDAR features, which are processed by some convolution layers to generate the fused features. Finally, the fused features are solely used as a complementation to the LiDAR features in the fusion module. The LiDAR features remain unchanged, decreasing the impact of image data caused by light and weather conditions. The authors [38] proposes a multi-modality fusion method for object detection in severe weather including fog, snow, and heavy rain. The deep learning network consists of four branches that deal with LiDAR, RGB camera, gated camera, and radar separately. To guide the

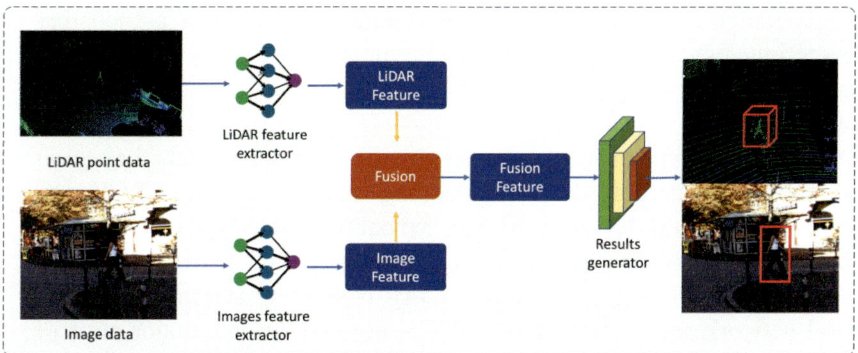

Figure 7.8 Typical global feature fusion structure.

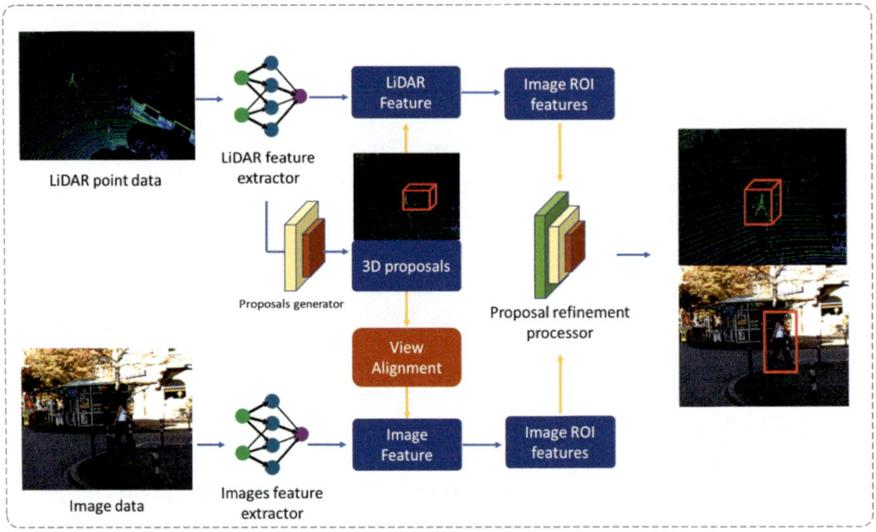

Figure 7.9 Typical ROI feature fusion structure.

fusion between sensors, feature exchange occurs in four branches at different scale levels using concatenation. The authors [39] detects vehicles in a variety of environments utilizing monocular images and radar scans. A radar branch and an image branch are included in the detector. The radar branch takes in range and Doppler data and uses ResNet to extract radar features, which are then concatenated with image branch features. RVNet [40] offered a similar solution, fusing features from radar and images by projecting the point cloud's range and Doppler information to the image coordinate system and utilizing the concatenation to fuse images features. The authors [41] proposes FusionNet, which implements two branches, a radar branch that processes Range-Azimuth-Doppler Tensor data and a camera branch that processes images captured by front-facing cameras. After the independent feature extractor branch, these feature maps are concatenated and fed to following layers.

There is also a type of feature fusion algorithm which extracts the region of interest before merging the corresponding features, as shown in Figure 7.9. MV3D [42] and AVOD [43] are typical works. MV3D proposes a multi-view 3D object detection network that implements 3D detection using multi-modality 2D data (LiDAR BEV, LiDAR front view, and images). A 3D proposal network and a region-based fusion network are the major parts of the MV3D. The 3D proposal network generates preliminary 3D proposals by using a BEV of the LiDAR point cloud. By projecting 3D proposals onto feature maps from different views, a multi-view fusion network extracts multi-regional features, and integrates them using concatenation and element-wise mean to refine the final proposals. AVOD takes the BEV map generated from LiDAR point cloud and RGB image as input, and then runs them separately through a feature extraction network to generate high-resolution full-scale feature maps. Two subnetworks, the region proposal network (RPN) and the second-stage detector network, share these feature maps. This RPN can leverage high-resolution features from multi-modalities input to generate preliminary proposals, and then crop ROI regions in the LiDAR and image features, respectively. It conducts the element-wise mean operation on these two regions' features to generate the fused ROI features. Finally, based on the fused features, accurate directional 3D bounding box can be refined. MMF [35] not only has the fusion of feature level, but also the refinement of proposals. The BEV feature map after point-wise fusion can predict many 3D detection results for each BEV voxel. After NMS and score filtering, a small number of high-quality 3D detection results and corresponding 2D detection results can be obtained. Then features within image ROIs and BEV-oriented ROIs are concatenated and processed to obtain more accurate 2D and 3D detection frames.

[44–51] are other feature-level fusion methods with similar feature fusion modules. Feature-level fusion extracts a particular information from the original data, which solves the problem of inconsistency between different sensors' original data. Higher-dimensional features offer better discrimination abilities, which improve fusion efficiency and overcomes the inherent flaws of a single sensor.

7.4.2.1.3 Object-level fusion

Object-level fusion is a method that integrates and optimizes the preliminary results of multi-modality, as shown in Figure 7.10. By fusing the primary results of multi-modalities, excellent object-level fusion algorithms can generate trustworthy and authentic target information, such as 3D position, category, or target speed, preventing false alarms and missed inspections. This type of work is usually based on rules to fuse different object-level information, such as using Kalman filtering [52] or particle filtering [53].

[54] detects and tracks objects by fusing LiDAR point cloud and radar signals at the object level. The LiDAR and radar signals are analyzed individually to estimate the obstacle status (location and velocity) on the road. The

information of nonlinear moving objects in multi-modalities can be fused by passing the obstacle state to the extended Kalman filter. The prediction and tracking of the object state can be realized at the same time. The authors [55] proposes a new Camera-LiDAR Candidate Object (CLOCs) fusion network, which is constructed using a deep learning network rather than traditional rule-based methods to directly fuse detected boxes. CLOCs fuses the combined output candidates before NMS of 2D and 3D detectors, rather than using ROI features to refine findings. CLOCs fusion network architecture converts individual 2D and 3D detection candidates into a collection of joint detection candidates, and then refines the preliminary candidates using a 2D CNN to process nonempty components in the sparse input tensor.

Since the methods for obtaining preliminary object information from each pipeline is independently, object-level fusion can be integrated with the state-of-the-art algorithms easily. In addition, each mode's influence is highly understandable, which makes it easier to analyze the outcome.

7.4.2.2 Sensor synchronization (solutions for challenge 4)

Sensor synchronization is the process of synchronizing numerous sensor data streams with varied frequencies and delays [56]. Cameras, for example, typically collect images at 30 FPS, but LiDAR sensors often scan at less than 10 Hz. Synchronizing numerous sensor data reduces collected scene

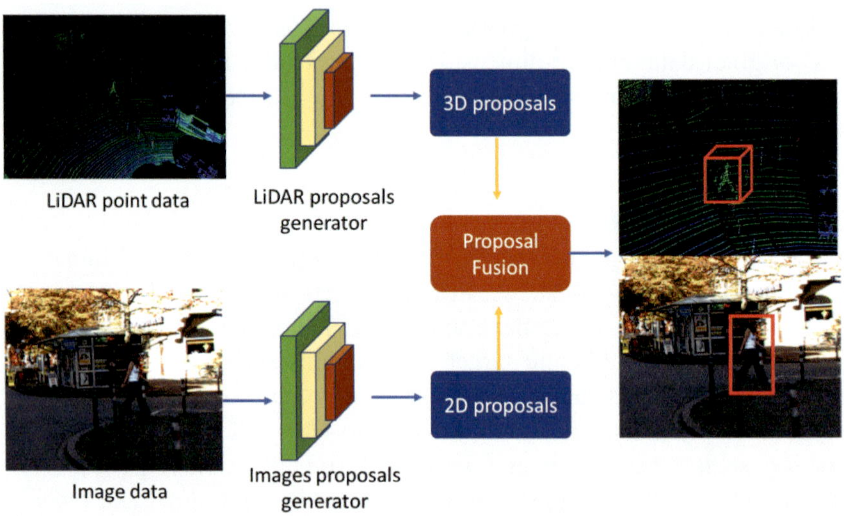

Figure 7.10 Typical object-level fusion structure.

differences and makes sensor fusion algorithms easier to construct. Obtaining timestamps from various sensors and constructing the closest match is one approach to synchronize sensor data. In theory, however, message timestamp synchronization is poor because sensors may have unpredictable delays, such as communication transmission delays or sensor processing delays. These unknown delays are difficult to predict and may differ from sensor to sensor [57].

Making a pulse generator, which is used to trigger all sensors, is another common solution. To synchronize the LiDAR and camera in the KITTI dataset [58], the timestamp of the Velodyne 3D LiDAR is used as a reference. A reed contact on the bottom of the spinning scanner triggers the camera when facing forward, limiting the range of dynamic objects and image viewing variations. The exposure of the camera is triggered when the top LiDAR sweeps over the center of the camera FOV for the nuScenes dataset [59]. The exposure trigger time is the timestamp of the image. The LiDAR scan timestamp is the time when the current LiDAR frame completes a full rotation. Given the camera's near-instantaneous exposure duration, this method frequently results in good data alignment. For Waymo Open Datasets [60], all camera synchronization errors are within [-6ms, 7ms] with 99.7% confidence and within [-6ms, 8ms] with 99.9995% confidence.

7.5 Multi-sensor Fusion in Detection and Tracking

3D object detection and multi-object tracking, which offer fundamental methods for autonomous driving perception and comprehension of scenes, can provide crucial information for later decision-making [65–67]. This section introduces multi-sensor fusion methods for 3D objects detection and multi-object tracking.

7.5.1 Data processing representation

Different sensors' data are handled differently because of the formats and characteristics. Fortunately, data processing forms are often uniform because they are mostly connected to the data type rather than tasks. For data processing, since the existing image-based perception algorithms are relatively advanced, the data processing forms of LiDAR and radar are principally described here.

7.5.1.1 LiDAR point cloud representation

The use of LiDAR point clouds for 3D detection and tracking becomes popular due to its inherent 3D structure and position measuring capabilities. A point cloud is a collection of data points that represent a LiDAR surface measurement of an item, usually determined by the reflection intensity and the x, y, and z coordinates [68]. Point clouds are continuous, sparse, irregular, and unordered data structures. Finding a proper point cloud representation is crucial for effective processing. The majority of approaches currently in use fall into three categories: 2D view-based representation, point-based representation, or voxel-based representation [69] [70].

2D view-based representation projects point clouds onto various 2D view planes to produce different view images [71, 72]. These produced multi-view images can directly be manipulated utilizing image processing methods. BEV is frequently utilized in autonomous driving because there are few occlusions between objects, and no change in object size from the BEV perspective. The range view (RV), also known as the range image, which is a special view of the spinning LiDAR sensor, is also used in the 3D object identification model [73]. The information is compact and dense in the RV perspective. However, the object's size will change, and it is obscured.

Voxel-based representation discretizes the 3D space into 3D voxels, converting irregular point clouds into compactly formed voxels and dividing the point cloud into a fixed-resolution 3D grid [74]. 3D convolution can be used directly to extract features from the voxel representation. However, it has performance issues due to information loss during voxelization such as the loss of spatial resolution and fine-grained 3D geometry.

Point-based representation does not transform the raw point cloud into an intermediate data representation. As a result, the original point cloud's geometry is kept as much as possible. Recent research has concentrated on defining specific convolution procedures for point-based representation, such as PointNet [75] and its derivatives or Graph Neural Networks (GNNs) [76]. However, point-based representation is computationally intensive, such as the widely used Velodyne LiDAR HDL-64E, which collects more than 100,000 points in a single scan in large scenes.

7.5.1.2 Radar data representation

Point clouds are typically used in MMW radar data. In theory, this is similar to the point cloud of LiDAR, but each point here contains different data. The points of LiDAR contain x, y, and z coordinates and the strength of the reflected signal, whereas the points of MMW radar contain range, angle, radar cross-section (the size and capability of the target reflected radar energy), and Doppler (object speed). If 4D imaging radar is used, each point will also contain elevation data. Consequently, LiDAR's representation also applies to MMW radar. However, MMW radar's point cloud is quite sparse compared to LiDAR, necessitating certain particular algorithmic solutions.

The majority of radar processing methods used today employ point cloud data as input, and process point clouds directly to produce results. However, performance of actual road conditions based on point data processing is generally not satisfied. There are two reasons. First, effective timing fusion is not adopted, the radar detection result of a single frame contains a lot of noise, and the traditional radar object algorithm also relies on timing information [77]. Second, the point cloud is too sparse, and a lot of useful information has been lost in the preprocessing stage. The original radar signal is processed by three times FFT to obtain a 3D signal with the coordinates of Range–Doppler–Azimuth. The 3D signal will be filtered using the constant false-alarm rate (CFAR) algorithm [78] in conventional radar signal processing, which will only keep relative strong responses. The dynamic selection of the sample threshold is the key of the CFAR algorithm, which eliminates a lot of noise, but it loses detailed information. Thus, the point cloud data acquired is the sparse data obtained following CFAR sampling.

Consequently, some researchers have started directly processing the raw 3D radar signal using techniques like Range–Doppler–Azimuth (RAD). RAD data is better for deep learning methods, despite having more information and requiring more computation. This has the advantage of fully utilizing the deep neural network's robust feature learning capabilities, automatically removing noise from the original signal, and preserving important data [79]. The design of this portion of the network is more difficult since it must take the features of the radar more into account.

7.5.2 3D object detection

To identify objects and scenes in images using statistical learning or mathematical models is a common problem in computer vision. However, as the image is the projection of 3D space on the 2D optical plane, it increases the difficulties to precisely predict the shape, location, and orientation of the object in 3D space in the field of autonomous driving. Locating, classifying, and estimating oriented bounding boxes in 3D space are the goals of 3D object detection. The perception and comprehension of scenes by autonomous vehicles depends heavily on 3D object detection. Although 3D object detection has made significant advancements, it is still a more difficult task than well-researched 2D object detection. In order to specify a bounding box that is 3D-oriented, 3D object detection needs more output parameters than 2D detection. To draw the 3D bounding box, the object's center 3D position, length, width, height, and deflection angle must be estimated. There are three basic categories of 3D object detection methods: image-based, point cloud-based, and multi-sensor-based 3D detection.

7.5.2.1 Image-based 3D object detection

Image-based 2D object detection is relatively mature; there are methods based on template matching [80], geometric properties [81], and pseudo-point cloud [20–22] to generate 3D detection boxes directly using images. These methods only take monocular/stereo images as input, providing color attributes and texture information. But they rely on a lot of domain expertise to compute the location of 2D detection in the 3D space. At the same time, due to the lack of depth information in 2D images, the accuracy of 3D object detection is not high.

7.5.2.2 Point cloud-based 3D object detection

Obtaining target information from point clouds, especially from dense LiDAR point clouds, is a relatively intuitive and accurate 3D target detection method. Since the depth of points can be directly measured, the 3D detection problem is essentially transformed into a 3D point classification problem. According to the different methods from point cloud representation, point cloud-based 3D object detection algorithms can be divided into four categories: multi-view based [71–72], voxel-based [82–83], point-based [84–85], and point-voxel-based methods [86–87]. The accuracy of 3D object detection based on point cloud has been greatly improved compared with those based on images. The best 3D detection algorithms at this stage are basically based on LiDAR point clouds.

7.5.2.3 Multi-sensor-based 3D object detection

Although multi-sensor fusion can theoretically integrate the different advantages of multiple sensors, so as to achieve more accurate 3D detection, it is not easy to exploit the synergistic effect of multiple sensors due to challenges such as perspective misalignment, time drift, etc. Current methods based on multi-sensor fusion have gradually surpassed point cloud-based methods. Here, the state-of-the-art 3D object detection algorithms based on multi-sensor fusion are primarily introduced and outlined.

There are primarily two sorts of competitive 3D object detection fusion algorithms: sequential approaches and parallel approaches. Sequential approaches have multiple stages. Multi-senor features are extracted in the first stage, and the final 3D detection result depends on the features. Parallel approaches process the multi-sensor data in parallel, gather the features of each modality, fuse them in the feature space, and then feed them into the supervised learning module.

Sequential approaches based on 2D images features, like PointPainting [15] and F-Convnet [88], typically produce better results. F-ConvNet, which is inspired by Frustum PointNets, extracts 2D regions using the traditional 2D object detection network and creates a frustum in the point cloud space for each proposed 2D region. A series of frustums are created by movement with same step length along the frustum axis. With the help of PointNet, each frustum's point is transformed into frustum-level features, which are organized as 2D feature maps and given to FCN for feature extraction to create end-to-end estimates of 3D-oriented boxes. However, F-ConvNet relies on the precision of 2D object detection. PointPainting, introduced in calibration-based data augmentation part, uses semantic segmentation data from images to assist point clouds in producing 3D detection boxes. All point cloud-based 3D object detection models can employ the PointPainting method since it produces enhanced point clouds by merging the LiDAR point with image segmentation result. However, PointPainting finds corresponding points between LiDAR and image based on sensor calibration. The performance of such hard association has a heavy reliance on high-quality calibration and can be severely degraded with low-quality image features, e.g., images under harsh lighting conditions. Transfusion [89] proposes a multi-modal detection framework to shift the fusion process from hard to soft associations. Two transformer decoding layers are used by Transfusion as detection heads. A strong and efficient soft-association fusion method is produced as a result of the attention mechanism of transformer [90], which enables the model to adaptively decide the position and information that should be gathered from the image. The first transformer layer creates initial 3D bounding boxes using a sparse set of object queries, which have been enhanced with location and category knowledge. The second-layer transformer closely associates and fuses the object query from the first stage with the image features, and it employs the attention mechanism to adaptively decide which aspect of the image features is more crucial of being transferred to the point cloud, thereby promoting the fusion of object queries and useful image information.

In parallel approaches, the performance of algorithms that fuse LiDAR BEV feature and image feature is significantly improved compared with classical algorithms such as MV3D and AVOD. ContFuse [91] proposes a BEV-based 3D object detector that projects image features into the LiDAR BEV space using continuous fusion layers. ContFuse uses two branches to extract the features of the image and the LiDAR BEV map, respectively. After that, continuous fusion layers locate the k nearest point clouds for each pixel in the LiDAR BEV feature and projects those point clouds onto the image plane to obtain the k corresponding image features. Then, MLP operation is used to encode image features and 3D geometry offsets between the k nearest LiDAR points and the target pixel in the LiDAR BEV feature to produce the final fusion. ContFuse converts image features to BEV views to create fused BEV features, which are capable of recording dense, exact geometric relationships of positions for the two modalities. Point-wise feature fusion, however, is sparse by nature of LiDAR observation when the LiDAR points are extremely sparse, such as for distant objects or small objects. MMF [35] is a multi-task framework that also uses a continuous fusion layer to combine LiDAR BEV features and image features. Along with 3D detection, multi-sensor data can be utilized to forecast dense depth, which can then be used to establish dense pixel correspondences between image features and LiDAR BEV features. Thus, it is possible to perform dense point-wise fusing of features, which addresses ContFuse issue with sparse point clouds. 3D-CVF [92] extracts LiDAR BEV features and multi-view images features, respectively, and uses auto-calibrated projection to transform the multi-view image features to BEV. A high-resolution camera voxel structure is built during auto-calibrated projection. The final voxel value is then calculated by interpolating using the image features of the associated position and surrounding positions after each voxel's center point is projected onto the camera-view plane. Finally, a gated adaptive fusion network is utilized to point-wise fusion of the image and LiDAR BEV features.

Even while 3D-CVF outperforms earlier fusion-based works, there is still a gap between LiDAR-based methods. It's important to delve deeper into the cause. In the generation-based augmentation, a class of pseudo-LiDAR-based algorithms is introduced. These algorithms, like PL++ [25], use LiDAR point to correct depth information and generate dense pseudo-point clouds from images, but their performance is not even close to that of the 3D object detection algorithm mentioned in this part. However, certain LiDAR-based

Table 7.3 3D object detection algorithm AP_{3D} (%) comparison on KITTI test split by submitting to official test server.

Method	Type	Time	Car		
			Easy	Moderate	Hard
F-ConvNet [88]		0.47 s	87.36	76.39	66.69
PointPainting	Data augment	0.4 s	82.11	71.70	67.08
PointRCNN [15]					
ContFuse [91]		0.06s	83.68	68.78	61.67
MMF [35]	Feature-level fusion	0.08 s	88.40	77.43	70.22
3D-CVF [92]		0.06 s	89.20	80.05	73.11
SFD [23]	Data and feature fusion	0.1 s	91.73	84.76	77.92
PointRCNN [85]	LiDAR only	0.1 s	86.96	75.64	70.70
PointPillars [82]		16ms	82.58	74.31	68.99

methods, such as SPG [93], which is far more efficient than pseudo-LiDAR algorithms, also concentrate on creating dense point clouds to enhance the effect of 3D detection. One explanation for this could be that the generated pseudo-LiDAR point still differs significantly from the true LiDAR point. Better outcomes might be obtained by directly guiding the pseudo-LiDAR point with the true LiDAR signal. In SFD [23], the generated RGB pseudo-LiDAR point and the original LiDAR signal are fused to obtain a better 3D detection performance than LiDAR-only based algorithms. As described in generation-based data augmentation, SFD first creates RGB pseudo-LiDAR points through fusion, recovers ROIs based on LiDAR signals, and then merges LiDAR and pseudo-LiDAR features in the ROI area. It is possible to fuse each pair of grid features separately because LiDAR and pseudo-LiDAR share the same data format. Therefore, each part of an object is improved by the appropriate pseudo-grid features. 2022 SFD is to convert images to pseudo-LiDAR, which prevents the feature fusion just using concatenation. Future research may focus on enhancing data directly from LiDAR and pseudo-LiDAR without limiting the subsequent detection network, which may also produce better results than point-based methods.

Meanwhile, the unified expression of BEV space is also a common and effective fusion direction. Because camera and LiDAR features have very different densities, BEVFusion [94] claims that enhancing LiDAR points with image features in point-wise manner does not effectively preserve the semantic information of the image. Less than 5% of camera features match LiDAR points for a 32-channel LiDAR scanner. Thus, BEVFusion fuses the LiDAR signal with the multi-view camera signal to generate a coherent

Table 7.4 3D object detection algorithm AP_{3D} (%) comparison on nuScenes test set.

	Type	mAP	Car	Pedestrian	Bicycle
PointPainted PointPillars+ [15]	Data augment	46.4	77.9	73.3	24.1
3D-CVF [92]		57.8	83.9	82.4	30.6
TransFusion [89]	Feature-level fusion	68.9	87.1	88.4	52.9
BEVFusion [94]		75.0	90.5	91.8	62.9
PointPillars+ [15]	LiDAR only	40.1	76.0	64.0	14.0
PointPillars [82]		30.5	68.4	59.7	1.1

expression in the BEV space that effectively maintains both geometric and semantic information. BEVFusion extracts features from multi-sensor. Then, all features are transformed into BEV space via view transformation. Features that have been converted to BEV are simple to combine using element-wise operators like concatenation. Finally, a fully convolutional BEV encoder is employed to correct the fused BEV features' local spatial misalignment. In order to support various target tasks, several task-specific headers are applied. BEVFusion ranked second in the nuScenes detection task until July 2022. Despite not having any KITTI test findings to compare with all mentioned algorithms, its performance on nuScenes is greater than 3D-CVF.

Tables 7.3 and 7.4 show the performance of the aforementioned algorithms on KITTI or nuScenes test datasets [95, 96]. Because PointPainting supports many LiDAR-based detection architectures, it employs PointRCNN [85] on KITTI and PointPillars+ [15] on nuScenes. The performance of PointRCNN and PointPillars [82] is also provided as a reference in the table.

7.5.3 Multiple object tracking (MOT)

MOT aims to follow the motion trajectories of multi-targets, which is essential to identify and detect multiple objects over various time periods of data [97]. Multi-object tracking is a crucial part for autonomous driving systems since it offers important data for a variety of on-board modules, including perception, prediction, and planning. However, MOT is a difficult work since it is challenging to recognize the same target at different times because the appearance of the target changes as a result of changes in light, viewing angle, and occurrence of excessive occlusion.

The vehicle-side and roadside tracking tasks are introduced individually here, because MOT task on the vehicle side uses a single group of sensors to track objects (single group of sensors are all on the vehicle), while the roadside sensors are made up of multiple groups of sensors in different positions (multiple groups of sensors are far away). Thus, the roadside MOT task involves tracking between multiple groups of sensors in addition to a single group of sensors.

7.5.3.1 Vehicle side

On the vehicle side, designing a suitable cost function to quantify similarity is essential for MOT systems in order to achieve data association. Earlier research used handcrafted features in cost functions, including color histograms and spatial distances. Motion models and appearance affinity are frequently used in modern methods. Commonly, 2D MOT systems and 3D MOT systems share the same parts. The distinction is that instead of 2D output, the tracking trajectories are in 3D space. Due to the importance of 3D coordinates for autonomous driving systems, this part primarily introduces the most recent research on 3D MOT.

Multi-sensor fusion contains multi-modality information, so some works in 3D MOT can extract multi-modality features to help subsequent data association. AlphaTrack [98] implements 3D MOT inspired by the LiDAR-only 3D detector CenterPoint. CenterPoint [99] detector, which predicts objects as points, associates objects using the greedy closest-point matching algorithm. AlphaTrack adds instance-aware appearance representation affinity in addition to position change association by increasing camera input. First, a PointPainting-like approach is used in 3D detection to connect LiDAR points with corresponding point-wise image features. The modified VoxelNet is then used to process both LiDAR features and camera features, producing both 3D detection results and appearance features. In the subsequent tracking process, the location information and appearance features generated are simultaneously associated with different objects, realizing high-accuracy 3D multi-target tracking. Based on the combination of radar and camera sensors, CFTrack [100] provides an end-to-end network for joint object recognition and tracking. CFTrack uses a greedy approach for object association along with a center-based radar-camera fusion network to detect objects. The greedy approach correlates objects with time using the depth, velocity, and displacement provided by the fusion network. The radar signal contains native depth and velocity data, which makes it easier to extract object features. The authors [101] uses LiDAR point clouds and camera images as input to compute appearance and motion affinity in 3D space. After detecting objects from LiDAR point, the features from 2D images and 3D LiDAR point clouds are extracted and fused to capture the appearance feature for each

Table 7.5 Multi-object tracking algorithm AMOTA (%) comparison on nuScenes test set.

	Overall	Car	Pedestrian	Bicycle
BEVFusion [94]	74.1	83.1	83.7	56.0
TransFusion [89]	71.8	82.1	79.6	53.9
AlphaTrack [98]	69.3	84.2	74.3	47.1
Probabilistic multi-modal [101]	65.5	83.0	74.1	46.9

object. Then, for data association, a combination of feature distances and Mahalanobis distances is trained using the fused features from time t and tracks from time t-1, and a greedy match technique is applied. Finally, the Kalman filter is used to refine the final object states after matched pairs are processed.

3D MOT based on multi-sensor fusion is currently superior to LiDAR-based 3D MOT. Thus, different features derived from multi-sensor inputs can in fact aid in data association and improve the performance. In addition, BEVFusion [94] and Transfusion [89], which rank first and second in nuScenes tracking task, both fuse LiDAR points with images to better realize 3D object detection, and employ the method described in Center-Point to achieve tracking on the basis of 3D detection. That is, objects are associated based on position affinity without the addition of other features such as appearance. Therefore, improving the detection performance and increasing the data association information are the two main focuses of 3D MOT. Table 7.5 depicts the effect of the aforementioned algorithms on nuScenes.

7.5.3.2 Roadside

Roadside MOT places emphasis on long-distance tracking using multiple groups of sensor modules, such as hundreds of LiDAR and thousands of cameras mounted on a highway to follow moving vehicles continuously. Numerous advanced inferences and predictions are supported by long-distance tracking, which has a wide range of applications. The majority of the research for this kind of study is multi-target multi-camera (MTMC) tracking or MOT across cameras. MTMC tracking is generally based on the tracking results of a single camera, and a strategy is formulated to achieve cross-camera tracking.

[102] proposes a solution for the track 3 multi-camera vehicle tracking task in the AI City challenge 2021. A MTMC vehicle tracking framework is proposed based on crossroad zones' guidance. To construct single-camera

trajectories and Re-ID features, [102] first detects object and extract appearance features using detection and vehicle re-identification models. The IDs between cameras are then synchronized to generate multi-camera trajectories. The crossroad is first separated into four sections before creating multi-camera trajectories, and a tracklet filter strategy is suggested to remove false positive tracklets such stationary traffic signs from the trajectory of a single camera. Then the constraints between tracks are matched using a Direction-Based Temporal Mask. The range of the Re-ID is significantly narrowed by the removal of objects that do not match the motion direction. Subclustering in adjacent cameras is a suggestion for multi-camera trajectory matching on the basis of this. Subclustering in adjacent cameras uses both inter-zone clustering and inter-camera clustering as subclustering techniques. Inter-zone clustering is used to cluster between the zones of different cameras, and inter-camera clustering is used to cluster between connected cameras. It can match as many trajectories as possible while maintaining accuracy using these two clustering.

Since MTMC's work is currently in 2D, estimating distance and speed from a vehicle is challenging. To acquire the 3D MOT under a single set of sensors, the camera data can be fused with LiDAR or radar (similar to the work done on the vehicle side), and then the 3D MOT results of numerous sets of sensors can be combined using the same method.

7.6 Conclusion

Multi-sensor fusion for autonomous driving perception has improved quickly in recent years, ranging from increasingly complex deep learning models to advanced feature representations. The advantages offered by various multi-sensor fusion algorithms still need to be investigated to guide effective algorithms' development. Despite the fact that there is still no consensus regarding the best data type for multi-sensor fusion algorithms or the best fusion technique for fusing features, we can get some inspiration from the aforementioned work.

1. Since the LiDAR-based approach performs good in 3D space tasks, approaches based solely on LiDAR point may be useful to increase point cloud information without disrupting its spatial structure and minimize the loss of spatial resolution, such as PointPainting. Thus, as noted in the 3D object detection, there is still much to learn about pseudo-LiDAR as a useful data representation.

2. Deep learning can be effectively used to extract abstract features, and the feature dimension can be self-adjusted. Consequently, the fusion processing of features is simpler than data augmentation. The multi-sensor information, however, cannot be effectively integrated if it is only a straightforward concatenation or element-wise process. The LiDAR BEV feature is improved in ContFuse, MMF, and 3D-CVF. Although they employ feature-level fusion, the fused features also contain a substantial amount of object space information, and hence these works still manage to portray the environment in a cohesive manner. BEV can be used for autonomous driving to provide unified information expression because it lacks occlusions and scale change. The viewing angle for roadside sensors is particularly near to BEV, and therefore the distortion following BEV transformation is small. BEV may thus perform better in tracking and detecting tasks.

This chapter has provided an in-depth introduction to the current multi-sensor fusion algorithms in detection and tracking. According to the general challenges faced by multi-sensor fusion, the solutions in the existing work are given. Simultaneously, the multi-sensor fusion algorithm's specific implementation for particular tasks is introduced, and some potential directions and solutions are suggested.

References

[1] S. Campbell, et. al., 'Sensor Technology in Autonomous Vehicles : A review', 2018 29th Irish Signals and Systems Conference (ISSC), pp. 1-4, 2018.

[2] E. Yurtsever, J. Lambert, A. Carballo and K. Takeda, 'A Survey of Autonomous Driving: Common Practices and Emerging Technologies', in IEEE Access, pp. 58443-58469, 2020.

[3] S. Jusoh and S. Almajali, 'A Systematic Review on Fusion Techniques and Approaches Used in Applications', in IEEE Access, pp. 14424-14439, 2020.

[4] M. B. Alatise and G. P. Hancke, 'A Review on Challenges of Autonomous Mobile Robot and Sensor Fusion Methods', in IEEE Access, pp. 39830-39846, 2020.

[5] G. Velasco-Hernandez, D. J. Yeong, J. Barry and J. Walsh, 'Autonomous Driving Architectures, Perception and Data Fusion: A Review', 2020 IEEE 16th International Conference on Intelligent

Computer Communication and Processing (ICCP), pp. 315-321, 2020.
[6] Z. Wang, Y. Wu and Q. Niu, 'Multi-Sensor Fusion in Automated Driving: A Survey', in IEEE Access, pp. 2847-2868, 2020.
[7] https://www.tesla.com/ownersmanual/modely/en_us/GUID-EDA77281-42DC-4618-98A9-CC62378E0EC2.html
[8] https://www.youtube.com/watch?v=j0z4FweCy4M
[9] T. Raj, F. H. Hashim, A. B. Huddin, M. F. Ibrahim, and A. Hussain, 'A Survey on LiDAR Scanning Mechanisms', Electronics, 2020.
[10] T. Zhou, M. Yang, K. Jiang, H. Wong, and D. Yang, 'MMW Radar-based Technologies in Autonomous Driving: A Review', Sensors, 2020.
[11] https://radar-scenes.com/dataset/examples/
[12] M. Stolz, M. Wolf, F. Meinl, M. Kunert and W. Menzel, 'A New Antenna Array and Signal Processing Concept for an Automotive 4D Radar', 2018 15th European Radar Conference (EuRAD), pp. 63-66, 2018.
[13] https://arberobotics.com/product/
[14] M. Pereira, D. Silva, V. Santos, and P. Dias, 'Self Calibration of Multiple LIDARs and Cameras on Autonomous Vehicles', Robotics and Autonomous Systems, 2016.
[15] S. Vora, et. al., 'Pointpainting: Sequential Fusion for 3D Object Detection', In Proceedings of the IEEE/CVF conference on computer vision and pattern recognition, pp. 4604-4612, 2020.
[16] L. Caltagirone, M. Bellone, L. Svensson, and M. Wahde, 'LIDAR–camera Fusion for Road Detection using Fully Convolutional Neural Networks', Robotics and Autonomous Systems, pp. 125-131, 2019.
[17] H. Gao, B. Cheng, J. Wang, K. Li, J. Zhao and D. Li, 'Object Classification Using CNN-Based Fusion of Vision and LIDAR in Autonomous Vehicle Environment', in IEEE Transactions on Industrial Informatics, pp. 4224-4231, Sept. 2018.
[18] Z. Li, W. Wang, H. Li, et. al., 'BEVFormer: Learning Bird's-Eye-View Representation from Multi-Camera Images via Spatiotemporal Transformers', arXiv preprint, 2022.
[19] L. Xie, C. Xiang, Z. Yu, G. Xu, Z. Yang, D. Cai, and X. He, 'PI-RCNN: An Efficient Multi-sensor 3D Object Detector with Point-based Attentive Cont-conv Fusion Module', In Proceedings of the AAAI conference on artificial intelligence, pp. 12460-12467, 2020.
[20] Y. Wang, W. Chao, D. Garg, B. Hariharan, M. Campbell, and K.Q. Weinberger, 'Pseudo-lidar from Visual Depth Estimation: Bridging the

Gap in 3D Object Detection for Autonomous Driving', In Proceedings of the IEEE/CVF Conference on Computer Vision and Pattern Recognition, pp. 8445-8453, 2019.

[21] X. Weng, and K. Kris, 'Monocular 3D Object Detection with Pseudo-LiDAR Point Cloud', In Proceedings of the IEEE/CVF International Conference on Computer Vision Workshops, 2019.

[22] R. Qian, G. Divyansh, W. Yan, Y. Yurong, et. al., 'End-to-end Pseudo-LiDAR for Image-based 3D Object Detection', In Proceedings of the IEEE/CVF Conference on Computer Vision and Pattern Recognition, pp. 5881-5890. 2020.

[23] X. Wu, L. Peng, H. Yang, L. Xie, C. Huang, C. Deng, H. Liu, and D. Cai, 'Sparse Fuse Dense: Towards High Quality 3D Detection with Depth Completion', In Proceedings of the IEEE/CVF Conference on Computer Vision and Pattern Recognition, pp. 5418-5427, 2022.

[24] J. Wang, M. Zhu, B. Wang, D. Sun, H. Wei, C. Liu, and H. Nie, 'KDA3D: Key-point Densification and Multi-attention Guidance for 3D Object Detection', Remote Sensing, 2020.

[25] Y. You, Y. Wang, W. Chao, et. al., 'Pseudo-LiDAR++: Accurate Depth for 3D Object Detection in Autonomous Driving', International Conference on Learning Representations, 2020.

[26] F. Negahbani, O.B. Töre, F. Güney, and B. Akgun, 'Frustum Fusion: Pseudo-LiDAR and LiDAR Fusion for 3D Detection', arXiv preprint, 2021.

[27] A. Haselhoff and A. Kummert, 'A Vehicle Detection System based on HAAR and Triangle Features', 2009 IEEE Intelligent Vehicles Symposium, pp. 261-266, 2009.

[28] C. Schuldt, I. Laptev and B. Caputo, 'Recognizing Human Actions: a Local SVM Approach', Proceedings of the 17th International Conference on Pattern Recognition, pp. 32-36, 2004.

[29] Li, L. Wang, and E. Sung, 'AdaBoost with SVM-based Component Classifiers', Engineering Applications of Artificial Intelligence, pp. 785-795, 2008.

[30] D. P. Solomatine and D. L. Shrestha, 'AdaBoost.RT: a Boosting Algorithm for Regression Problems,' 2004 IEEE International Joint Conference on Neural Networks, pp. 1163-1168, 2004.

[31] J. Fayyad, M. A. Jaradat, D. Gruyer, and H. Najjaran, 'Deep Learning Sensor Fusion for Autonomous Vehicle Perception and Localization: A Review', Sensors, 2020.

[32] K. Muhammad, A. Ullah, J. Lloret, J. D. Ser and V. H. C. de Albuquerque, 'Deep Learning for Safe Autonomous Driving: Current Challenges and Future Directions', in IEEE Transactions on Intelligent Transportation Systems, pp. 4316-4336, July 2021.

[33] Y. Li, et. al., 'Deep Learning for LiDAR Point Clouds in Autonomous Driving: A Review', in IEEE Transactions on Neural Networks and Learning Systems, pp. 3412-3432, Aug. 2021.

[34] D. Xu, D. Anguelov, and A. Jain, 'Pointfusion: Deep Sensor Fusion for 3D Bounding Box Estimation', In Proceedings of the IEEE/CVF Conference on Computer Vision and Pattern Recognition, pp. 244-253, 2018

[35] M. Liang, B. Yang, Y. Chen, R. Hu, and R. Urtasun, 'Multi-task Multi-sensor Fusion for 3D Object Detection', In Proceedings of the IEEE/CVF Conference on Computer Vision and Pattern Recognition, pp. 7345-7353, 2019.

[36] T. Huang, Z. Liu, X. Chen, and X. Bai, 'EPNet: Enhancing Point Features with Image Semantics for 3D Object Detection', In European Conference on Computer Vision, pp. 35-52, August, 2020.

[37] Z. Zhuang, R. Li, K. Jia, Q. Wang, Y. Li, and M. Tan, 'Perception-aware Multi-sensor Fusion for 3D LiDAR Semantic Segmentation', In Proceedings of the IEEE/CVF International Conference on Computer Vision, pp. 16280-16290, 2021.

[38] M. Bijelic, T. Gruber, F. Mannan, F. Kraus, W. Ritter, K. Dietmayer, and F. Heide, 'Seeing Through Fog without Seeing Fog: Deep Multimodal Sensor Fusion in Unseen Adverse Weather', In Proceedings of the IEEE/CVF Conference on Computer Vision and Pattern Recognition, pp. 11682-11692, 2020.

[39] S. Chadwick, W. Maddern and P. Newman, 'Distant Vehicle Detection Using Radar and Vision', 2019 International Conference on Robotics and Automation (ICRA), pp. 8311-8317, 2019.

[40] V. John, and S. Mita, 'RVNet: Deep Sensor Fusion of Monocular Camera and Radar for Image-based Obstacle Detection in Challenging Environments', In Pacific-Rim Symposium on Image and Video Technology, pp. 351-364, November, 2019.

[41] T.Y. Lim, A. Ansari, B. Major, D. Fontijne, M. Hamilton, R. Gowaikar, and S. Subramanian, 'Radar and Camera Early Fusion for Vehicle Detection in Advanced Driver Assistance Systems', In Machine Learning for Autonomous Driving Workshop at the 33rd Conference on Neural Information Processing Systems, December, 2019.

[42] X. Chen, H. Ma, J. Wan, B. Li, and T. Xia, 'Multi-view 3D Object Detection Network for Autonomous Driving', In Proceedings of the IEEE conference on Computer Vision and Pattern Recognition, pp. 1907-1915, 2017.

[43] J. Ku, M. Mozifian, J. Lee, A. Harakeh and S. L. Waslander, 'Joint 3D Proposal Generation and Object Detection from View Aggregation,' 2018 IEEE/RSJ International Conference on Intelligent Robots and Systems (IROS), pp. 1-8, 2018.

[44] V. A. Sindagi, Y. Zhou and O. Tuzel, 'MVX-Net: Multimodal VoxelNet for 3D Object Detection', 2019 International Conference on Robotics and Automation (ICRA), pp. 7276-7282, 2019.

[45] C. Chen, L. Z. Fragonara and A. Tsourdos, 'RoIFusion: 3D Object Detection From LiDAR and Vision', in IEEE Access, pp. 51710-51721, 2021.

[46] X. Cheng, Y. Zhong, Y. Dai, P. Ji, and H. Li, 'Noise-aware Unsupervised Deep LiDAR-stereo Fusion', In Proceedings of the IEEE/CVF Conference on Computer Vision and Pattern Recognition, pp. 6339-6348, 2019.

[47] R. Théodose, D. Denis, T. Chateau, V. Frémont, and P. Checchin, 'R-AGNO-RPN: A LIDAR-Camera Region Deep Network for Resolution-Agnostic Detection', arXiv preprint, 2020.

[48] L. -H. Wen and K. -H. Jo, 'Fast and Accurate 3D Object Detection for Lidar-Camera-Based Autonomous Vehicles Using One Shared Voxel-Based Backbone', in IEEE Access, pp. 22080-22089, 2021.

[49] G. Wang, B. Tian, Y. Zhang, L. Chen, D. Cao, and J. Wu, 'Multi-view Adaptive Fusion Network for 3D Object Detection', arXiv preprint, 2020.

[50] C. Wang, C. Ma, M. Zhu, and X. Yang, 'Pointaugmenting: Cross-modal Augmentation for 3D Object Detection', In Proceedings of the IEEE/CVF Conference on Computer Vision and Pattern Recognition, pp. 11794-11803, 2021.

[51] Z. Wang, W. Zhan, and M. Tomizuka, 'Fusing Bird View LiDAR Point Cloud and Front View Camera Image for Deep Object Detection', arXiv preprint, 2017.

[52] G. Welch, and G. Bishop, 'An Introduction to the Kalman Filter', 1995.

[53] F. Gustafsson, 'Particle Filter Theory and Practice with Positioning Applications', in IEEE Aerospace and Electronic Systems Magazine, pp. 53-82, July 2010.

[54] B. Shahian Jahromi, T. Tulabandhula, and S. Cetin, 'Real-time hybrid multi-sensor fusion framework for perception in autonomous vehicles', Sensors, 2019.

[55] S. Pang, D. Morris and H. Radha, 'CLOCs: Camera-LiDAR Object Candidates Fusion for 3D Object Detection', 2020 IEEE/RSJ International Conference on Intelligent Robots and Systems (IROS), pp. 10386-10393, 2020.

[56] D.J. Yeong, G. Velasco-Hernandez, J. Barry, and J. Walsh, 'Sensor and Sensor Fusion Technology in Autonomous Vehicles: A Review', Sensors, 2021.

[57] M. Bouain, K.M. Ali, D. Berdjag, N. Fakhfakh, and R.B. Atitallah, 'An Embedded Multi-Sensor Data Fusion Design for Vehicle Perception Tasks', J. Commun., pp. 8-14, 2018.

[58] A. Geiger, P. Lenz, C. Stiller, and R. Urtasun, 'Vision Meets Robotics: The KITTI dataset', The International Journal of Robotics Research, pp. 1231-1237, 2013.

[59] H. Caesar, V. Bankiti, A.H. Lang, S. Vora, V.E. Liong, Q. Xu, A. Krishnan, Y. Pan, G. Baldan, and O. Beijbom, 'nuScenes: A multimodal Dataset for Autonomous Driving', In Proceedings of the IEEE/CVF conference on computer vision and pattern recognition, pp. 11621-11631, 2020.

[60] P. Sun, H. Kretzschmar, X. Dotiwalla, A. Chouard, V. Patnaik, P. Tsui, J. Guo, Y. Zhou, Y. Chai, B. Caine, and V. Vasudevan, 'Scalability in Perception for Autonomous Driving: Waymo Open Dataset', In Proceedings of the IEEE/CVF conference on computer vision and pattern recognition, pp. 2446-2454, 2020.

[61] O. Schumann, et. al., 'RadarScenes: A Real-World Radar Point Cloud Data Set for Automotive Applications', 2021 IEEE 24th International Conference on Information Fusion (FUSION), pp. 1-8, 2021.

[62] J. Rebut, A. Ouaknine, W. Malik, and P. Pérez, 'Raw High-Definition Radar for Multi-Task Learning', In Proceedings of the IEEE/CVF Conference on Computer Vision and Pattern Recognition, pp. 17021-17030, 2022.

[63] A. Zhang, F. E. Nowruzi and R. Laganiere, 'RADDet: Range-Azimuth-Doppler based Radar Object Detection for Dynamic Road Users', 2021 18th Conference on Robots and Vision (CRV), pp. 95-102, 2021.

[64] L. Zheng, Z. Ma, X. Zhu, B. Tan, S. Li, K. Long, W. Sun, S. Chen, L. Zhang, M. Wan, and L. Huang, 'TJ4DRadSet: A 4D Radar Dataset for Autonomous Driving', arXiv preprint, 2022.

[65] E. Arnold, O. Y. Al-Jarrah, M. Dianati, S. Fallah, D. Oxtoby and A. Mouzakitis, 'A Survey on 3D Object Detection Methods for Autonomous Driving Applications', in IEEE Transactions on Intelligent Transportation Systems, pp. 3782-3795, Oct. 2019.

[66] D. Feng, A. Harakeh, S. L. Waslander and K. Dietmayer, 'A Review and Comparative Study on Probabilistic Object Detection in Autonomous Driving', in IEEE Transactions on Intelligent Transportation Systems, 2021.

[67] F. Leon, and M. Gavrilescu, 'A Review of Tracking and Trajectory Prediction Methods for Autonomous Driving', Mathematics, 2021.

[68] S. Royo, and M. Ballesta-Garcia, 'An Overview of LiDAR Imaging Systems for Autonomous Vehicles', Applied sciences, 2019.

[69] Y. Wang, Q. Mao, H. Zhu, Y. Zhang, J. Ji, and Y. Zhang, 'Multimodal 3D Object Detection in Autonomous Driving: a Survey', arXiv preprint, 2021.

[70] R. Qian, X. Lai, and X. Li, '3D Object Detection for Autonomous Driving: a Survey', Pattern Recognition, 2022.

[71] B. Li, T. Zhang, and T. Xia, "Vehicle Detection from 3D LiDAR using Fully Convolutional Network," in Robotics, 2016.

[72] B. Yang, W. Luo, and R. Urtasun, 'PIXOR: Real-time 3D Object Detection from Point Clouds', In Proceedings of the IEEE conference on Computer Vision and Pattern Recognition, pp. 7652-7660, 2018.

[73] Y. Gu, Y. Huang, C. Xu, and H. Kong, 'MaskRange: A Mask-classification Model for Range-view based LiDAR Segmentation', arXiv preprint, 2022.

[74] Y. Zhou, and O. Tuzel, 'VoxelNet: End-to-end Learning for Point Cloud based 3D Object Detection', In Proceedings of the IEEE conference on computer vision and pattern recognition, pp. 4490-4499, 2018.

[75] C.R. Qi, H. Su, K. Mo, and L.J. Guibas, 'PointNet: Deep Learning on Point Sets for 3D Classification and Segmentation', In Proceedings of the IEEE conference on computer vision and pattern recognition, pp. 652-660, 2017.

[76] W. Shi, and R. Rajkumar, 'Point-GNN: Graph Neural Network for 3D Object Detection in a Point Cloud', In Proceedings of the IEEE/CVF conference on computer vision and pattern recognition, pp. 1711-1719, 2020

[77] X. Li, Y. Yang, Z. Sun, G. Cui and T. S. Yeo, 'Multi-Frame Integration Method for Radar Detection of Weak Moving Target', in

IEEE Transactions on Vehicular Technology, pp. 3609-3624, April 2021.

[78] R. Nitzberg, 'Constant-False-Alarm-Rate Signal Processors for Several Types of Interference', in IEEE Transactions on Aerospace and Electronic Systems, pp. 27-34, Jan. 1972.

[79] Y. Wang, Z. Jiang, Y. Li, J. -N. Hwang, G. Xing and H. Liu, 'RODNet: A Real-Time Radar Object Detection Network Cross-Supervised by Camera-Radar Fused Object 3D Localization', in IEEE Journal of Selected Topics in Signal Processing, pp. 954-967, June 2021.

[80] X. Chen, K. Kundu, Y. Zhu, A. G. Berneshawi, H. Ma, S. Fidler, and R. Urtasun, '3D Object Proposals for Accurate Object Class Detection', in NeurIPS, 2015.

[81] A. Mousavian, D. Anguelov, J. Flynn, and J. Kosecka, '3D Bounding Box Estimation using Deep Learning and Geometry', In Proceedings of the IEEE conference on Computer Vision and Pattern Recognition, pp. 7074-7082, 2017.

[82] A.H. Lang, S. Vora, H. Caesar, L. Zhou, J. Yang, and O. Beijbom, 'PointPillars: Fast Encoders for Object Detection from Point Clouds', In Proceedings of the IEEE/CVF conference on computer vision and pattern recognition, pp. 12697-12705, 2019.

[83] Y. Yan, Y. Mao, and B. Li, 'SECOND: Sparsely Embedded Convolutional Detection', Sensors, 2018.

[84] Z. Yang, Y. Sun, S. Liu, and J. Jia, '3DSSD: Point-based 3D Single Stage Object Detector', In Proceedings of the IEEE/CVF conference on computer vision and pattern recognition, pp. 11040-11048, 2020.

[85] S. Shi, X. Wang, and H. Li, 'PointRCNN: 3D Object Proposal Generation and Detection from Point Cloud', In Proceedings of the IEEE/CVF conference on computer vision and pattern recognition, pp. 770-779, 2019.

[86] Y. Chen, S. Liu, X. Shen and J. Jia, 'Fast Point R-CNN', 2019 IEEE/CVF International Conference on Computer Vision (ICCV), pp. 9774-9783, 2019.

[87] C. He, H. Zeng, J. Huang, X.S. Hua, and L. Zhang, 'Structure Aware Single-stage 3D Object Detection from Point Cloud', In Proceedings of the IEEE/CVF conference on computer vision and pattern recognition, pp. 11873-11882, 2020.

[88] Z. Wang and K. Jia, 'Frustum ConvNet: Sliding Frustums to Aggregate Local Point-Wise Features for Amodal 3D Object Detection', 2019 IEEE/RSJ International Conference on Intelligent Robots and Systems (IROS), pp. 1742-1749, 2019.

[89] X. Bai, Z. Hu, X. Zhu, Q. Huang, Y. Chen, H. Fu, and C.L. Tai, 'Transfusion: Robust LiDAR-camera fusion for 3D Object Detection with Transformers', In Proceedings of the IEEE/CVF Conference on Computer Vision and Pattern Recognition, pp. 1090-1099, 2022.

[90] K. Han, A. Xiao, E. Wu, J. Guo, C. Xu, and Y. Wang, 'Transformer in Transformer', Advances in Neural Information Processing Systems, 2021.

[91] M. Liang, B. Yang, S. Wang, and R. Urtasun, 'Deep continuous fusion for multi-sensor 3D Object Detection', In Proceedings of the European conference on computer vision (ECCV), pp. 641-656, 2018.

[92] J.H. Yoo, Y. Kim, J. Kim, and J.W. Choi, '3D-CVF: Generating Joint Camera and LiDAR Features using Cross-view Spatial Feature Fusion for 3D Object Detection', In European Conference on Computer Vision, pp. 720-736, August, 2020.

[93] Q. Xu, Y. Zhou, W. Wang, C.R. Qi, and D. Anguelov, 'SPG: Unsupervised Domain Adaptation for 3D Object Detection via Semantic Point Generation', In Proceedings of the IEEE/CVF International Conference on Computer Vision, pp. 15446-15456, 2021.

[94] Z. Liu, H. Tang, A. Amini, X. Yang, H. Mao, D. Rus, and S. Han, 'BEVFusion: Multi-Task Multi-Sensor Fusion with Unified Bird's-Eye View Representation', arXiv preprint, 2022.

[95] http://www.cvlibs.net/datasets/kitti/eval_object.php?\obj_benchmark =3d

[96] https://www.nuscenes.org/object-detection?\externalData=all&map Data=all&modalities=Any

[97] W. Luo, J. Xing, A. Milan, X. Zhang, W. Liu, and T.K. Kim, 'Multiple Object Tracking: A literature review', Artificial Intelligence, 2021.

[98] Y. Zeng, C. Ma, M. Zhu, Z. Fan and X. Yang, 'Cross-Modal 3D Object Detection and Tracking for Auto-Driving', 2021 IEEE/RSJ International Conference on Intelligent Robots and Systems (IROS), pp. 3850-3857, 2021.

[99] T. Yin, X. Zhou, and P. Krahenbuhl, 'Center-based 3D Object Detection and Tracking', In Proceedings of the IEEE/CVF conference on computer vision and pattern recognition, pp. 11784-11793, 2021.

[100] R. Nabati, L. Harris and H. Qi, 'CFTrack: Center-based Radar and Camera Fusion for 3D Multi-Object Tracking', 2021 IEEE Intelligent Vehicles Symposium Workshops (IV Workshops), pp. 243-248, 2021.

[101] H. Chiu, J. Li, R. Ambruş and J. Bohg, 'Probabilistic 3D Multi-Modal, Multi-Object Tracking for Autonomous Driving', 2021 IEEE

International Conference on Robotics and Automation (ICRA), pp. 14227-14233, 2021.
[102] C. Liu, Y. Zhang, H. Luo, J. Tang, W. Chen, X. Xu, F. Wang, H. Li, and Y.D. Shen, 'City-scale Multi-camera Vehicle Tracking Guided by Crossroad Zones', In Proceedings of the IEEE/CVF Conference on Computer Vision and Pattern Recognition, pp. 4129-4137, 2021.

Authors' Biography

Li Liu received the B.S. degree in Automation from Xi'an Jiaotong University, Shanxi, China in 2015 and the M.S. degree in Automation from Tsinghua University, Beijing, China in 2018. She is currently a computer vision researcher in AI Lab, Lenovo Research. Her research interests include image/video processing, signal processing, 3D object detection, medical image analysis, etc.

Dr. Jiang Tian, an expert in Machine Learning and Computer Vision, is currently a Principal Researcher in AI Lab, Lenovo Research, focusing on health and 3D geometry-related technologies. He got his Ph.D. degree in Computer Science from Iowa State University, US. He participates many social services, such as serving as session co-chair at IROS. He is a reviewer for many top conferences and journals worldwide.

Zhongchao Shi received the M.S. degree from Harbin Institute of Technology, Heilongjiang, China in 2002, and the Ph.D. degree from the Institute of Automation, Chinese Academy of Sciences (CASIA), Beijing, China in 2005. He is currently working in AI Lab of Lenovo Research as a distinguished researcher. His research interests include computer vision, machine learning, deep learning, etc.

Dr. Jianping Fan is a well-known technology leader in artificial intelligence, especially in computer vision, machine learning, and privacy protection. He has published 200+ peer-reviewed papers, and is a recipient of the China national natural science advancement award (3rd class), Shanghai municipal science and technology advancement award (1st class), Shan'xi provincial science and technology advancement award (1st class), and ACM Multimedia best paper runner up (2004). He joined Lenovo as the VP and Head of AI Lab in 2020.

Dr. Yong Rui is the Chief Technology Officer and Senior Vice President of Lenovo Group. He directs Lenovo's corporate technical strategy, research and development directions, and leads the Lenovo Research that investigates AI, intelligent devices, cloud/edge, 5G, AR, blockchain, as well as smart vertical solution development in smart manufacturing, education, healthcare, etc. A Fellow of ACM, IEEE, IAPR, SPIE, AAAS, and a Foreign Member of Academia Europaea, he is a world-renowned technologist.

8

Teleoperation for Complementing and Enhancing Autonomous Vehicles

Sidi Lu[1], Ren Zhong[2], and Weisong Shi[3]

[1]Department of Computer Science, William and Mary, USA
[2]Department of Computer Science, Wayne State University, USA
[3]Department of Computer & Information Sciences, University of Delaware, USA

Abstract

Although autonomous driving technologies are becoming more mature, the arrival of fully autonomous vehicles still requires a long way to go. Today, teleoperation is widely considered as a key and effective solution for advancing the arrival of autonomous vehicles. In the event that the vehicle's autonomous driving abilities are unable to safely continue execution, the vehicle can be monitored, assisted, and partially controlled from a distance by a remote operator. This chapter first provides a detailed definition of teleoperation and then specifically discusses the impact of teleoperation on autonomous vehicles. Then, after introducing the regulations of teleoperation, this chapter summarizes useful technical considerations and improvement techniques, as well as important evaluation criteria based on the current mainstream work. Next, this chapter summarizes the major research efforts of well-known automotive teleoperation participants and diverse teleoperation case studies. In the end, research challenges are presented, and this chapter also points out some possible and promising future research directions in the teleoperation field.

Keywords: Autonomous driving, teleoperation, remote assistance.

8.1 Background

The increased availability of robotics, sensing technologies, communication, and artificial intelligence has led to the wide adoption of autonomous vehicles [42]. There has been a surge in research and development efforts aimed at realizing the vision of fully autonomous driving. Diverse brands of autonomous vehicles, such as Waymo [14], Super Cruise [8], Autopilot [15], and Apollo [47], are currently gaining high momentum in the automotive industry. Since 2015, the global autonmous vehicle market has sustained an average annual growth rate of 12.7%. It is estimated that such a massive market will reach $818.6 billion in 2019, with projections to surpass $3000 billion in 2030 [37].

8.1.1 Human intervention

SAE International, formerly known as the Society of Automotive Engineers, is a US-based professional organization with global operations serving engineering professionals across multiple industries. It has established a six-level classification system for automated driving, ranging from level 0 to level 5, based on the extent of human intervention necessary [43, 26]. This classification system is summarized in Table 8.1.

Today, a portion of level 3 autonomous vehicles are available for consumeial usages, such as Super Cruise and Autopilot. These vehicles are capable of autonomous driving in limited, specific scenarios but require a human driver to be prepared to assume control at any time [53]. Leading automotive companies such as Waymo, General Motors, and Argo AI are testing and operating level 4 autonomous vehicles in defined locations and under favorable conditions, such as clear weather and daytime. Although the robotaxi pilot program of Waymo in Arizona employs human remote operators, a human in-vehicle driver is still generally present in the vehicle [6].

8.1.2 Level 5 fully autonomous driving is still far away

Due to ongoing technical challenges and unpredictable driving conditions, even though autonomous driving technologies have made impressive advancements, there remains a substantial distance to fully realizing autonomous driving [51]. Currently, most level 3 and level 4 test vehicles are limited to predetermined paths or optimized under limited and specific scenarios, such as highway cruising and autonomous parking [50]. However,

Table 8.1 Six levels of driving automation.

	Human drives the vehicle		
	Level 0 (no self-driving)	**Level 1 (human assistance)**	**Level 2 (partially autonomous)**
Driver	Performs all self-driving tasks	Manually control vehicle	Continuous engagement and alertness: must sustain an active involvement and remain vigilant.
Vehicle	Responds to driver's operation	Provide basic driving assist features: e.g., emergency braking	Perform steering, acceleration, and braking in specific situations
	Advanced driver-assistance system (ADAS) drives the vehicle		
	Level 3 (conditionally autonomous driving)	**Level 4 (highly autonomous driving)**	**Level 5 (fully autonomous driving)**
Driver	The human operator needs to assume control of the vehicle at any given moment	The operator may take on the role of a passenger and assume control of the vehicle when the autonomous driving capabilities encounter limitations	No driver required
Vehicle	Assumes full control over the vehicle's acceleration, steering, and braking functions	Perform autonomous driving under most of driving scenarios	Perform autonomous driving under all conditions

when testing autonomous vehicles in unfamiliar settings, some of these vehicles have to stop in the middle of the road due to insufficient environmental perception or issues of control decisions [52].

Within this context, teleoperation presents a promising interim solution for maximizing the potential of current autonomous driving capabilities while preserving driving safety. By incorporating human intelligence, teleoperation enhances the capabilities of autonomous vehicles and enables them to navigate challenging scenarios that they are unable to tackle independently. The evolution of teleoperation technologies is likely to play a pivotal role in the eventual attainment of fully autonomous driving.

As a reminder, the structure of this chapter is as follows. In Section 8.2, teleoperation is defined, and its significance for autonomous vehicles is explored in Section 8.3. The teleoperation regulations are summarized in Section 8.4, while Section 8.5 focuses on technical considerations, enhancement techniques, and evaluation criteria of teleoperation performance, which are described in Section 8.6. The major research efforts of key players in the automotive teleoperation industry and the use cases of teleoperation are outlined in Section 8.7. Section 8.9 addresses ongoing research challenges and introduces potential research directions for future research. In the end, Section 8.10 presents the conclusion of this chapter.

8.2 Definition of Teleoperation

The term "teleoperation" is not a novel word that has emerged recently. Instead, this term refers to operating the machine or robot from a remote location, with roots dating back to the invention of remotely operated weapons in the nineteenth century [11]. Currently, there are numerous machines that can be remotely controlled, from toys to a growing number of Unmanned aerial vehicles, allowing the extension of human perception, assistance, and control capabilities to far-off locations via teleoperation [12].

8.2.1 Teleoperation in connected mobility

According to Section 8.1, the deployment timeline of fully autonomous vehicles remains uncertain and can span anywhere from a few years to several decades. In order to address this issue, teleoperation has emerged as a potential interim solution, which involves a remote operator communicating and potentially intervening manually in a vehicle's operations through wireless channels. This approach is currently being explored by several leading companies in the automotive industry.

To be concrete, the teleoperator (also called remote operator) collects information about the driving scenario from a location outside of the vehicle or a remote office. For example, by observing the vehicle's camera feed in real time or reviewing LiDAR data. They then provide suggestions or directions to the vehicle in regards to tasks that the vehicle may struggle with, such as decoding sensor data or communicating directly with other humans near the vehicle. The vehicle then carries out the task based on the inputs from the remote operator, who may be supported by the vehicle's automation and cognitive abilities to varying degrees.

The key ideas behind remote operation for an autonomous vehicle are illustrated in Figure 8.1. Human remote operators help guide autonomous vehicles to make the right decisions, enabling them to perform efficiently and accurately in challenging or unforeseen circumstances. This approach lessens

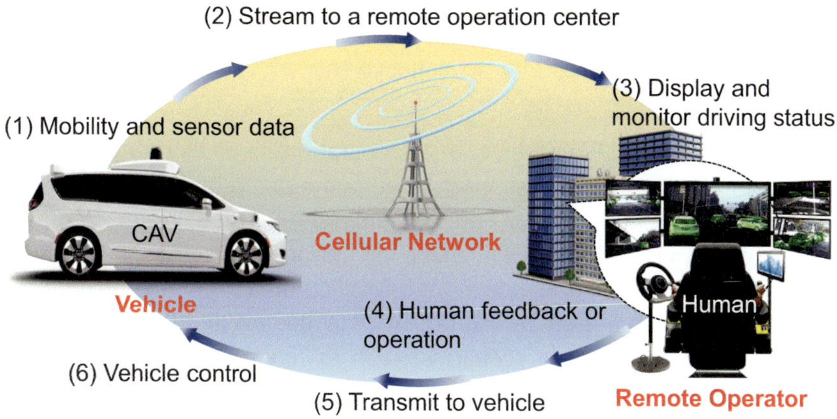

Figure 8.1 A schematic representation of teleoperation [26]. The vehicle transmits data to a remote office via various communication channels, such as satellite connections and cellular networks. The remote operators then interact with the vehicle by sending feedback or commands through a user interface specifically designed for this purpose.

the risk associated with relying solely on wireless networks for full remote operation, as the vehicle operates in collaboration with its own automated features and AI capabilities, instead of surrendering complete control to the remote operator.

8.2.2 Diverse forms of teleoperation

Currently, it is widely acknowledged that the teleoperation of autonomous vehicles can be classified into three distinct forms:

1. **Remote surveillance/monitoring:** This category of teleoperation encompasses the real-time remote surveillance of autonomous vehicles as they operate. Nowadays, autonomous vehicles are equipped with an array of diverse sensors, including cameras, LiDAR systems, GPS, and radars, which generate a significant volume of data during regular driving. The generated data will be transmitted to a remote office through a secure, low-bandwidth wireless network, typically involving data encoding and compression approaches. The remote office will be responsible for decoding the received information and synchronizing and aggregating the diverse data sources, and finally, storing vehicular data in a database. Besides, web pages will be designed for remote operators to monitor the driving status of vehicles, and an edge device

will be used to analyze historical data to support further improvement of involved methods. Through this mechanism, remote surveillance can provide valuable insights to enhance the safety and efficiency of autonomous driving.

2. **Remote assistance**: This aspect of teleoperation involves indirect control methods that provide support to autonomous vehicles in making decisions and executing given computation tasks. This generally encompasses either presentation of a menu of options for selection by the remote operator or the issuance of high-level commands by the operator, who leverages their human intelligence, in the event of unexpected or low-confidence driving scenarios, such as navigating through construction zones, encountering obstacles, dealing with congested parking spaces, traversing challenging off-road terrain, or being stuck at an intersection due to the nonmovement of the right-of-way owner. Ultimately, the autonomous vehicle independently determines the most practical action.
3. **Remote control**: This form of teleoperation entails real-time, direct control over the vehicle, which is typically performed over an extended period. Remote operators are able to manipulate the steering, brakes, and accelerator of the vehicle through the delivery of low-level commands with precise instructions on how to perform the operation (e.g., "turn 20 degrees right" and "increase speed by 10%"). In this scenario, the teleoperated vehicles are not entirely autonomous, as it still involves human control. In this case, the remote control is frequently viewed as a supplementary tool to address obstacles, unfavorable driving conditions, sensor malfunctions, and challenging or hazardous driving scenarios, such as navigating off-road terrain, by involving human intelligence in the loop.

8.3 Importance of Teleoperation

Within the autonomous vehicle ecosystem, teleoperation is regarded as a vital constituent, which assists autonomous vehicles in managing challenging scenarios, thereby ensuring a wide range of autonomous vehicle deployments and secure autonomous driving. The following subsection outlines the comprehensive motivations behind the development of teleoperation.

8.3.1 Challenges of Fully Autonomous Driving

8.3.1.1 Failures in environment perception and recognition

The effectiveness of an autonomous vehicle's ability to understand and respond appropriately to the driving environment heavily relies on the implementation of deep learning-powered systems and involved data analysis algorithms. Nevertheless, as autonomous vehicles become increasingly prevalent, reports of vehicle collisions have also emerged. Recently, the National Highway Traffic Safety Administration released its first-ever report on autonomous vehicle crashes, revealing that there were nearly 400 accidents involving partially or fully autonomous vehicles in the US during a 10-month period [7].

The aforementioned reports clearly indicate that a significant proportion of autonomous vehicle accidents are attributed to the failures of environmental perception and recognition. This is primarily due to the fact that the inference performance of the vehicle's machine-learning models is highly dependent on the particular training datasets. When autonomous vehicles are deployed to challenging scenarios with substantial occlusions or extreme lighting conditions (such as night-time or rainy conditions), their systems and applications have a high possibility of experiencing an unavoidable degradation in performance.

Using lane marking detection as an example, it can be observed that the performance of state-of-the-art (SOTA) methods deteriorates significantly from normal scenarios to challenging scenarios. For instance, SCNN [35] and ENet-SAD [19], which are among the SOTA methods, can achieve over 96% accuracy on the TuSimple dataset [13] collected under normal weather conditions. However, their inference accuracy drops sharply to only 36.56% and 35.79%, respectively, when tested on the BDD100K dataset [49]. BDD100K dataset includes highly complex scenarios, including but not limited to unexpected weather conditions. Therefore, in the practical deployment and application of autonomous vehicles, it is imperative to incorporate teleoperation to leverage remote human intelligence in addressing the open problems of driving environment perception and recognition in challenging situations.

8.3.1.2 Prolonged downtime due to the unfamiliar environmental interpretation

In the context of autonomous vehicle deployment, unfamiliar environments such as exotic cities, construction zones, boxed-in vehicles, and chaotic parking lots pose significant challenges for environment interpretation. Current methods for handling such scenarios are often limited to a shallow level of information, such as color changes and position shifting [18, 21, 40]. Encountering unacquainted environments can result in prolonged downtime for autonomous vehicles that are unable to navigate such scenarios.

Recently, unsupervised learning has been considered a promising technique to leverage unacquainted environment data for domain adaptation (e.g., feature alignment [24, 41, 54], and style transfer [48, 16, 46]). However, the current state-of-the-art approaches still fall into limited applications due to insufficient data interpretation, which makes it difficult for autonomous vehicles to operate effectively and autonomously in the unexpected environment [45].

A common approach in the industry is to use brute-force methods, which involve collecting extensive amounts of training data to cover as many uncertain environments as possible. However, even with exhaustive data collection and annotation efforts, autonomous vehicles will likely encounter unknown situations. In fact, even a hypothetical autonomous vehicle that can handle 99% of driving scenarios independently would still experience an average downtime of 8 min during a 12-h driving day. Such prolonged downtime is unacceptable from both a safety and customer experience standpoint [32].

Thus, in this scenario, teleoperation is required to address unfamiliar driving situations, commonly referred to as edge cases. This approach is to transform these edge cases into known driving events that the autonomous vehicle can confidently handle when confronted with them during driving.

8.3.2 Unforeseen sensor malfunction and invalid sensor data

Autonomous vehicles may encounter unanticipated sensor failures and invalid sensor data, which can be caused by various factors such as mechanical conditions, weather factors, insufficient maintenance, and deficiencies in examinations and evaluations. Sensor malfunctions are normal due to electrochemical reactions and aging processes. Even if the sensors are healthy, the generated data may still not accurately report the real situations and lead to incorrect or irrelevant information. For example, during vehicle operation, the camera may be obstructed by unknown objects such as leaves, mud, or insects, and the radar may deviate from its fixed position due to wind forces. In such situations, autonomous vehicles may not be able to navigate themselves by analyzing invalid sensor data, which necessitates the use of teleoperation as an effective solution to ensure safe and unobstructed vehicle travel.

8.4 Teleoperation Legislation and Standards

Until now, legislative measures concerning autonomous vehicles have been implemented in over 41 states across the United States. As explained in Section 8.3, the incorporation of teleoperation is vital for enabling the widespread integration of autonomous vehicles. Therefore, establishing comprehensive teleoperation regulations can play a pivotal role in clearing the path for autonomous vehicles.

8.4.1 Teleoperation legislation

8.4.1.1 Regulations in US states

In 2018, California included teleoperation in its regulations pertaining to autonomous vehicles. Several other states, such as Ohio, Arizona, Florida, Michigan, and Texas, have mandated the use of teleoperation as part of their autonomous vehicle regulations to facilitate vehicle testing [4, 11]. Additionally, regulations in other states, including Georgia, North Carolina, Nebraska, Nevada, and Tennessee, allow for the use of teleoperation, though it is not explicitly mandated [11].

8.4.1.2 Other countries' regulations

Apart from the United States, numerous other countries have integrated teleoperation into their regulations concerning autonomous vehicles, including the United Kingdom, Canada, Finland, Sweden, and the Netherlands. China also included teleoperation in its autonomous vehicle regulations [4].

8.4.1.3 Illustration: Regulations on teleoperation in Germany

Germany has taken a leading role in the implementation of teleoperation. In 2021, Germany approved fresh regulations for level 4 autonomous vehicles that do not require human driver involvement while the vehicle is in operation, making commercial deployment of autonomous vehicles on German roads feasible. In line with this groundbreaking initiative, Germany has integrated teleoperation into its legal framework, specifying that humans must possess the ability to intervene from a remote site. To prevent ambiguity and misuse, the new law unambiguously lays down that:

(i) Direct remote driving is prohibited, and only indirect control (termed "technical supervision" to clarify role differentiation) is allowed for teleoperation.

(*ii*) The automated driving system (ADS) is responsible for determining when to initiate a teleoperation session. Teleoperation is carried out only in low-risk situations and involves low-risk activities, such as bringing autonomous vehicles to a safe stop and turning on hazard lights. Only under exceptional circumstances (e.g., in the event of a cyber security breach) can a teleoperation session be initiated directly from the remote operations center.

(*iii*) The ADS retains control and responsibility for driving at all times, and it determines the safe timing and method for executing any commands received from a remote operator.

(*iv*) The remote operator is not required to provide continuous monitoring of autonomous vehicles, but only as needed.

(*v*) Continuous data transmission over cellular networks must be ensured.

This illustration highlights Germany's pioneering role in teleoperation regulation. Despite teleoperation legislation lagging behind autonomous vehicle technology, it is gradually catching up.

8.4.2 Teleoperation Standards

At present, government institutions in the United States, including NIST (National Institute of Standards and Technology), recognize the necessity of remote operation for autonomous vehicles, but clear standards for teleoperation are still lacking. The next version of SAE J3016 is likely to provide an official definition of teleoperation, which will serve as a foundation for the development of teleoperation standards. Nonprofit organizations such as the NIST Vehicle Teleoperation Forum [31] and the Teleoperation Consortium [38] are expected to play a significant role in the creation and advancement of teleoperation standards.

NIST Vehicle Teleoperation Forum is a virtual platform that aims to promote teleoperation concepts and standardization in autonomous vehicles [31]. The forum is designed to bring together experts from academia, industry, and government agencies to discuss potential development directions and explore opportunities for teleoperation in different types of autonomous vehicles.

Teleoperation Consortium (TC) aims to provide a communication and collaboration platform for the teleoperation and automotive industry. It has more than 30 industry members, including dozens of leading automotive companies such as Aptiv, Baidu Appollo, Xpeng Motors, and Real-Time

Innovations (autonomous systems software framework companies) [38]. These members are working to develop important industry guidelines.

8.5 Technical Considerations and Improvement Techniques

After discussing the importance of teleoperation and related regulations, we list critical technical considerations and improvement approaches for teleoperations in this section.

8.5.1 Technical considerations

Ensuring safe and reliable teleoperation requires careful consideration of key factors such as connectivity, system architecture, and the remote display of critical driving environments.

(i) **Reliable and high-capacity connectivity with low latency** is crucial to support teleoperation of autonomous vehicles. The timely and accurate delivery of high-resolution video, audio, and sensor data from the vehicle to the remote operations center depends on a stable and fast connection. This is particularly important for video transmission, which typically requires a minimum of 4–10 megabits per second [3]. However, the unstable nature of cellular networks poses a challenge for consistently achieving this level of performance in a moving vehicle.

(ii) **Advanced system architecture:** Integrating the vehicle software and teleoperation hardware into a single unit requires an advanced system architecture that considers the level of teleoperation involved. The corresponding control interfaces must be carefully designed and implemented to handle vehicle steering, acceleration, and braking, among others. An efficient system architecture must coordinate the interaction between hardware and software to achieve higher resource utilization and reduced system response time. Additionally, the architecture should minimize the addition of new hardware to achieve lower space, weight, and power requirements.

(iii) **Remote display of critical driving environments** is a critical aspect of teleoperation, but it also poses significant challenges. To ensure safe and effective teleoperation, ultra-low latency is necessary to provide remote operators with real-time situational awareness of the distant driving environment. In addition, it is important to present only relevant and critical information to remote operators to prevent information overload, which can potentially lead to errors. DriveU.auto recommends that the information presented should be carefully selected to provide operators with the most critical and relevant data [3].

8.5.2 Improvement approaches

Various research domains, including control, communication, sensors, artificial intelligence, edge computing, security, and more, contribute to improving teleoperation. In this section, we categorize teleoperation improvement methods into four main groups, namely operator perception enhancement, interface improvement, control system improvement, and latency compensation. The details of each approach are discussed below.

8.5.2.1 Perception enhancement for operators

To facilitate rapid and informed decision-making, accurate perception of the driving environment, space, and motion is crucial for remote operators. As highlighted in [30], researchers have explored several areas to improve perception accuracy, including:

- **Automated view adjustment** is a technique that involves changing the perspective of the camera from a fixed position to a more flexible one, allowing for automatic prediction and adjustment of the desired viewing direction.
- **Vibro-tactile feedback** involves generating nonvisual sensor information, such as force feedback, auditory cues, and haptic signals, to remote operators in order to provide extra information about the remote driving environment. This approach can be integrated into the teleoperation process to enhance situational awareness and improve operator performance.
- **Viewpoint shifting** refers to the technique of changing the egocentric view of the captured environment, which is a first-person perspective, to an exocentric view, which is a third-person perspective. This technique is used to obtain a broader perspective.
- **Stereoscopic vision** involves building a 3D scenario by utilizing visual inputs from different vision sensors. This approach can provide remote operators with improved environmental perception through depth estimation.

- **Map merging** refers to the process of combining visual information gathered from one or multiple autonomous vehicles to obtain a more comprehensive and accurate view of the driving environment.

8.5.2.2 Immersive virtual driving interface

The term "virtual driving environment" describes the reconstruction or simulation of the real-world driving experience. Current user interfaces for remote operators enable the use of specialized devices such as joysticks and steering wheels. However, these interfaces are limited in providing feedback such as motion and vibration to enhance the remote operator's perception of the environment.

Augmented reality (AR) is a technology that enables an interactive and immersive experience by overlaying digital information in the real-world environment. It has been identified as an enabling technology that can improve communication between autonomous vehicles and remote operators. By designing an immersive interface that triggers a 3D virtual reconstruction and using commercially available AR glasses and headsets (if necessary), remote operators can be fully immersed in the vehicle environment, allowing for spatial dialogue and intuitive visual signals.

8.5.2.3 Enhancement directions for control systems

Enhancing the control systems of autonomous vehicles to facilitate teleoperation has been a longstanding area of research. These approaches can be broadly classified into three categories, as outlined below.

- **Enabling automatic tracking** can alleviate the challenges associated with the remote operation of multiple vehicles by a single operator. By ensuring that the control of the lead vehicle automatically tracks subsequent vehicles, the workload involved in the teleoperation of a fleet of autonomous vehicles can be significantly reduced.
- **Enhancing local autonomy** is considered an effective means to improve teleoperation performance, as full autonomous vehicle development is still a long-term future research topic. Improving autonomous capabilities for decision-making, such as automatic collision avoidance, real-time diagnostics, and trajectory planning, can be implemented to increase the vehicle's ability to operate independently, reducing the reliance on teleoperation.
- **Improvement of control policies** is a research area focused on developing intelligent control strategies to enhance the controllability of

teleoperated vehicles. One such approach is to leverage compensation and adaptation algorithms to determine speed references by integrating speed commands from the remote operator, sensor feedback, and the vehicle's current state with time delays.

8.5.2.4 Compensation for latency

Communication delays between autonomous vehicles and remote operators are an inevitable challenge that can significantly impact the effectiveness of teleoperation. In challenging environments, such as those with long and time-varying delays, effective teleoperation can become impossible. The following section discusses two potential methods for latency compensation.

- **Parallel virtual vehicle control**: To mitigate the effects of communication delays in teleoperated autonomous vehicles, one potential solution is to implement parallel virtual vehicle control. This involves creating a parallel virtual environment that displays the immediate responses of the autonomous vehicles to the remote operator on a virtual interface. The state of the calibrated virtual vehicle is represented on the user interface, allowing the remote operator to generate action commands prior to receiving actual visual feedback from the real environment. By using this virtual environment to simulate the control of autonomous vehicles, it can reduce the impact of communication delays and improve the overall teleoperation performance.
- **Predictive display** is a method developed to compensate for the communication and control loop delays. It displays the current and future positions of all observed vehicles, enabling remote operators to anticipate the response of the autonomous vehicle system before it actually occurs. This approach helps avoid potential collisions and mitigate the impact of high- latency delays in challenging environments.

8.6 Assessment Standards for Teleoperation Performance

Teleoperation is a multifaceted research domain spanning across multiple disciplines. Due to the variability in vehicle and interface types, control technologies, driving scenarios, and diverse mobile services, a standardized set of evaluation metrics may not be applicable to all situations, and developing a unified evaluation approach for teleoperation of autonomous vehicles can be challenging. This section outlines a collection of possible evaluation techniques that can be employed.

Based on [30], remote operators can assess the effectiveness of teleoperation by conducting simple tasks with autonomous vehicles, such as driving the vehicle along a predetermined track, recognizing objects, or maneuvering the vehicle to halt near an obstacle. The quantitative assessment criteria may involve measuring the completion time of multiple tasks, the velocity of the vehicle, the frequency of collisions with obstacles, the actual vehicle location compared to the intended location, the distance traveled by the vehicle over a set period, the number of executed commands, the number of unfulfilled tasks, and the actual trajectory compared to the intended trajectory. One possible future research direction in the field of teleoperation could be to develop a more comprehensive tool for evaluating teleoperation performance.

8.7 Distinguished Participants in Automotive Teleoperation

Automotive manufacturers, startups, ride-sharing firms, food delivery companies, and renowned cloud service providers have been making strides in the development of cutting-edge teleoperation solutions. This section delves into the research endeavors of prominent automotive teleoperation players, as outlined in Tables 8.2 and Table 8.3.

8.7.1 Automotive company and startup

Despite Tesla's position as an early innovator in autonomous vehicles, and its highly-regarded Autopilot system, the company does not seem to publicly acknowledge the use of teleoperation [1].

Jaguar Land Rover, a luxury vehicle and sport utility vehicle manufacturer and a subsidiary of Tata Motors Limited, has showcased how a remote-controlled Land Rover Sport research vehicle can be directed by a driver outside the vehicle via a smartphone app to navigate obstacles or exit a parking space [20]. Specifically, remote operators can control the vehicle's steering, brakes, and accelerator through the smartphone application to safely guide the vehicle out of challenging situations, such as difficult off-road terrain. By walking alongside the vehicle, the driver can continuously assess the angle at which the vehicle should approach and depart, especially when crossing streams or slippery roads caused by rain, mud, or snow. Additionally, if the vehicle is parked too closely, the driver can use the smartphone app to reverse the vehicle out of the parking space, serving as their own off-road spotter. The teleoperation feature is only accessible if the driver is within 10

Table 8.2 An overview of major participants in the automotive industry.

Company	General infomation	Research efforts in teleoperation
Jaguar Land Rover [20]	• Automotive company • Founded date: 2008 • Headquater: Whitley, Coventry, United Kingdom	• Remote operation of a vehicle can be performed by drivers through a smartphone application, enabling them to safely steer, accelerate, and brake their vehicle out of difficult scenarios. • Teleoperation should only be activated when the user is located within 10 m of the vehicle, and the smart key can be detected.
Waymo [17]	• Automotive company • Founded date: 2009 • Headquater: Mountain View, California, United States	• Remote operators monitor a constant stream of video data from eight cameras that are installed on each vehicle. • Remote operators do not manipulate vehicles but instead answer specific questions under challenging situations, relying on their human understanding.
Uber [2]	• Mobility-as-a-service provider • Founded date: 2009 • Headquater: San Francisco, California, United States	• Develop teleoperation technology in-house, incorporating industry-standard protocols, to maintain human involvement in the loop without direct control. • Remote operators receive training in remote assistance and autonomous systems, but are not trained specifically in steering, braking, or other driving maneuvers.
Cruise [8, 22, 32]	• Automotive company • Founded date: 2013 • Headquater: San Francisco, California, United States	• Cruise aims to commercialize its fleets through a ride-sharing platform, enabling passengers to communicate with remote operators via a button press. • Cruise acquired Voyage, an autonomous vehicle startup that developed a customized workstation for remote operators called the Telelessist Pod.
Toyota [39]	• Automotive company • Founded date: 1937 • Headquater: Toyota, Aichi, Japan	• Transmit a selected portion of data collected by the vehicle's multiple cameras, radar, and LiDar to the remote operators. • Remote operators may take direct control of the vehicle or provide instructions for the vehicle to perform specific actions.
Motional [33]	• Automotive startup • Founded date: 2020 • Headquater: Boston, Massachusetts, United States	Utilize teleoperation technology to provide remote vehicle assistance while operating SAE level 4 autonomous vehicles.
Aurora [1]	• Automotive startup • Founded date: 2017 • Headquater: Pittsburgh, Pennsylvania, United States	The developed system, called "Teleassist," is designed to notify remote operators of situations requiring assistance, enabling autonomous testing fleets in the Bay Area and Pittsburgh to navigate various road conditions.

Table 8.2 *Continued*

Company	General infomation	Research efforts in teleoperation
Drive.AI [34]	• Automotive company • Founded date: 2015 • Headquater: Mountain View, California, United States	• Remote operators respond to specific queries within a unique driving scenario grounded in human comprehension.• The decisions of remote operators are also integrated into the vehicle's software to enhance its capacity for decision-making.
Aptiv [23, 29]	• Automotive supplier • Founded date: 1994 • Headquater: Troy, Michigan, United States	Create a proprietary teleoperation solution (potentially from the Las Vegas operations center).
Zoox[27]	• Automotive startup • Founded date: 2014 • Headquater: Foster City, California, United States	Develop TeleGuidance, a critical constituent of the autonomous driving stack, to integrate human guidance into the motion planning algorithms of the vehicle.

m of the vehicle and the related Smart Key can be detected. If the driver exceeds the specified range or gets too close to the vehicle, the system will stop the vehicle.

Waymo, a well-known autonomous driving company that provides a commercial autonomous driving taxi service, is a subsidiary of Alphabet Inc., the parent company of Google. Waymo has a team of remote operators stationed at its facilities in suburban Phoenix who monitor and assist its ride services, though they do not operate vehicles directly [17]. Instead, they observe real-time video feeds from the vehicle's eight cameras via a graphical user interface. In situations where the vehicle encounters low-confidence scenarios, such as construction zones, a flock of birds, or an unexpected turn, the remote operators can provide assistance by answering specific questions about the ambiguous situation using relevant buttons. With the help of human comprehension and knowledge of the surrounding environment, the vehicles can safely navigate through such situations.

Toyota is a multinational automotive company that has developed a control system for autonomous and semi-autonomous vehicles that supports teleoperation. In situations where the vehicles encounter unfamiliar driving environments that are not suitable for autonomous operation, such as road construction or obstructions, the system sends vehicle data captured by various cameras, radar, and LiDAR sensors to a remote operator. The remote operator can manually operate the vehicle remotely or issue commands to the vehicle systems. To optimize bandwidth, the captured data can be transmitted

Table 8.3 An overview of major teleoperation technology providers.

Company	General infomation	Research efforts in teleoperation
Phantom Auto [44]	• Teleoperation company • Founded date: 2017 • Headquater: Mountain View, California, United States	• Augment vehicle autonomy through the remote monitoring of numerous vehicles, with the ability to focus on specific vehicles for precise remote control when necessary. • Enable remote operation of material handling vehicles (e.g., forklifts, yard trucks, and tuggers in intermodal shipping areas) by humans situated thousands of miles away. • Teleoperators can delineate a route for a Postmate (an on-demand small-size delivery vehicle) to follow.
Designated Driver [10, 11]	• Teleoperation company • Founded date: 2018 • Headquater: Portland, Oregon, United States	• Provide teleoperation-as-a-service for fleets, which includes certified and trained remote operators, as well as hardware and software kits. • Offer remote driving and assistance services. • Deliver teleoperation solutions for fixed-route shuttles (in collaboration with Texas A&M University).
Scotty Labs [28]	• Teleoperation startup • Founded date: 2017 • Headquater: Menlo Park, California, United States	• Utilizing teleoperation to address self-driving difficulties, while enabling the company to concentrate on improving their autonomous driving technology. • Collaborated with Voyage, a firm dedicated to developing a fully autonomous driving taxi platform.
Ottopia [11, 36]	• Teleoperation company • Founded date: 2018 • Headquater: Tel Aviv, Israel	• Provide remote assistance and remote driving support to solve the main challenges in the field of teleoperation • Produce vehicle remote control modules and remote control center consoles • Established partnerships with Denso, Via, Bestmile, BMW, Innoviz, Telekom, Deutsche, Gaussin, May Mobility, and EasyMile.
RoboAuto [9]	• Teleoperation startup • Founded date: 2017 • Headquater: MBrno-Královo Pole, Czechia	Providing teleoperation support for different industrial vehicles, including excavators, forklifts, harvesters, and so on.
DriveU.auto [3]	• Teleoperation platform provider • Founded date: 2019 • Headquater: Kfar Saba, HaMerkaz, Israel	• Develop a teleoperation platform called DriveU, which has been used on public roads • The SDK and open API of the product support fast and direct integration • DriveU platforms could be deployed on various NVIDIA Jetson platforms.

to the remote operator in a limited portion. This control system is disclosed in a Toyota patent on remote control technology [39].

Cruise, a subsidiary of General Motors focused on autonomous driving technology, is working toward the commercialization of its Cruise fleets through a ride-sharing platform [22]. Passengers will be able to use mobile apps to request rides and customize their ride experience through climate and radio settings. Upon entering the vehicle, passengers will have access to real-time status information via touchscreen tablets, which will also remind them to fasten their seat belts and close all doors. Additionally, passengers will be able to communicate with remote operators by pressing a button. Cruise has also acquired Voyage [8], an autonomous vehicle startup known for its Telessist Pod product, which provides a proprietary workstation designed for remote operators [32]. The Telessist Pod uses automotive-grade components for steering, acceleration, and braking.

Aptiv is an automotive technology supplier headquartered in Dublin, Ireland, and the United States. The company is currently developing its own teleoperation solution, which may potentially be operated from its operations center located in Las Vegas [29]. Aptiv is also a member of the Teleoperation Consortium (TC), a nonprofit business league dedicated to fostering collaboration among companies, organizations, and governmental bodies involved in bidirectional vehicle communication development [29]. In addition to their work on teleoperation in vehicles, Aptiv researchers have also focused on modeling the teleoperation of Unmanned Ground Vehicles (UGVs) [23].

Uber, a US-based mobility-as-a-service provider that offers on-demand taxi services, views teleoperation as an essential intermediary stage toward autonomy. As such, the company is actively developing teleoperation technology with industry-standard protocols. The primary aim is to keep humans in the loop, enabling remote operators to monitor the actions of vehicles in real time and provide operational recommendations to ensure safe and efficient operations [2]. These remote operators are trained in the use of autonomous vehicle systems and remote assistance but are not specifically trained in driving maneuvers such as steering or braking.

Motional is an American autonomous vehicle start up that operates one of the most established public fully autonomous robotaxi services, and is a joint venture between automaker Hyundai Motor Group and auto supplier Aptiv. In order to support its SAE level 4 autonomous vehicles, Motional has partnered with Ottopia, a global leader in teleoperation, to provide remote assistance to its robo-taxi fleet [33]. Motional has also announced a partnership with Lyft to launch an autonomous taxi service in multiple U.S. markets, which will support teleoperation and provide remote assistance to millions of Lyft riders starting in 2023.

Aurora, an autonomous driving technology startup backed by Amazon, is recognized for its development of the Aurora Driver, an integrated computer system for autonomous driving in vehicles [1]. The company acknowledges that developing a teleoperation system is crucial to ensure public acceptance of its Aurora Driver. To that end, Aurora has created a remote control system named "teleassist," which empowers its testing fleets in the Bay Area and Pittsburgh to handle various road conditions. Trained remote operators can access a vehicle's sensors and provide guidance and advice in exceptional scenarios. Aurora's vehicles will move to the side of the road safely to make such communications for guidance.

Drive.AI, a technology company owned by Apple, specializes in developing autonomous driving systems for autonomous vehicles using AI. The company leverages teleoperation technologies to guide autonomous vehicles through challenging or ambiguous situations [34]. For example, if pedestrians have the right-of-way but are not crossing, the vehicle may communicate with a remote operator for guidance on how to proceed. The remote operator can provide guidance to the vehicle based on their expertise, and the vehicle can learn from this experience. Feedback from remote operators is incorporated into Drive's learning software, allowing the autonomous vehicle to make better decisions on its own in similar situations in the future.

Zoox, an autonomous driving startup that was acquired by Amazon, is working on developing autonomous vehicles for Mobility-as-a-Service. Zoox has created TeleGuidance, which is a significant part of its autonomous driving stack. This technology enables Zoox's vehicles to incorporate human contextual guidance into algorithms such as path planning and object recategorization, allowing the vehicles to navigate complex scenarios by utilizing real-time guidance from remote operators [27]. For example, in path planning, a remote operator can assist the vehicle in finding a new path that avoids an obstruction such as a double-parked truck. Furthermore, Zoox has demonstrated how TeleGuidance can aid vehicles by reclassifying objects with the help of remote operators. For instance, a stationary vehicle, such as a truck whose driver is doing road construction work, can be appropriately classified as a stationary object rather than an active vehicle waiting to move forward by using teleoperation.

8.7.2 Companies offering teleoperation services

Designated Driver is a leading teleoperation provider that offers comprehensive teleoperation technology solutions for both autonomous and nonautonomous vehicles. The company provides teleoperation-as-a-service for fleets, including hardware and software kits, as well as trained and certified remote operators. Designated Driver offers both remote driving and remote assistance for teleoperation [10]. Its products are available through AutonomouStuff, which offers a wide range of autonomous vehicle products. Additionally, in August 2019, Designated Driver partnered with Texas A&M University to provide teleoperation for fixed-route shuttles [11].

Phantom Auto is an industry pioneer specializing in teleoperation, offering software solutions that are not tied to specific vehicles, and have low-latency communications and APIs for both remote driving and remote assistance [44]. Their technical products and training services enable human operators to remotely operate material handling vehicles, such as forklifts, tuggers, stackers, counterbalances, pallet jacks, and yard trucks, manufactured by a range of companies including UniCarriers and KION Group. Alternatively, they can complement vehicle autonomy by remotely monitoring multiple vehicles, such as the autonomous sidewalk vehicles used by Postmates, and quickly take control of individual vehicles when needed. Operators can draw paths for the vehicles to follow, and even take direct control when necessary. Phantom Auto's technology is currently being used by major companies such as Uber, ITS ConGlobal, and Postmates to support a range of use cases, including last-mile delivery, material handling, and yard operations.

Ottopia is a leading provider of automotive-grade teleoperation, with investment from Hyundai Motor Company. Its products include both remote assistance and remote driving solutions that aim to tackle key challenges in teleoperation, such as network connectivity, safety, and cybersecurity [36]. For example, its teleoperation software platform is AI-powered and can predict the quality of cellular carriers along a vehicle's route. Additionally, it employs video compression techniques to minimize the amount of video transmitted during teleoperation. Ottopia offers an in-vehicle teleoperation module and a teleoperation center console [11]. The in-vehicle module comprises advanced remote assistance systems with safety algorithms, and its advanced teleoperation APIs provide scalable control methods for different application scenarios. The teleoperation center console incorporates off-the-shelf hardware and specialized software that can fuse real-time video streams and vehicle sensor data. Ottopia has established partnerships with BMW, Denso, Bestmile, EasyMile, Innoviz, Deutsche Telekom, Gaussin, Via, and May Mobility to date.

Scotty Labs is a Google-backed teleoperation company, which focuses on providing teleoperation services for autonomous vehicles. It asserts that the greatest potential for the future of autonomous driving is the ability of humans and machines to collaborate and tackle the most challenging problems together [28]. The first company to partner with Scotty Labs is Voyage, which is dedicated to developing a fully autonomous driving taxi platform. Scotty Labs has previously collaborated with Voyage to provide teleoperation services for autonomous vehicles in retirement communities.

DriveU.auto is a software developer that specializes in creating a cutting-edge connectivity platform called DriveU for teleoperating autonomous vehicles. The platform features an SDK and open APIs, which enable seamless integration and transmission of high-resolution video, sensor data, audio, and custom applications [3]. The DriveU platform is designed to run on the main computing units of vehicles powered by NVIDIA Jetson platforms, such as Jetson AGX Xavier, Jetson Xavier NX, and Jetson Nano GPU-accelerated systems-on-module. DriveU has already been deployed and utilized on public roads and sidewalks to support the large-scale deployment of autonomous vehicles.

RoboAuto is a teleoperation startup that specializes in the development and integration of comprehensive teleoperation services for a variety of industrial vehicles used in areas such as mining, logistics, demolition, and agriculture. Its focus is on developing solutions for a range of industrial vehicles, including forklifts, excavators, and harvesters [9]. For instance, in the case of building demolition, RoboAuto's teleoperation services provide a means of removing human operators to a safe distance. Additionally, its teleoperated vehicles are capable of transporting materials and samples across large areas such as factories, chemical facilities, and refineries without requiring human intervention.

8.8 Teleoperation Case Studies

Teleoperation is a crucial component of autonomy across various industries, including but not limited to robotaxis, shuttles, trucks, electric scooters, mining equipment, flying vehicles, military systems, and medical robotics. At present, teleoperation is used not only in teleoperated autonomous vehicles, such as Waymo, Cruise, and Zoox robotaxis, but also in sidewalk vehicles, such as Kiwibot and Postmates, and in various industrial vehicles, including forklifts, trucks for port and loading/unloading operations, combines and harvesters in agriculture, and excavators in mining.

(a) **Safety mechanism during autonomous vehicle trials**: To ensure safety during autonomous vehicle trials, a safety mechanism is commonly implemented, which typically involves the presence of a safety driver, either physically or remotely. This requirement may be stipulated by regulatory bodies or self-imposed by the testing organization. In the event of an emergency or hazardous situation, the remote operator or safety driver is responsible for promptly intervening and taking necessary measures to ensure the safe operation of the vehicle.

(b) **The provision of a reliable fallback mechanism for autonomous vehicles in challenging and unanticipated scenarios** presents a major obstacle to realizing full autonomy in driving. The deployment of numerous decision-making algorithms capable of handling diverse driving scenarios, including those at the fringes, poses a formidable challenge. Nonetheless, it is practically unattainable to cover every conceivable real-world situation. Consequently, teleoperation plays an essential role in providing guidance to autonomous vehicles when they face novel or intractable situations, such as intricate off-road terrain and crowded parking lots, thereby guaranteeing the safe accomplishment of their objectives.

(c) **Package and food delivery**: Teleoperation is currently extensively employed in the delivery of food and packages, particularly for last-mile delivery robots and autonomous trucks that manage the entire delivery process, from pick-up to handover.

(d) **Vehicle ride-sharing services** enable passengers to utilize mobile applications to request rides and personalize their ride experience by adjusting climate control and radio station settings. Furthermore, passengers can conveniently communicate with remote operators by pressing a button to make special requests and report any issues or concerns.

(e) **Electric scooters utilized in shared mobility services**: Teleoperation is also utilized in the management of shared electric scooters, allowing for the remote repositioning of scooters to charging stations or designated drop-off locations after rental usage, thereby enhancing operational efficiency.

(f) **Logistics operations within chemical factories or refineries**: Teleoperated vehicles are commonly employed in logistics operations in chemical factories or refineries to transport materials within vast areas. Given that all operations are remotely performed, and one operator can oversee or manage multiple vehicles simultaneously, teleoperation can ensure operator safety while enhancing work efficiency.

(g) **House demolition and building dismantling** are inherently risky tasks. Teleoperation represents a novel approach that enables human workers to be positioned at a safe distance, thereby enhancing work safety and optimizing worker efficiency.

(h) **Mowing tasks in challenging terrain**, such as steep hills, along highways, or in hot climates, pose significant challenges for operators. However, teleoperation presents a viable solution that simplifies and enhances the safety of such tasks.

(i) **Minning operations** frequently require excavators to either be actively engaged in digging or waiting for the next task. However, with teleoperation, remote operators can eliminate the waiting period by quickly disconnecting from one vehicle and connecting to another to seamlessly continue excavation work.

8.9 Open Problems and Potential Future Directions

Enabling effective and efficient teleoperation for maneuvering autonomous vehicles on public roads represent a highly intricate task, given the need to ensure that the vehicle can accurately detect and interact with all conceivable objects on roads with varying traffic, weather, and visibility conditions. This section highlights the principal challenges that are particularly relevant to teleoperation.

8.9.1 Open problems in teleoperation

8.9.1.1 Bandwidth limitations

To enable teleoperation, real-time data transmission and display of high-quality sensor data, video, and diverse commands are essential. Insufficient bandwidth can result in delays in video and information transmission, which may impair spatial awareness and potentially cause the operator to miss important environmental cues. However, maintaining adequate signal strength can be challenging due to factors such as distance, electronic interference, or physical obstructions. Thus, ensuring consistent and reliable bandwidth for remote vehicle operation continues to be a significant challenge in the field of teleoperation.

8.9.1.2 Fluctuating and excessive latency

The overall latency in teleoperation encompasses both the latency of control commands sent from the remote operator to the autonomous vehicle and the latency of feedback received from the vehicle back to the remote operator.

(1) **Fluctuating latency** results in a more pronounced degradation of teleoperation performance than constant latency [25]. The greater the delay's standard deviation, the more it impairs the remote operator's control of the vehicle. Thus, a crucial challenge is to ensure sufficient teleoperation performance and safety of the vehicle in situations where latency varies significantly.

(2) **Excessive latency**: Moreover, prior research has demonstrated that a total latency of less than approximately 170 ms has only a minor impact on teleoperation performance [5]. Conversely, when the total latency exceeds 300 ms, remote operators are more likely to wait for feedback on the previous command before issuing the next command, which becomes increasingly noticeable as the latency increases [5]. Additionally, some studies have found that teleoperation performance significantly deteriorates when the total latency exceeds 700 ms, rendering vehicle teleoperation is largely infeasible [25].

(3) **Taking appropriate precautions**: Remote operators or vehicles should promptly initiate necessary proactive measures in advance, such as reducing driving speed, pulling over or stopping the vehicle, or avoiding complex maneuvers.

(4) **Anticipating disconnections**: Remote operators and autonomous vehicles need to be capable of accurately forecasting the timing and duration of disconnections.

(5) **Self-rescue in the event of unforeseen disconnections**: In the event of an unexpected disconnection where the remote operator loses control of the autonomous vehicle, the vehicle should be able to initiate a self-rescue maneuver to transition into a safe mode until the remote operator is able to regain control.

8.9.1.3 Other open problems in teleoperation

The following presents the remaining four open issues in teleoperation [30].

(1) **Overwhelmed information**: Remote operating systems usually generate an overwhelming amount of monitoring information, making it challenging for remote operators to comprehend and process the data in real-time. Thus, it is crucial to design an interface and system that can present the driving environment data in an easy-to-understand format.
(2) **Trade-off among various factors**: Achieving an optimal balance among diverse factors, such as vehicle characteristics, mobility tasks, teleoperation methods, vehicle speed, video frame rates, pixel resolutions per frame, and bit depths per frame, is a significant challenge in teleoperation.
(3) **Determining optimal parameters**: One of the key challenges in the development of a visual assistive user interface for remote operators is to determine the optimal parameters for a given scene, such as the appropriate field of view, number of video feeds, and acceptable resolution.
(4) **Uncertain operation of human remote operators**: The abundance of high-latency teleoperation tasks can be stressful for remote operators, and they may experience fatigue and over-correction, leading to erroneous operations. Nevertheless, current systems do not account for the uncertainty associated with the remote operator, including the absence of real-time monitoring of the remote operator's physical condition.

8.9.2 Potential future directions

Autonomous teleoperation: Currently, autonomous vehicle teleoperation is solely dependent on human remote operators. However, according to Zhang *et al.* [53], the teleoperation system is expected to become more intelligent as AI-powered cloud or edge devices can offer advanced assistance to human remote operators. These devices can predict network conditions and send warnings to human remote operators, which they might overlook, leading to the emergence of autonomous teleoperation.

In this section, we provide an overview of several teleoperation topics that are of interest to both industry and academia, including but not limited to: safety of teleoperation, latency and bandwidth reduction in teleoperation, operational design of teleoperation, interface design for vehicles, and teleoperation systems, open and customizable teleoperation systems, and multimodal teleoperation systems. Furthermore, a potential avenue for future research could be the development of a comprehensive tool for evaluating the overall performance of teleoperation. Additionally, a collaborative effort in defining teleoperation standards will be instrumental in advancing the industry.

8.10 Conclusion

In spite of the impressive advancements in autonomous vehicles, fully autonomous driving still remains a distant goal. This chapter presents teleoperation as a promising interim solution that can bridge the gap between the current state of autonomous driving and the widespread adoption of autonomous vehicles. Teleoperation complements autonomous driving systems by leveraging human intelligence. This can make autonomous vehicles more capable and enable them to navigate through tricky situations that they cannot handle independently. Teleoperation of autonomous vehicles is generally classified into three levels: (i) remote monitoring, which involves real-time remote monitoring of how autonomous vehicles are driving; (ii) remote assistance, which refers to an indirect control method that assists autonomous vehicles in the decision-making process; and (iii) remote control, which enables real-time and direct control (such as remote driving) over a significant period of time (as described in Chapter 8.2).

Why is teleoperation important? Teleoperation is essential for autonomous vehicles due to the challenges of achieving fully autonomous driving, which include the following: (1) Failures in environment perception and recognition under complex scenarios: Autonomous vehicle systems rely heavily on machine-learning models trained on specific datasets, and when deployed to challenging scenarios with severe occlusions or extreme lighting conditions, the inference performance of these models may degrade, making teleoperation necessary. (2) Long downtimes in unacquainted environments that autonomous vehicles cannot handle: Despite vigorous data collection and annotation, autonomous vehicles will still encounter unknown situations. Even an ideal autonomous vehicle capable of handling 99% of driving situations on its own would still be stuck for 8 min during a 12-h driving day. This downtime is unacceptable from both a safety and customer experience perspective, making teleoperation a necessary solution. (3) Malfunctioning sensors and invalid sensor data: Autonomous vehicles may not be able to navigate themselves by analyzing invalid sensor data. Teleoperation is an effective and necessary solution to ensure safe vehicle trip progression and unblock the vehicle in the event of unexpected sensor malfunctions or invalid data (as described in Section 8.3).

Teleoperation regulations and standards: As of now, over 41 states in the United States have established regulations for autonomous vehicles. In 2018, California included teleoperation in its autonomous vehicle regulations. Several other states, including Michigan, Arizona, Ohio, Florida, and Texas, have mandated teleoperation as a component of their autonomous vehicle regulations to facilitate testing of autonomous vehicles [4, 11]. Although teleoperation is not explicitly mandated, other states, such as Georgia, Nebraska, North Carolina, Nevada, and Tennessee, allow it through their enacted regulations [11]. Germany has been a pioneer in the implementation of teleoperation, and the law stipulates that humans must be capable of intervening from a distance (as discussed in Section 8.4).

Technical considerations: Next, we outline critical technical considerations that must be taken into account to ensure safe and reliable teleoperation. These considerations include: (1) reliable, high-capacity, and low-latency connectivity, which is essential to support the timely delivery of high-resolution video, audio, and data from multiple vehicle sensors to remote operators; (2) advanced system architecture that integrates the vehicle software and teleoperation hardware into a single unit with well-designed control interfaces; and (3) remote display of critical driving environments, presenting only relevant and critical information as remote operators can become overwhelmed with unnecessary data, which may result in more harm than good (as discussed in Section 8.5.1).

Enhancement methods for teleoperation: The enhancement methods of teleoperation involves various research domains such as control, communication, sensor technology, artificial intelligence, edge computing, security, among others. We categorize the methods for improving teleoperation into four main categories: operator perception enhancement, interface improvement, control system improvement, and latency compensation. These approaches are described in detail in Section 8.5.2.

Teleoperation performance evaluation criteria: Teleoperation is a complex research field that spans across multiple disciplines. Given the variety in vehicle and interface types, driving environment, control technologies, and types of mobile services, a predefined set of evaluation metrics may not be suitable for all scenarios, and developing a unified evaluation technique for autonomous vehicle teleoperation is a challenge. In Section 8.6, we discuss a set of potential evaluation approaches that could be applied.

Prominent players in the automotive industry: Automotive startups, ride-sharing and food delivery companies, and established cloud providers, have been actively engaged in the development of advanced teleoperation solutions. In Section 8.7, we explore the research endeavors of major automotive teleoperation players, such as Waymo, Cruise, Jaguar Land Rover, Uber, Motional, Toyota, Aptiv, Zoox, Designated Driver, Drive.AI, Ottopia, Scotty Labs, Phantom Auto, RoboAuto, and DriveU.auto.

Teleoperation case studies: Furthermore, teleoperation has potential applications in various fields such as robotaxis, trucks, shuttles, electric scooters, flying vehicles, mining equipment, medical robotics, military systems, and more. At present, in addition to conventional autonomous vehicles, teleoperation is widely utilized in sidewalk vehicles and industrial vehicles such as forklifts and trucks (for port and loading/unloading operations), combines and harvesters (in agriculture), and excavators (in mining). We summarize the most promising use cases of teleoperation, which include (1) serving as a fallback mechanism for autonomous vehicles in challenging and unforeseen scenarios, (2) acting as a safety mechanism during autonomous vehicle testing, (3) facilitating food and package delivery, (4) supporting shared electric scooters, (5) enabling vehicle ride-sharing, (6) assisting in housebreaking and building demolition, (7) optimizing logistics in chemical factories or refineries, (8) mowing around highways or in challenging terrains, and (9) enhancing mining operations (Section 8.8).

Research challenges and potential future directions: Finally, we address major challenges that are critical to teleoperation, such as limited bandwidth, high and varying latency, and ensuring safety in poor network conditions. In the future, teleoperation of autonomous vehicles may provide advanced support to remote human operators. Additionally, we summarize several teleoperation topics that may be of interest to industry and academia, including teleoperation safety, reducing latency and bandwidth requirements, operational design, interface design for vehicles and teleoperation systems, customizable and open teleoperation systems, and multimodal teleoperation systems. Future research could focus on developing a comprehensive tool for evaluating overall teleoperation performance, as well as collaboratively defining teleoperation standards to drive industry progress (Section 8.9).

References

[1] Alan Ohnsman. Amazon-backed Aurora readies an air-traffic control system for self-driving cars. https://www.forbes.com/sites/alanohnsman/2019/12/16/amazon-backed-aurora-readies-air-traffic-control-system-for-self-driving-cars/?sh=111bb6c01a16, 2019.

[2] Alejandra Sarmiento. Smooth teleoperator: The rise of the remote controller. https://venturebeat.com/2020/08/17/smooth-teleoperator-the-rise-of-the-remote-controller/, 2022.

[3] Alon Podhurst. What's teleoperation got to do with it? the connection between remote driving and autonomous vehicles. https://driveu.auto/blog/the-complete-guide-to-av-teleoperation/, 2021.

[4] Amit Rosenzweig. Regulators know teleoperation is key for self-driving vehicles to succeed. https://venturebeat.com/2021/06/17/regulators-know-teleoperation-is-a-must-have-for-self-driving-vehicles-to-succeed/, 2021.

[5] Oscar Bodell and Erik Gulliksson. Teleoperation of autonomous vehicle with 360° camera feedback. *Department of Signals and Systems, Chalmers University of Technology*, 2016.

[6] I Boudway and J Brustein. Waymo sees human drivers in autonomous cars for foreseeable future. 7. *Verfügbar unter: https://www.ttnews.com/articles/waymos-long-term-commitment-safety-drivers-autonomous-cars-0, veröffentlicht,* 2020.

[7] David Kroman. First-ever self-driving vehicle crash report released. Nearly all the WA wrecks involved Teslas. https://www.seattletimes.com/seattle-news/transportation/first-ever-self-driving-vehicle-crash-report-released-nearly-all-involved-teslas-in-washington/, June 15, 2022.

[8] David Welch. Gm-backed Cruise buys self-driving startup Voyage. https://www.ttnews.com/articles/gm-backed-cruise-buys-self-driving-startup-voyage, 2021.

[9] David Welch. Control all your vehicles remotely from one spot. https://roboauto.tech/, 2022.

[10] Alex Davies. The war to remotely control self-driving cars heats up. *Retrieved October*, 13:2020, 2019.

[11] Egil Juliussen. Avs will need teleoperation: Here's why. https://www.eetimes.com/avs-will-need-teleoperation-heres-why/, 2021.

[12] Terrence Fong and Charles Thorpe. Vehicle teleoperation interfaces. *Autonomous robots*, 11(1):9–18, 2001.

[13] Mohsen Ghafoorian, Cedric Nugteren, Nóra Baka, Olaf Booij, and Michael Hofmann. El-gan: Embedding loss driven generative adversarial networks for lane detection. In *Proceedings of the European Conference on Computer Vision (ECCV)*, pages 0–0, 2018.

[14] Samuel Gibbs. Google sibling waymo launches fully autonomous ride-hailing service. *The Guardian*, 7, 2017.

[15] Samineh Gillmore and Nathan L Tenhundfeld. The good, the bad, and the ugly: Evaluating tesla's human factors in the wild west of self-driving cars. In *Human Factors and Ergonomics Society Annual Meeting*, 2020.

[16] Shirsendu Sukanta Halder, Jean-François Lalonde, and Raoul de Charette. Physics-based Rendering for Improving Robustness to Rain. In *Proceedings of the IEEE/CVF International Conference on Computer Vision (CVPR)*, pages 10203–10212, 2019.

[17] A Hawkins. Waymo's driverless car: ghost-riding in the back seat of a robot taxi. *The Verge*, 2019.

[18] Dan Hendrycks and Thomas Dietterich. Benchmarking Neural Network Robustness to Common Corruptions and Perturbations, 2019.

[19] Yuenan Hou, Zheng Ma, Chunxiao Liu, and Chen Change Loy. Learning lightweight lane detection CNNs by self attention distillation. In *Proceedings of the IEEE International Conference on Computer Vision (ICCV)*, pages 1013–1021, 2019.

[20] Jaguar Land Rover. Jaguar Land Rover showcases new technologies including a remote control range rover sport. https://media.landrover.com/en-gb/news/2015/06/jaguar-land-rover-showcase-remote-control-range-rover-sport-controlled-driver, 2015.

[21] Christoph Kamann and Carsten Rother. Benchmarking the robustness of semantic segmentation models with respect to common corruptions. *International Journal of Computer Vision*, pages 1–22, 2020.

[22] Jamie L LaReau. How General Motors is leading the race for self-driving cars, 2018.

[23] Chen Li, Yue Tang, Yingshi Zheng, Paramsothy Jayakumar, and Tulga Ersal. Modeling human steering behavior in teleoperation of unmanned ground vehicles with varying speed. *Human factors*, 64(3):589–600, 2022.

[24] Haoliang Li, Sinno Jialin Pan, Shiqi Wang, and Alex C Kot. Domain Generalization with Adversarial Feature Learning. In *Proceedings of the IEEE/CVF Conference on Computer Vision and Pattern Recognition (CVPR)*, pages 5400–5409, 2018.

[25] Ruilin Liu, Daehan Kwak, Srinivas Devarakonda, Kostas Bekris, and Liviu Iftode. Investigating remote driving over the LTE network. In *Proceedings of the 9th international conference on automotive user interfaces and interactive vehicular applications*, pages 264–269, 2017.

[26] Sidi Lu, Ren Zhong, and Weisong Shi. Teleoperation technologies for enhancing connected and autonomous vehicles. In *2022 IEEE 19th International Conference on Mobile Ad Hoc and Smart Systems (MASS)*, pages 435–443. IEEE, 2022.

[27] Mario Herger. How teleguidance helps zoox vehicles to navigate difficult traffic scenarios. https://thelastdriverlicenseholder.com/2020/11/12/how-teleguidance-helps-zoox-vehicles-to-navigate-difficult-traffic-scenarios/, 2020.

[28] Megan Rose Dickey. Scotty Labs raises 6$ million for remote-controlled autonomous car platform. https://techcrunch.com/2018/03/28/scotty-labs-raises-6-million-for-remote-controlled-autonomous-car-platform/, 2018.

[29] Moderation Team. Aptiv raise stakes on self-driving. https://www.selfdrivingcars360.com/aptiv-raises-stakes-on-self-driving/amp/, 2020.

[30] MD Moniruzzaman, Alexander Rassau, Douglas Chai, and Syed Mohammed Shamsul Islam. Teleoperation methods and enhancement techniques for mobile robots: A comprehensive survey. *Robotics and Autonomous Systems*, page 103973, 2021.

[31] NIST. Nist vehicle teleoperation forum. https://www.nist.gov/news-events/events/2020/11/nist-vehicle-teleoperation-forum, 2020.

[32] Oliver Cameron. Introducing Voyage Telessist. https://news.voyage.auto/introducing-voyage-telessist-a085e4c1f691, July 8, 2020.

[33] Ottopia. Motional selects Ottopia to enable teleoperation of its mass robotaxi fleets. https://www.prnewswire.com/news-releases/motional-selects-ottopia-to-enable-teleoperation-of-its-mass-robotaxi-fleets-301343097.html, 2021.

[34] Ottopia Ltd. Drive.ai launches robot car pilot in Texas with a focus on humans. https://spectrum.ieee.org/driveai-launches-dallas-robot-car-pilot-with-a-focus-on-humans, 2018.

[35] Xingang Pan, Jianping Shi, Ping Luo, Xiaogang Wang, and Xiaoou Tang. Spatial as deep: Spatial CNN for traffic scene understanding. In *32nd AAAI Conference on Artificial Intelligence (AAAI)*, 2018.

[36] Paul Sawers. Ottopia's remote assistance platform for autonomous cars combines humans with AI. https://venturebeat.com/2018/12/21/ottopia s-remote-assistance-platform-for-autonomous-cars-combines-humans -with-ai/, 2018.

[37] PolicyAdvice. 25 Astonishing Self-Driving Car Statistics for 2021. https://policyadvice.net/, 2021.

[38] Real-Time Innovations (RTI). RTI joins the teleoperation consortium to collaboratively advance autonomous vehicle adoption. https://www.gl obenewswire.com/en/news-release/2021/06/02/2240375/0/en/RTI-Join s-the-Teleoperation-Consortium-to-Collaboratively-Advance-Autono mous-Vehicle-Adoption.html, 2021.

[39] Research Briefs. Remote control: Companies researching teleoperation for autonomous vehicles. https://www.cbinsights.com/research/autono mous-vehicle-teleoperation-patents/, 2017.

[40] Evgenia Rusak, Lukas Schott, Roland S Zimmermann, Julian Bitterwolf, Oliver Bringmann, Matthias Bethge, and Wieland Brendel. A Simple Way to Make Neural Networks Robust against Diverse Image Corruptions. In *European Conference on Computer Vision*, pages 53–69, 2020.

[41] Swami Sankaranarayanan, Yogesh Balaji, Carlos D Castillo, and Rama Chellappa. Generate to Adapt: Aligning Domains using Generative Adversarial Networks. In *Proceedings of the IEEE/CVF Conference on Computer Vision and Pattern Recognition (CVPR)*, pages 8503–8512, 2018.

[42] Weisong Shi, Jie Cao, Quan Zhang, Youhuizi Li, and Lanyu Xu. Edge computing: Vision and challenges. *IEEE internet of things journal*, 3(5):637–646, 2016.

[43] Erik Stayton and Jack Stilgoe. It's time to rethink levels of automation for self-driving vehicles [opinion]. *IEEE Technology and Society Magazine*, 39(3):13–19, 2020.

[44] Steve Crowe. Scaling teleoperation of next-gen material handling equipment. https://www.therobotreport.com/phantom-auto-scaling-teleoper ations-logistics-equipment/, 2021.

[45] Wen Sun, Shiyu Lei, Lu Wang, Zhiqiang Liu, and Yan Zhang. Adaptive Federated Learning and Digital Twin for Industrial Internet of Things. *IEEE Transactions on Industrial Informatics*, 2020.

[46] Georg Volk, Stefan Müller, Alexander von Bernuth, Dennis Hospach, and Oliver Bringmann. Towards Robust CNN-based Object Detection through Augmentation with Synthetic Rain Variations. In *Proceedings of the IEEE Intelligent Transportation Systems Conference (ITSC)*, pages 285–292, 2019.

[47] Kecheng Xu, Xiangquan Xiao, Jinghao Miao, and Qi Luo. Data driven prediction architecture for autonomous driving and its application on apollo platform. In *2020 IEEE Intelligent Vehicles Symposium (IV)*, pages 175–181. IEEE, 2020.

[48] Zhiyuan Xu, Xiaoming Liu, and Xiaonan Chen. Fog removal from video sequences using contrast limited adaptive histogram equalization. In *Proceedings of the International Conference on Computational Intelligence and Software Engineering (CISE)*, pages 1–4. IEEE, 2009.

[49] Fisher Yu, Haofeng Chen, Xin Wang, Wenqi Xian, Yingying Chen, Fangchen Liu, Vashisht Madhavan, and Trevor Darrell. BDD100K: a diverse driving dataset for heterogeneous multitask learning. In *Proceedings of the IEEE/CVF Conference on Computer Vision and Pattern Recognition (CVPR)*, pages 2636–2645, 2020.

[50] Ming-Yuan Yu, Ram Vasudevan, and Matthew Johnson-Roberson. Occlusion-aware Risk Assessment for Autonomous Driving in Urban Environments. *IEEE Robotics and Automation Letters*, 4(2):2235–2241, 2019.

[51] Ekim Yurtsever, Jacob Lambert, Alexander Carballo, and Kazuya Takeda. A survey of autonomous driving: Common practices and emerging technologies. *IEEE access*, 8:58443–58469, 2020.

[52] Qingyang Zhang, Hong Zhong, Jie Cui, Lingmei Ren, and Weisong Shi. Ac4av: A flexible and dynamic access control framework for connected and autonomous vehicles. *IEEE Internet of Things Journal*, 8(3):1946–1958, 2021.

[53] Tao Zhang. Toward automated vehicle teleoperation: Vision, opportunities, and challenges. *IEEE Internet of Things Journal*, 7(12):11347–11354, 2020.

[54] Xiaoqin Zhang, Jinxin Wang, Tao Wang, Runhua Jiang, Jiawei Xu, and Li Zhao. Robust feature Learning for Adversarial Defense via Hierarchical Feature Alignment. *Information Sciences*, 560:256–270, 2021.

Authors' Biography

Sidi Lu is a Ph.D. candidate at Wayne State University under the guidance of Prof. Weisong Shi. Her research interests include edge computing and applied science intending to make connected and mobile computer systems more reliable, scalable, secure, and efficient. Specifically, she has accomplished a series of projects based on the vehicle-edge-cloud closed-loop framework. Then, she further implemented a series of designs targeting system support for vehicles, edge/IoT devices, and cloud, respectively, contributing to a range of edge-enabled applications, such as autonomous driving, temporal compressive sensing, collaborative learning, multimodal signal processing, and anomaly prediction.

9

Hybrid Edge/Cloud Solutions for Supporting Autonomous Vehicles

Loris Belcastro, Fabrizio Marozzo, and Alessio Orsino

Department of Informatics, Modeling, Electronics and Systems (DIMES), University of Calabria, Italy
E-mail: lbelcastro@dimes.unical.it, fmarozzo@dimes.unical.it, aorsino@dimes.unical.it

Abstract

Researchers and IT companies have proposed in recent years the use of new hybrid edge/cloud solutions to efficiently process the huge amounts of data produced by IoT devices. In fact, edge combined with cloud computing is used in different application scenarios, such as that of autonomous vehicles, which often require low latency, energy saving, privacy protection, and scalable services. Specifically, edge computing is useful in managing tasks that require real-time analysis and low response times, such as driving assistance, collision avoidance and road sign recognition. Instead, the use of the cloud is convenient for tasks that require a lot of resources and access to large data sets, such as diagnostic data collection and analysis, routing, and targeted advertising. Designing and testing edge/cloud architectures to support autonomous vehicles are still open issues due to their large-scale, heterogeneity, and complexity. In this chapter, we analyze how edge/cloud solutions can be exploited for efficiently managing tasks related to autonomous vehicle driving. In particular, through a simulation-based approach, we demonstrate that these solutions are capable of providing great performance benefits in support of autonomous vehicle systems, especially as the number of vehicles and computing nodes increase.

Keywords: Edge/cloud architecture, IoT infrastructure, modeling and Simulation, autonomous vehicles, urban computing, smart cities.

9.1 Introduction

In the last few years, huge volumes of data have been generated by and collected from several sources, such as sensors, cameras, smart meters, mobile devices, and wearables, which are commonly referred as Internet of Things (IoT) devices [13]. Such huge volumes of data, coupled with the speed with which they are generated, poses new research challenges in collecting, storing and analyzing them. To efficiently extract useful information and produce helpful knowledge for science, industry and public services, novel technologies, architectures, and algorithms have been developed to capture and analyze this data [3].

Existing applications used for processing data from IoT devices are usually highly centralized and rely on the cloud for all operations related to data management, such as data collection, integration, and analysis. However, fully relying on the cloud may result ineffective in terms of network traffic management, response times and energy consumption, particularly for some critical applications (e.g., medical and security) where it is essential to implement low-latency services to avoid serious problems, including fatal accidents [2]. Because of this, researchers and IT companies have proposed in recent years the adoption of the edge computing paradigm and the use of novel hybrid edge/cloud solutions for processing data closer to where it is generated. In this way, edge computing and cloud computing complement each other so that tasks that require real-time analysis and low response times can be run at the edge, while big data applications that benefit from data aggregation and compute-intensive analytics will run in the cloud.

Even in the field of autonomous vehicles, the use of edge/cloud solutions can prove to be extremely effective in managing the different tasks that are generated. For example, tasks such as driver assistance, collision avoidance and traffic sign recognition, which require real-time analysis and low response times, can take advantage of edge computing. Differently, tasks such as diagnostic data collection and analysis, route calculations and targeted advertising, which require a lot of computing resources and access to large data sets, can benefit from the use of cloud computing. When working with advanced machine learning tasks that require real-time analytics with minimal latency, edge-to-cloud cooperation turns out to be even more essential [16].

Designing and testing large-scale and multi-layer edge/cloud architectures are still open issues. In particular, designing and testing a distributed and heterogeneous infrastructure, composed of several components using different technologies and software stacks could be very expensive and hard to manage [19]. For these reasons, simulation-based approaches are a powerful and flexible tool for reproducing and testing edge/cloud architectures, by avoiding the risks, costs, and failures associated with extensive field experimentation [4].

In this chapter, we analyze how edge/cloud architectures can be exploited as an effective solution for managing tasks related to autonomous vehicle driving. In particular, we defined a hybrid edge/cloud architecture that distributes the workload between the cloud and the edge nodes in order to optimize application performance and managing a large number of autonomous vehicles. Furthermore, through a simulation-based approach, we also evaluated different design choices for improving the performance of autonomous vehicles tasks in terms of processing time, network delay, task failure and computing resource utilization.

The structure of this chapter is as follows. Section 9.2 discusses related work. Section 9.3 describes the hybrid edge/cloud architecture. Section 9.4 presents a case study and a performance evaluation. Finally, Section 9.5 concludes the chapter.

9.2 Related Work

With the widespread availability of IoT devices, there was a growing demand for new solutions for extracting useful information from large amounts of data produced by such devices. These solutions are necessary for big data analytics in IoT environments in order to process and analyze huge volumes of structured, unstructured, or semi-structured data. Machine learning algorithms are used to identify patterns, trends, and correlations in data [1, 14], but it is important to consider that these algorithms are run by devices having limited resources such as memory, processing, bandwidth, and energy [23]. As a result, it is essential to strike the correct balance between performance (e.g., the accuracy of the machine learning model) and the amount of resources required for computing.

To analyze and validate these solutions, thorough testing on IoT environments with an ever-increasing number of devices is necessary. Modeling and simulation techniques can help in the design and validation of IoT environments. The primary issues in testing big IoT applications were described by Bosmans et al. [6]. The authors, in particular, proposed a novel simulation-based testing technique capable of facilitating interactions between real-life and virtual local IoT entities. Instead, D'Angelo et al. [7] presented the main issues of modeling IoT systems and how to improve scalability using simulation approaches. Through a performance analysis related to the use of services in a smart city, the same authors in [8] analyzed the feasibility of using simulation to construct and compose heterogeneous IoT simulation scenarios.

In terms of tools and software solutions, different simulators, such as iFog-Sim [9], EdgeCloudSim [21], IoTSim [24], and FogNetSim++ [17], have been proposed in recent years to simulate IoT environments [15]. Among them, EdgeCloudSim results are well-suited for modeling complex IoT systems thanks to the support of architectures of different types and sizes. Moreover, it also supports mobility, which is a typical aspect to be considered when modeling autonomous vehicles scenarios.

These simulation tools are frequently used in the field of autonomous vehicles to perform in-depth tests that take into account the features of the vehicular applications (e.g., upload/download sizes, task description, network models, and mobility) in order to demonstrate how task-offloading schemes perform in a realistic IoT scenario. For example, Sonmez et al. [21] proposed a two-stage machine learning-based vehicular edge orchestrator, which considers both task completion success and service time, and used EdgeCloudSim for a detailed performance evaluation. Hossain et al. [10] proposed an efficient dynamic task offloading approach based on a noncooperative game (NGTO) and assessed its performance on different scenarios via EdgeCloudSim simulator.

Other research effort is devoted to design architectures and models for autonomous vehicles. As an example, Ibn-Khedher [11] designed an end-to-end architecture that enables the allocation of compute-intensive autonomous driving services to edge computing servers, ensuring reliability and low latency. Sasaki [20] presented an edge/cloud computing model for autonomous vehicles based on the Autoware software platform.

9.3 System Architecture

Although cloud computing provides highly scalable and dynamically allocable computing resources, it could have performance issues due to the centralization of the operations of data collection and processing [12, 18]. Using a hybrid edge/cloud architecture could address these issues, enabling a distributed and efficient management of data produced by IoT devices. Even in the context of autonomous vehicles, these architectures may improve computation times and scalability, reduce network congestion, and task failure rate. Figure 9.1 shows a three-layer edge/cloud architecture that we propose for supporting the management of autonomous vehicles.

The *device layer* includes the components that are exploited by vehicles to share information during their movements across different urban cells, which define a partitioning of an urban area. These vehicles produce a very high volume of data by using the embedded components (e.g., GPS, infotainment devices, and on-board cameras) and send it to the edge server of the current cell. This data can be combined with personal data of the users (e.g., preferences and behaviors) and information about the surrounding environment, so as to offer advanced, customized, and context-aware services.

The *edge layer* includes different types of hardware (e.g., Arduino, FPGA, and Raspberry Pi) that make up the infrastructure for gathering the raw data generated at the device layer. The edge layer processes incoming data as long as it has sufficient computing and storage resources. When such resources are no longer sufficient, data is forwarded to the cloud for further processing by the *Edge Orchestrator (EO)*. In particular, the EO is a component that can be configured to apply different orchestration policies in order to improve the performance of the whole architecture. These policies can take into account different parameters, such as the level

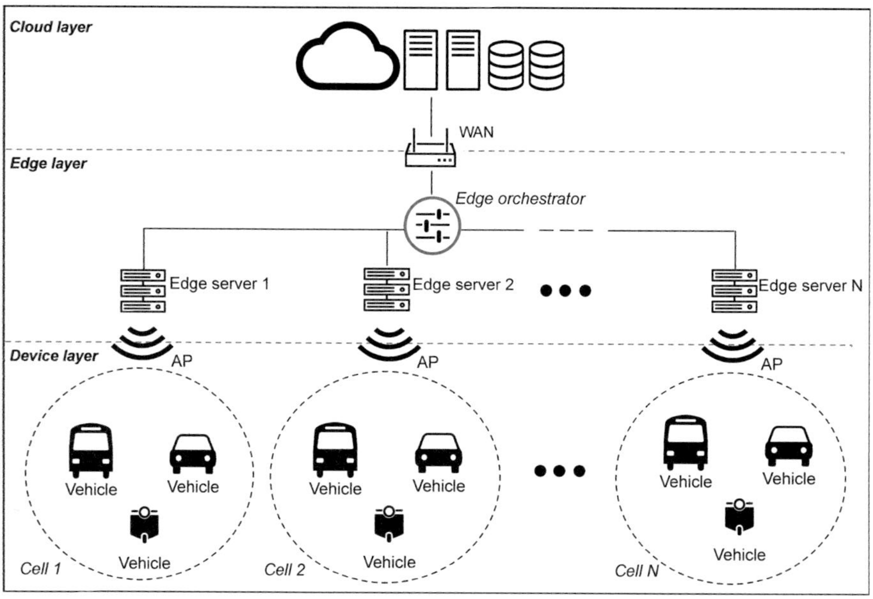

Figure 9.1 The three-layer edge/cloud architecture.

of network congestion, volume of data to be processed, and workload of edge and cloud. With reference to the specific case study presented in this chapter, two orchestration policies have been used, i.e., *Network Based (EO-NB)* and *Utilization Based (EO-UB)*, capable of efficiently and effectively managing the data produced by autonomous vehicles (for more details see Section 9.4).

Finally, the *cloud layer* represents a large set of computing and storage resources, which can be dynamically allocated, for executing tasks that cannot be performed on behalf of edge servers. From a client perspective, the cloud is an abstraction for remote, infinitely scalable provisioning of computation and storage resources, which has emerged as an effective computing platform to face the challenge of processing big data repositories in limited time, as well as to provide efficient data analysis environments to both researchers and companies [5].

9.4 Performance Evaluation

To evaluate the performance of our hybrid edge/cloud architecture for supporting autonomous vehicle driving, we considered three common tasks:

- *Collision avoidance*, which consists in alerting the driver and/or initiating an automated emergency braking action to avoid an accident with a very low-latency response.
- *Route calculation*, which consists in finding routes from a starting location to a given destination.
- *Targeted advertising*, which consists in providing drivers with the most appropriate products or services, such as rest stops, hotels, and restaurants, based on their personal interests and behaviors. Differently from the other two tasks, this results to be a latency-tolerant task.

Specifically, we used the EdgeCloudSim simulator, a Java-based, open-source, and discrete event-based simulator designed for modeling IoT devices, applications, and hybrid edge/cloud architectures. Following the approach proposed in [22], the tasks are generated through a Poisson distribution with different active/idle task generation patterns and interarrival times. A large number of experiments have been carried out using an architecture composed of a cloud and 20 edge servers, each of which provides services to the vehicles that are located on a specific cell of an urban area. In particular, the cloud has been configured as a set of 10 virtual machines (VMs) having 2 cores, 8 GB of RAM and 512 GB of storage memory each. Instead, edge servers have been configured as VMs having 4 cores, 8 GB of RAM and

256 GB of storage memory. In addition, we considered a variable number of autonomous vehicles, ranging from 100 to 1500.

The experiments are used to assess the behavior of the hybrid edge/cloud solution compared to centralized ones that exploit only cloud or edge resources. Specifically, the three configurations we evaluated are the following:

- *Cloud-only*: the tasks are performed exclusively on the cloud.
- *Edge-only*: the tasks are performed directly on the edge.
- *Edge/cloud*: the tasks are performed locally on edge servers; however, if the edge servers are unable to complete the tasks (e.g., due to the lack of resources), the edge orchestrator offloads them to the cloud.

The edge orchestrator was configured to distribute the workload between the cloud and the edge using two policies, namely *EO-UB* and *EO-NB*. Specifically, *EO-UB* schedules tasks based on the average utilization of edge nodes. If the average edge utilization is greater than a fixed threshold, the incoming task is offloaded to the cloud; otherwise, it is assigned to a generic edge device. On the other hand, *EO-NB* measures the network congestion from the edge device to the cloud to decide where incoming tasks must be performed.

Table 9.1 reports the main simulation parameters along with their description. The actual values used for configuring the EdgeCloudSim simulator are listed in Table 9.2. Instead, Table 9.3 reports the parameter values related to the three tasks described above.

Table 9.1 Description of the main EdgeCloudSim simulation parameters.

Parameter	Description
Simulation time	Duration of the simulation in seconds.
Mobile devices	Number of mobile devices used in the simulation scenarios.
Edge servers	Number of edge servers.
MIPS for edge server VM	Computing processor's speed of edge servers in terms of Million Instructions Per Second.
MIPS for cloud VM	Computing processor's speed of cloud in terms of Million Instructions Per Second.
Poisson interarrival	Mean interarrival time between two tasks.
Active period	The active period of the task.
Idle period	The idle period of the task.
Upload data size	Mean input file sizes to upload.
Download data size	Mean output file sizes to download.
Task length	Mean number of instructions to execute the emerging task.

Table 9.2 EdgeCloudSim simulation parameters.

Parameter	Value
Simulation time (min)	36
Mobile devices (i.e., vehicles)	100âĂŞ1500
Edge servers	20
MIPS for edge server VM	10,000
MIPS for cloud VM	37,500

Table 9.3 Parameter values of the three tasks.

Parameter	Task		
	Collision avoidance	Route calculation	Targeted advertising
Poisson interarrival	5	3	15
Active period	3600	3600	3600
Idle period	1	1	15
Upload data size	40 KB	20 KB	20 KB
Download data size	20 KB	80 KB	20 KB
Task length	10,000 MIPS	3000 MIPS	20,000 MIPS

Figure 9.2 reports the performance of the different configurations as the number of vehicles increases. Specifically, Figures 9.2(a)-(d) report the average processing time, percentage of failed tasks, network delay and VM utilization, respectively. As a general consideration for evaluating results, low percentage of failed tasks and low network delay are essential to ensure a high quality of services (e.g., in terms of reliability and responsiveness). Also reducing VM utilization is a crucial aspect in large-scale applications, as it enables cost and energy consumption optimization, as well as handling any unexpected workload spikes.

About the results obtained, the edge-only solution achieved the best results in terms of network delay but with high processing times, failure rate and VM utilization. In fact, the average processing time ranges from 0.86 seconds with 100 vehicles to 1.67 seconds with 1,500 vehicles, the percentage of failed tasks reaches up to 13%, and the VM utilization goes up to 68%. The cloud-only solution achieved a very low average processing time and VM utilization, but it drastically increased the percentage of failed tasks (25% for 1500 vehicles). Regarding the network usage, data transfer directly to the cloud caused a significant increase in delay (80% vs. edge-only). Instead, the hybrid edge/cloud solution turned out to provide the best compromise among all the performance metrics considered. In fact, both

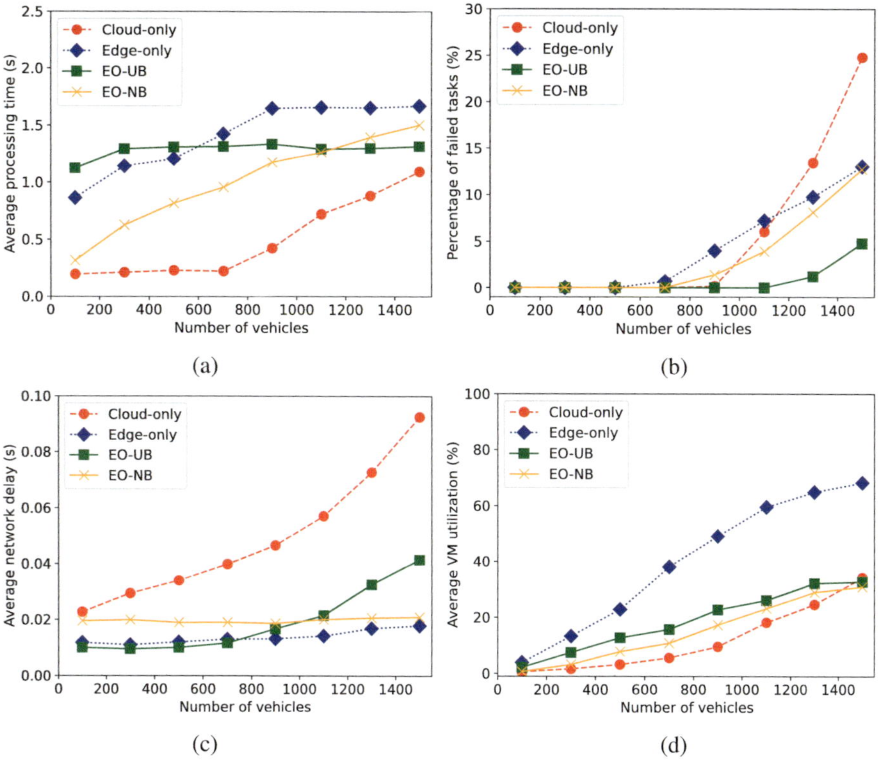

Figure 9.2 Performance results of the tasks related to autonomous vehicle driving obtained for the different simulation scenarios.(a) Processing time. (b) Failed tasks. (c) Network delay. (d) VM utilization.

orchestration policies ensured good processing times, low percentages of failed tasks and VM utilization, and limited network delays. However, if the number of vehicles is high, the EO-UB policy turned out to be the best choice, as it reached lower processing times and percentage of failed tasks.

In order to understand how computing resources are exploited by the hybrid edge/cloud solution with the two orchestration policies, Figure 9.3 shows the percentage of VM utilization for the cloud and edge. As shown, the two orchestration policies attempt to make the most of the edge layer's resources, employing the cloud-only when necessary, and as a result are able to reduce both latency and percentage of failed tasks.

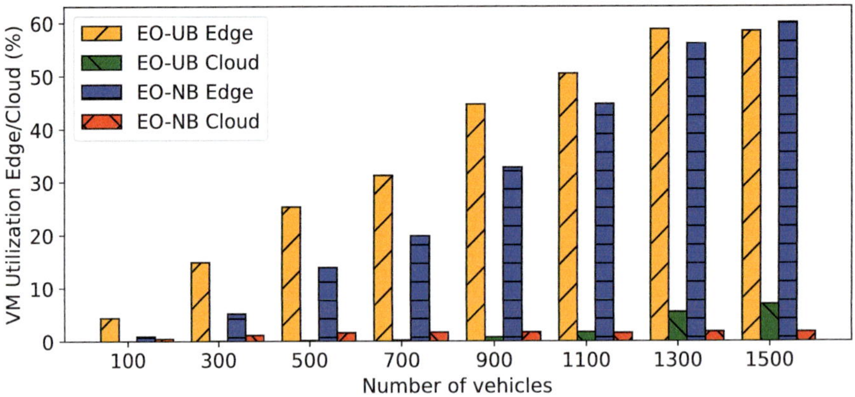

Figure 9.3 Comparison of EO-UB and EO-NB policies about the VM utilization on the edge and cloud.

Overall, the presented architecture and orchestration policies outperformed the conventional cloud- or edge-only approaches, considerably improving processing times and percentage of failed tasks. Furthermore, they ensure better management of workload spikes because of a more efficient resource utilization.

9.5 Conclusions

In recent years, the use of IoT solutions enabled the processing of data closer to where it is generated, so as to reduce network traffic and task failure rates. Generally, IoT infrastructures are composed of many heterogeneous components that interact with each other to pursue common goals. Designing, testing and deploying modern IoT infrastructures are going to be a great challenge in the next few years.

The chapter presented a hybrid edge/cloud solution to efficiently manage tasks concerning autonomous vehicles driving in a large-scale IoT computing environment. In particular, two orchestration policies were used for better balancing the workload between the edge and the cloud. Numerous experiments have been carried out to evaluate the behavior of the proposed solution compared to the centralized ones that exploit only cloud or edge resources. The results showed that our solution ensures a good compromise between processing time, network delay, and task failure rate compared to conventional centralized architectures.

References

[1] Ejaz Ahmed, Ibrar Yaqoob, Ibrahim Abaker Targio Hashem, Imran Khan, Abdelmuttlib Ibrahim Abdalla Ahmed, Muhammad Imran, and Athanasios V Vasilakos. The role of big data analytics in internet of things. *Computer Networks*, 129:459–471, 2017.

[2] Alessandro Barbieri, Fabrizio Marozzo, and Claudio Savaglio. Iot platforms and services configuration through parameter sweep: a simulation-based approach. In *2021 IEEE International Conference on Systems, Man, and Cybernetics (SMC)*, pages 1803–1808, 17-20 October 2021.

[3] Loris Belcastro, Riccardo Cantini, Fabrizio Marozzo, Alessio Orsino, Domenico Talia, and Paolo Trunfio. Programming big data analysis: Principles and solutions. *Journal of Big Data*, 9(4), 2022.

[4] Loris Belcastro, Alberto Falcone, Alfredo Garro, and Fabrizio Marozzo. Evaluation of large scale roi mining applications in edge computing environments. In *25th International Symposium on Distributed Simulation and Real Time Applications (DS-RT)*, pages 1–8, 2021.

[5] Loris Belcastro, Fabrizio Marozzo, Domenico Talia, and Paolo Trunfio. Big data analysis on clouds. In Albert Zomaya and Sherif Sakr, editors, *Handbook of Big Data Technologies*, pages 101–142. Springer, December 2017.

[6] Stig Bosmans, Siegfried Mercelis, Joachim Denil, and Peter Hellinckx. Testing iot systems using a hybrid simulation based testing approach. *Computing*, 101(7):857–872, 2019.

[7] Gabriele D'Angelo, Stefano Ferretti, and Vittorio Ghini. Simulation of the internet of things. In *2016 International Conference on High Performance Computing & Simulation (HPCS)*, pages 1–8. IEEE, 2016.

[8] Gabriele DâĂŹAngelo, Stefano Ferretti, and Vittorio Ghini. Distributed hybrid simulation of the internet of things and smart territories. *Concurrency and Computation: Practice and Experience*, 30(9):e4370, Nov 2017.

[9] Harshit Gupta, Amir Vahid Dastjerdi, Soumya K Ghosh, and Rajkumar Buyya. ifogsim: A toolkit for modeling and simulation of resource management techniques in the internet of things, edge and fog computing environments. *Software: Practice and Experience*, 47(9):1275–1296, 2017.

[10] Md. Delowar Hossain, Tangina Sultana, Md. Alamgir Hossain, Md. Abu Layek, Md. Imtiaz Hossain, Phoo Pyae Sone, Ga-Won Lee, and Eui-Nam Huh. Dynamic task offloading for cloud-assisted vehicular edge computing networks: A non-cooperative game theoretic approach. *Sensors*, 22(10), 2022.

[11] Hatem Ibn-Khedher, Mohammed Laroui, Mouna Ben Mabrouk, Hassine Moungla, Hossam Afifi, Alberto Nai Oleari, and Ahmed E. Kamal. Edge computing assisted autonomous driving using artificial intelligence. In *2021 International Wireless Communications and Mobile Computing (IWCMC)*, pages 254–259, 2021.

[12] Niloofar Khanghahi and Reza Ravanmehr. Cloud computing performance evaluation: issues and challenges. *Comput*, 5(1):29–41, 2013.

[13] Somayya Madakam, Vihar Lake, Vihar Lake, Vihar Lake, et al. Internet of things (iot): A literature review. *Journal of Computer and Communications*, 3(05):164, 2015.

[14] Mohsen Marjani, Fariza Nasaruddin, Abdullah Gani, Ahmad Karim, Ibrahim Abaker Targio Hashem, Aisha Siddiqa, and Ibrar Yaqoob. Big iot data analytics: architecture, opportunities, and open research challenges. *ieee access*, 5:5247–5261, 2017.

[15] Andras Markus and Attila Kertesz. A survey and taxonomy of simulation environments modelling fog computing. *Simulation Modelling Practice and Theory*, 101:102042, 2020.

[16] Fabrizio Marozzo, Alessio Orsino, Domenico Talia, and Paolo Trunfio. Edge computing solutions for distributed machine learning. In *The 8th IEEE International Conference on Cloud and Big Data Computing*, pp. 1-8, 2022.

[17] Tariq Qayyum, Asad Waqar Malik, Muazzam A Khan Khattak, Osman Khalid, and Samee U Khan. Fognetsim++: A toolkit for modeling and simulation of distributed fog environment. *IEEE Access*, 6:63570–63583, 2018.

[18] V Krishna Reddy, B Thirumala Rao, and LSS Reddy. Research issues in cloud computing. *Global Journal of Computer Science and Technology*, 2011.

[19] Maria Salama, Yehia Elkhatib, and Gordon Blair. Iotnetsim: A modelling and simulation platform for end-to-end iot services and networking. In *Proceedings of the 12th IEEE/ACM International Conference on Utility and Cloud Computing*, pages 251–261, 2019.

[20] Yu Sasaki, Tomoya Sato, Hiroyuki Chishiro, Tasuku Ishigooka, Satoshi Otsuka, Kentaro Yoshimura, and Shinpei Kato. An edge-cloud computing model for autonomous vehicles. In *In 11th IROS Workshop on Planning, Perception, Navigation for Intelligent Vehicle*, 2019.

[21] Cagatay Sonmez, Atay Ozgovde, and Cem Ersoy. Edgecloudsim: An environment for performance evaluation of edge computing systems. *Transactions on Emerging Telecommunications Technologies*, 29(11):e3493, 2018.

[22] Cagatay Sonmez, Can Tunca, Atay Ozgovde, and Cem Ersoy. Machine learning-based workload orchestrator for vehicular edge computing. *IEEE Transactions on Intelligent Transportation Systems*, 2020.

[23] Saniya Zahoor and Roohie Naaz Mir. Resource management in pervasive internet of things: A survey. *Journal of King Saud University-Computer and Information Sciences*, 2018.

[24] Xuezhi Zeng, Saurabh Kumar Garg, Peter Strazdins, Prem Prakash Jayaraman, Dimitrios Georgakopoulos, and Rajiv Ranjan. Iotsim: A simulator for analysing iot applications. *Journal of Systems Architecture*, 72:93–107, 2017.

Authors' Biography

Loris Belcastro is a Research Fellow of Computer Engineering at the University of Calabria, Italy. He received a Ph.D. in Information and Communication Engineering at the University of Calabria. In 2012, he held a scholarship at the Institute of High-Performance Computing and Networking of the Italian National Research Council (ICAR-CNR). His research interests include cloud computing, social media and big data analysis, distributed and parallel computing, and edge computing.

Fabrizio Marozzo is an Assistant Professor of Computer Engineering at the University of Calabria. He received a Ph.D. in Systems and Computer Engineering at the University of Calabria. In 2011âĂŞ2012 he visited the Barcelona SuperComputing Center for a research internship with the Grid Computer Research group in Computer Sciences Department. He is a member of the editorial boards of IEEE Access, Journal of Big Data, SN Computer Science, Scalable Computing: Practice and Experience, Big Data and Cognitive Computing, Algorithms, and Heliyon. His research focuses on big data analysis, social media analysis, parallel and distributed computing, cloud and edge computing, and machine learning.

Alessio Orsino is a Ph.D. student in Information and Communication Technologies at the University of Calabria, Italy. His research interests include big data analysis, parallel and distributed computing, cloud and edge computing, and machine learning.

10

Game-theoretic Decision-making for Autonomous Driving Vehicles

Mingfeng Yuan and Jinjun Shan

Department of Earth and Space Science, Lassonde School of Engineering, York University, Canada
E-mail: mfyuan@yorku.ca, jjshan@yorku.ca

Abstract

In the next 50 years, self-driving cars will coexist with human-driven vehicles on roads, resulting in constant interactions between the two types of vehicles. It is challenging to solve decision-making problems for autonomous driving vehicles in dynamic and interactive environments. Predefining rules via hard coding is almost impossible to address these issues. In this research, we aim to utilize a combination of game theory and deep reinforcement learning to address the interaction challenge faced by autonomous vehicles. Our approach involves leveraging learning vehicles to explore complex interactions that human drivers experience in real-life situations using a realistic simulator. As the policy evolves, autonomous driving vehicles can drive more successfully by collecting data through interactions with multiple vehicles from the virtual environments. To enable autonomous driving vehicles to make decisions in traffic scenarios with varying levels of reasoning, we have integrated the learning capability of deep q networks with the Level-k reasoning approach. Therefore, vehicles can learn complex driving behaviors via a self-play training mode. The trained policies were tested in different traffic scenarios including unsignalized intersections and urban lane-change scene with different driving models, and the experimental results show that the autonomous driving vehicles can still drive safely in such complex traffic environments, thus verifying the effectiveness of our proposed method.

Keywords: Level-k game theory, LSTM network, deep reinforcement learning, autonomous driving vehicles, sim2real, decision-making.

10.1 Introduction

Autonomous driving vehicles (ADVs) and human-driven vehicles (HDVs) will coexist in traffic scenarios within the next 50 years, leading to frequent interactions between the two types of vehicles. Current level-4 ADVs in the mainstream industry tend to restrict these interactions rather than enhance them. For instance, in complex scenarios where interactions are involved, the ADVs tend to decelerate and come to a halt rather than proactively seek alternative routes [1]. Addressing decision-making challenges for ADVs in dynamic and interactive traffic scenarios is a complex task, as relying solely on predefined codes or rules is nearly impossible. Furthermore, according to [2], ADVs must have a track record of driving 275 million miles without any fatal accidents to match the reliability of HDVs. Undoubtedly, the road-testing phase is inherently expensive and requires a significant amount of energy, not to mention the potential danger it poses to pedestrians. Hence, it is highly significant to model diverse driving behaviors and evaluate decision-making algorithms in high-fidelity simulation environments for enhancing the safety of ADVs and reducing their reliance on real-world road tests. Motivated by the need to develop simulation tools for verifying and validating autonomous driving systems operating in traffic with ADVs and HDVs, our main focus is on modeling vehicle interactions using game theory (GT) and deep reinforcement learning (DRL). Learning vehicles can effectively learn such complex interactions automatically via exploration, resulting in more successful driving experiences with data collected through interactions in multi-vehicle scenarios over time in virtual environments.

The behavior of ADVs in actual traffic scenarios displays distinct traits. To begin with, the decisions made by ADVs are influenced by the actions of nearby vehicles, and vice versa. Second, vehicles engage in rich dynamic interaction behaviors as they not only cooperate to avoid collisions, but also compete based on their diverse driving strategies. Third, ADVs have limited knowledge of the driving policies of HDVs and must rely on observing the actions of their counterparts to make estimations.

Several methodologies have been reported in literature for modeling interactions between multiple vehicles, which encompass decision trees [3,4], partially observable MDP [5], dynamic Bayesian networks [6], MPC [7], and data-driven methods [8], that mainly serve as high-level controllers. Game theory offers mathematical models to analyze the strategic interactions between rational agents, which can be employed to depict the strategic reasoning of numerous vehicles and simulate their interaction behavior. In conventional methods, the interaction behavior among vehicles is typically represented by a one-shot normal-form game, where each vehicle chooses driving actions (i.e. "move" and "wait") based on a payoff matrix, without taking into account the dynamic features of the vehicles. Moreover, as the number of vehicles increases, the computational complexity grows exponentially, making it increasingly difficult to achieve real-time decision-making. In the existing studies, Refs. [9, 10] utilized a game-theoretic methodology to model vehicle interactions in scenarios involving highways or intersections. Ref. [11] employed the leader-follower game to simulate the interactions among vehicles at unsignalized intersections. Ref. [12] utilizes a game-theoretic approach, along with MPC and an imitation learning algorithm, to derive control policies. Ref. [13] introduced an adaptive control approach that considers uncertainties in the estimation of the vehicle dynamic model and driver model. Our approach diverges significantly from prior research that often depends on simple simulations, disregarding the presence of noise in observations or dynamics in physical ADVs. As a result, these methods face significant limitations in their capacity to be applied to actual vehicles. The decision-making framework presented in this study depends entirely on the onboard sensors, without presuming any cooperation, communication, or joint regulation with the neighboring vehicles. Additionally, the vehicle's actions and movements are executed on a high-idelity simulator, mimicking real-life scenarios. An alternative idea of studies explore driving behavior through simulation, which is particularly suitable for reinforcement learning (RL) due to the ability to learn from failure cases in virtual environments.

Moreover, RL has proved its effectiveness in crafting controllers for single-agent as well as multi-agent systems, including wireless communication [14], and navigation [15]. Utilizing these methods to acquire autonomous driving policies is quickly emerging as a highly promising approach [16]. The unapologetic nature of the trial-and-error process in RL exacerbates the challenge of ensuring functional safety. These adversities call for learning that first takes place in virtual environments before transferring to the corresponding real applications [17], which is referred as sim2real. It presents a challenge owing to the gap between simulation and real-world traffic scenarios, encompassing factors like vehicle dynamics and sensor information" [18]. Simultaneously, the collision and near-collision cases are imperative to the learning process, as it facilitates the incorporation of these vital experiences into future policy rollout. How can we offer secure, multi-agent learning experiences without sacrificing the authenticity of realistic training data? There is a shortage of study that addresses this challenge.

In this chapter, we adopt a combination of DRL and GT to model decision makers with varying levels of reasoning in an unsignalized intersection scenario. Our approach stands out from previous research as it involves all drivers in a multi-move situation making strategic decisions simultaneously, rather than relying on the ADV to act as a decision-maker and assuming predetermined actions for other drivers [1]. This is accomplished by merging the cognitive hierarchy theory, also known as level-k reasoning, with a reinforcement learning method called Dueling Double Deep Q-Network with Prioritized Experience Replay (D3QN PER). Previous research [5, 9, 20–22] has aimed to merge RL and level-k theory, enabling vehicles to independently develop policies that incorporate rich interaction behaviors. They emphasized the importance of investigating how to extend to a more general setting where a level-k car determines its optimal response to the actions of another car, which is reasoning based on a distribution over lower levels, rather than only at level-(k-1) [20]. This chapter presents the integration of LSTM network with a RL framework, in order to efficiently tackle the issue of hidden state. The primary objective is to develop a policy that can adapt to a complicated and varied multi-strategy in mixed driving environments, for enabling multi-car interactions.

The arrangement of this chapter is as follows: Section 10.2 outlines two decision-making problems that will be addressed. Section 10.3 details the utilization of level-k game theory, LSTM network, and RL algorithm to construct a decision-making model for an intersection scenario involving four vehicles, and an adaptive cruise control system (ACCs) is developed using the same learning framework. In Section 10.4, we introduce the details of low-level control incorporating lateral control, longitudinal control, and generating of fifth-order polynomial trajectories. Trained driving policies

---

[1]This chapter is an extension of the work described in our paper published in IEEE Robotics and Automation Letters [19].

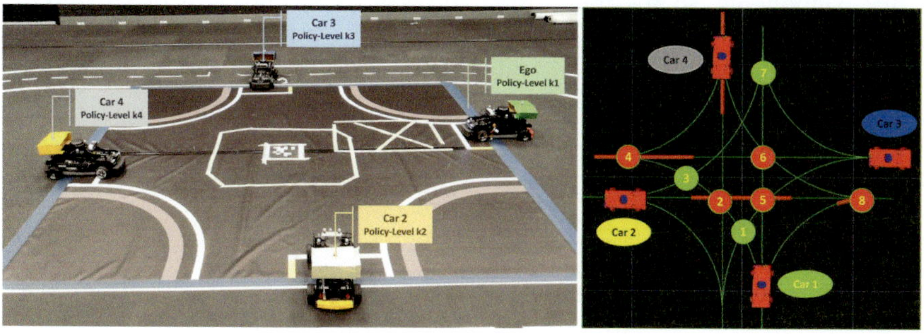

Figure 10.1 Four-way intersection, car 1 is the ego car while cars 2–4 are opponents with varying levels of reasoning. The collision areas, numbered from 1 to 8, are viewed from the perspective of the ego car.

are tested in Section 10.5 by both simulator and hardware. The chapter is summarized and concluded in Section 10.6.

10.2 Problem Definition

10.2.1 Modeling scenarios for unsignalized intersection

The unsignalized intersection was selected as the traffic scene of our study, owing to its complexity compared to other traffic scenarios. In this scenario, every vehicle has to maneuver through the intersection on its own, and drivers must engage in ongoing interaction with other nearby vehicles on the road to ensure a safe and smooth driving.

In this section, we will introduce a scenario that comprises of 12 paths and four cars, with car 1 being the ego vehicle and the remaining vehicles serving as opponents, as shown in Figure 10.1. To present the results more clearly, the problem can be simplified in the following manner: the number of opponent vehicles that the ego vehicle encounters at the intersection is generated randomly and can be either 0, 1, 2, or 3. Ego is tasked with three maneuvers: turning left, proceeding straight, and turning right. While all opponents proceed straight, they possess two trained policies – conservative and aggressive driving behaviors – which is unknown to the Ego vehicle. The specifics regarding the training of diverse driving policies are outlined in Section 10.3. It is noteworthy that the method proposed in this chapter can be extended to more intricate scenarios, where all vehicles have no constraints on their path selection. In this work, a total of 81 scenarios were identified. This is because there are 24 possible combinations when an ego car is faced with three opponents, 36 possible combinations when an ego car is faced with two opponents, and 18 possible combinations when an ego car is faced with one opponent. There are three possible combinations when the ego passes intersection without opponent vehicles. Throughout the training period, the probability of encountering interaction scenarios involving one car, two cars, three cars, and four cars will be 10%, 20%, 30%, and 40%, respectively, with the aim of highlighting intricate situations.

10.2.1.1 Observation space

Lidar's ability to adjust to various lighting conditions and its resilience to the surroundings make it one of the most vital sensors in advanced autonomous vehicles. To handle the long sequence of information generated by LiDAR, an LSTM network with 512 cell units is employed in this chapter to effectively process the raw data, especially in partially observed environments. Opponent cars adopt various strategies that may confuse the ego car if it is heavily rewarded for an action in one state but penalized for the same action in the same situation during the next attempt. Such inconsistency can cause instability in the training process. Thus, the ego car's action depends on not only the current observation but also a fixed number of frames of past observations. The point cloud data, with a frequency of 50 Hz and a detection range of 0.1 – 3 m, is restricted to 360 dimensions. The intersection scenario involves eight collision areas where both the ego vehicle and the opponent vehicle paths intersect, as depicted in Figure 10.1.

To address the limitations of relying solely on 2D Lidar, the distance of each vehicle to potential collision points was factored in. Moreover, the array $[v_1, v_2, v_3, v_4, p_1, p_2, p_3, p_4]$ captures the state information of vehicles, where each element of the array corresponds to the speed of cars 1–4 and the path selected by each car. This information enhances the observation space, and its generality can be maintained even in complex scenarios.

10.2.1.2 Action space

The action space comprises five feasible actions that each vehicle can undertake:

- Maintain: Keep the current speed of the vehicle.
- Accelerate: Increase the speed of the vehicle by 1.5 m/s^2, unless the maximum velocity has already been reached.
- Fast Accelerate: Increase the speed of the vehicle by 3 m/s^2, unless the maximum velocity has already been reached.

Table 10.1 Setting of reward function.

Conditions	Values
Collision status	−5
Dangerous zone status and car approaching	$-0.1*\Delta(t)$
Changes in acceleration	$-0.05*(\|a_t - a_{t-1}\|)$
Reach destination	2
Reach each checkpoint	0.1

- Brake: Reduce the speed of the vehicle by 1.5 m/s^2, unless the vehicle is already stationary.
- Hard Brake: Reduce the speed of the vehicle by 3 m/s^2, unless the vehicle is already stationary.

10.2.1.3 Reward function

The reward function illustrated in Table 10.1 is utilized to assess the driving performance of the ego vehicle, thereby motivating it to acquire effective human-like driving patterns. Reward function was designed to discourage collisions or hazardous states, and promote efficient behaviors, including reaching the destination and making progress. The reward function dictates that the ego car will be rewarded 0.1 for successfully reaching each checkpoint. If the ego successfully reaches its destination by following a straight path, it will receive a reward of 2 points for passing all 20 checkpoints along the way. In case of a collision, 5 points will be deducted from the reward. If the ego vehicle and opponent cars are stuck in a deadlock, 0.1 points will be subtracted for each time step. This will result in a maximum of −1.5 points deduction in each episode.

10.2.2 Modeling scenarios for urban lane-change

In a typical highway scenario, as depicted in Figure 10.2, the ego vehicle must possess the capability to adapt its actions to deal with the constantly changing traffic conditions. This work proposes a hierarchical decision-making architecture, as shown in Figure 10.3, which is mainly divided into two levels. The first is the high-level decision-making module, which mainly considers travel efficiency, comfort and driving safety, and other macro factors. This module will tell AVs the tasks to be performed at the current stage after observing the surrounding driving environment, and the output commands are car following, preparing for lane changing, and executing lane change tasks.

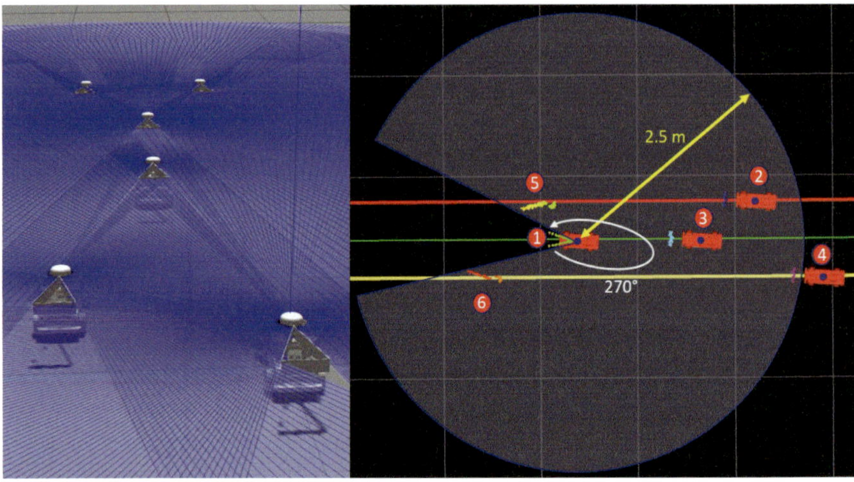

Figure 10.2 Simulation environment with multiple cars.

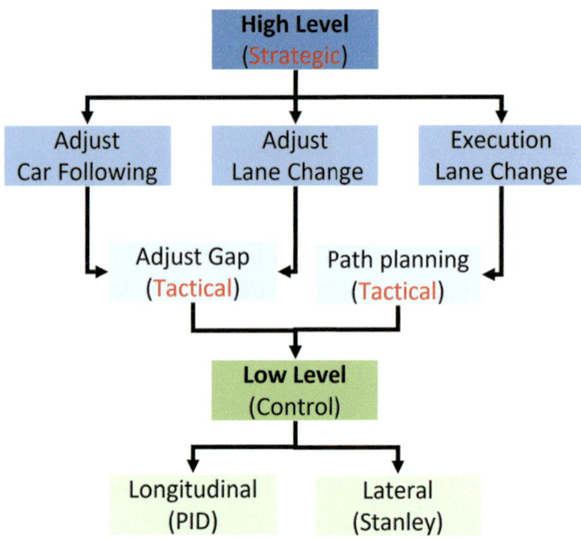

Figure 10.3 Hierarchical decision-making.

The main task of our ACCs is to adjust the throttle to keep a relative distance and a relative speed with the car in front. Preparing for lane changing is divided into preparation for left lane change or right lane change, which is to adjust the throttle to adjust their relative position with the front car in the

Table 10.2 Action set.

Action	Acceleration(m/s^2)
Maintain	0
Accelerate	0.15
Decelerate	−0.15
Hard accelerate	0.35
Hard brake	−0.35

current lane, as well as the front car and the rear car in the target lane. The requirement of the execution of a lane-changing task is to generate a smooth, safe trajectory. The above three tasks are still at the command level, but compared with the top level, the command is more specific, and its content can be acceleration, deceleration, maintenance, and the desired speed and steering angle. However, the specific instructions must require good execution ability, which is the task of the low-level module. It will be subdivided into throttle control and steering control. Desired control performance can be guaranteed by implementing controllers. For high-level modules, Level-2 and Level-3 AVs on the market are usually asking human drivers to undertake high-level tasks, which is also called the human-in-the-loop. In simple scenarios, it is also possible to define high-level policies in a rules-based manner. This work mainly focuses on decision-making tasks below the high-level module, that is, how to learn a driving policy with similar distribution to that of a HDV through RL algorithms.

10.2.2.1 Observation space

Imagine that when we are driving a car, we make decisions primarily based on the information captured by our eyes. However, the driving policies of the surrounding opponent vehicles are unknown to us, and we have to observe them for a period of time to approximately estimate their driving behaviors. Therefore, in our study, we prohibited the use of accurate speed and position information from other vehicles, and all observations were obtained from onboard 2D LiDAR plus ego vehicle speed which can be directly obtained from the vehicle dashboard. RL algorithms usually require observation information to be complete in order to ensure algorithm convergence. However, the driving environment mentioned above belongs to a partially observable environment.

Similar to the idea for the intersection decision-making problem, the LSTM network is used to preprocess the point cloud data. Hence, the decision-making of the ego vehicle will not only consider the current observations but also consider the history of the observations to predict the motion state of surrounding vehicles. Since we are using a 2D LiDAR, which is mounted on the top of the car, to enhance the point cloud data from the LiDAR detection of surrounding cars, we mounted an isosceles triangle frame behind LiDAR. The effective scanning region of LiDAR is [135°, −135°], and the scanning range is [0.1, 2.5] m. The scanning frequency is 12 Hz with 1160 points. The point cloud data is shown in Figure 10.2.

10.2.2.2 Action space

The action space of learning car consists of five basic actions, as shown in Table 10.2. The discrete representation of the action space has its foundation in the distribution of vehicle accelerations, which were procured through US101 data analysis. This approximation serves as a rational estimation of human drivers' actions at highway scenarios.

10.2.2.3 Reward function

According to US101 data, the intervehicle distance is mainly maintained between 11m and 25m above about 50% of the time. Because the size ratio of the car used in our study and the real vehicle is 1:10. Therefore, the values for both the distance and the acceleration will be divided by 10 to define our reward function and action space.

The stage reward function for car following, $R(t)$, is defined as

$$R(t) = \mathbf{w}_i^\top \mathbf{\Phi}(t) \tag{10.1}$$

where $\mathbf{\Phi}(t) = [\phi_1(t), \phi_2(t), \phi_3(t), \phi_4(t)]^\top$ represents feature vector and $\mathbf{w}$ is weight vector. If a collision occurs, feature ϕ_1 equals 1, otherwise 0. The corresponding weight $w_1 = -1000$; According to the distance, Dis, between Ego and the car in front, we divide them into three zones: danger zone ($0.5 < \text{Dis} < 1.1$ m) interaction zone ($1.1 < \text{Dis} < 2.5$ m) and safety zone ($\text{Dis} > 2.5$ m). When $\phi_2 = 1$, it indicates that the car in front is in its own danger zone, and is zero otherwise; The corresponding weight $w_2 = (\frac{1.1}{\text{Dis}})^3$; When $\phi_3 = 1$, which indicates that the car in front is in its interaction area, and we encourage the car to keep relatively stationary from the car in front. The weight $w_3 = |v_{\text{ego}} - v_{\text{front}}|$. When $\phi_4 = 1$, the car in front is outside the interaction area, and we encourage the Ego to keep the speed above the average speed. And the corresponding weight can be calculated as $w_4 = v_{\text{ego}} - (v_{\max} + V_{\min}/2)$ where $v_{\max} = 1.5$ m/s, and $v_{\min} = 0.2$ m/s.

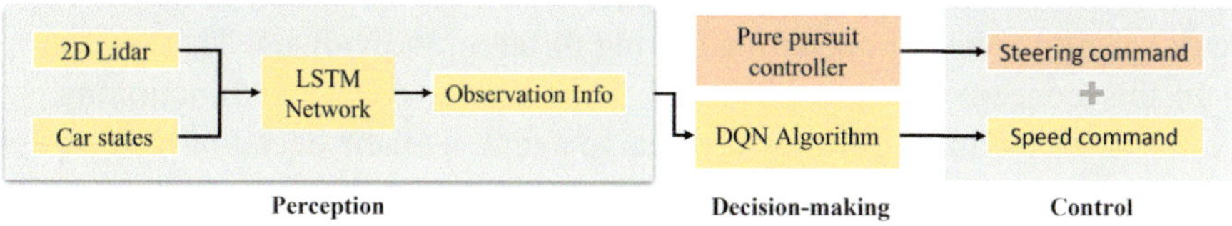

Figure 10.4 End-to-end control scheme.

10.2.3 End-to-end control scheme

The aim of this study is to develop an end-to-end plan that effectively models the driving policies of self-driving cars at an uncontrolled intersection. The first step involves processing the features of 2D Lidar point cloud data and car states via LSTM to generate observation information. The next step is to use the dueling structure to generate the Q value through the full connection layer.

We present a novel algorithm, D3QN PER, that incorporates the strengths of the double DQN, conventional deep Q-network (DQN), and Prioritized Experience Replay (PER) techniques. By combining these approaches, we have achieved significant enhancements in driving performance and learning efficiency. The optimal action for each state can be determined by selecting the Q value with the highest magnitude. As there are a total of 12 potential paths within the intersection, we implemented the pure pursuit controller to generate the steering commands. The visual representation of the research methodology is shown in Fig. 10.4.

10.3 Driver Interaction Model

The driver interaction model, developed in this chapter, facilitates the modeling of interactions between vehicles through the implementation of level-k reasoning and D3QN PER algorithm. The model represents a "policy" that functions as a stochastic map, taking the driver's observation space to their corresponding action space (refer to Section 10.2). In other words, this map presents a probability distribution for each observation that assigns possible actions. In the following sections, we will elaborate on how this model is generated.

10.3.1 level k reasoning

To depict the strategic decision-making process of human drivers, a game-theoretical concept named "level-k reasoning" is employed. The Level-K

game theory is a hierarchical decision-making concept that assumes the existence of varying levels of reasoning among different individuals. The lowest level in this concept is known as level 0 reasoning. An agent functioning at level 0 is deemed a nonstrategic or naive agent, as their decision-making process is not influenced by the potential actions of other agents but rather relies on preestablished moves. At a higher level, there exists a strategic agent operating at level-1, whose actions are determined by assuming that the reasoning levels of other agents are at level 0. Therefore, the actions of a level-1 agent represent the most optimal response to level 0 actions. Similarly, a level 2 agent regards other agents as level-1 and formulates its decisions based on this forecast. This pattern persists for subsequent levels, applying the same logic. In certain experiments, it has been observed that humans exhibit a maximum level of reasoning at level 3. It is important to note, however, that this may vary depending on the particular game being played. In general, all agents who are at level-k, with the exception of those at level 0, assume that the other agents are at level(k-1) and make their decisions based on this assumption. However, this belief may not always be accurate, which results in the agents exhibiting bounded rationality [23, 24].

10.3.2 Dueling double deep Q-network with prioritized experience replay algorithm (D3QN PER)

Our study employs the D3QN PER algorithm, which integrates the benefits of Dueling DQN, Double DQN, and PER, to enhance the effectiveness of our ego vehicle training. We evaluate and compare the learning efficacy of various algorithms in Section 10.5.

In the DQN algorithm [25], the deep Q-network is provided with labeled samples using Q-learning. The network's parameters θ are estimated by optimizing a sequence of loss functions at iteration i that minimize the expression $Q(s,a;\theta)$

$$L_i(\theta_i) = \mathbb{E}\left[\left(y_i^{DQN} - Q(s,a;\theta_i)\right)^2\right], \tag{10.2}$$

with

$$y_i^{DQN} = r + \gamma \max_{a'} Q\left(s',a';\theta^-\right), \tag{10.3}$$

where r denotes the reward gained by performing action a in state s. The action, a', is taken in the subsequent state, s'. The discount factor is represented by γ. The parameters of the target network, $Q\left(s',a';\theta^-\right)$, are frozen for a certain number of iterations while updating the online network, $Q(s,a;\theta_i)$, via gradient descent. The gradient update can be described as:

$$\nabla L_i(\theta_i) = \mathbb{E}\left[\left(y_i^{DQN} - Q(s,a;\theta_i)\right)\nabla_{\theta_i} Q(s,a;\theta_i)\right]. \tag{10.4}$$

Although Q-learning and DQN both suffer from overestimation problems due to the same values being used for action selection and evaluation by the max operator, Double DQN offers a solution by utilizing an alternative target in place of eqn (10.3).

$$y_i^{\text{Double}} = r + \gamma Q\left(s', \underset{a'}{\arg\max} Q\left(s', a'; \theta_i\right); \theta^-\right). \tag{10.5}$$

In order to speed up the training process, the Dueling structure aims to minimize the need for estimating the value of every possible action choice in various states. This is because certain states can cause the task to fail regardless of which action is taken by the agent. The Dueling structure can effectively decompose the estimator into a sum of:

$$Q(s,a;\theta,\alpha,\beta) = V(s;\theta,\beta) + A(s,a;\theta,\alpha), \tag{10.6}$$

There are two streams of fully-connected layers. One stream produces a scalar value, denoted as $V(s;\theta,\beta)$, while the other stream produces a vector of dimension $|\mathscr{A}|$, denoted as $A(s,a;\theta,\alpha)$. The parameters for these two streams of fully-connected layers are denoted as α and β. However, it should be noted that eqn (10.6) is not identifiable because V and A cannot be uniquely determined given Q. To solve this problem, the advantage function estimator can be modified to have zero advantage for the chosen action

$$Q(s,a;\theta,\alpha,\beta) = V(s;\theta,\beta) + \tag{10.7}$$

$$\left(A(s,a;\theta,\alpha) - \frac{1}{|\mathscr{A}|}\sum_{a'} A\left(s,a';\theta,\alpha\right)\right). \tag{10.8}$$

The rationale for using PER is to prioritize crucial experiences for effective training. Rather than selecting experiences randomly, those with significant discrepancies between the deep Q-network and target network are given priority. A probability is then calculated to determine which experiences are chosen for replay, using

$$P(i) = \frac{p_i^\alpha}{\sum_k p_k^\alpha} \quad \text{and} \quad p_i = |\delta_i| + e. \tag{10.9}$$

The absolute Temporal Difference Error is denoted as $|\delta_i|$. A small value of constant e is added to ensure that no experience has a probability of 0 to be selected. When $\alpha = 1$, the experiences with the highest priorities are chosen, while $\alpha = 0$ implies pure random selection. Priority sampling is employed, leading to bias toward high-priority samples. To rectify this bias, importance sampling weights (ISW) are utilized to adjust updating by reducing the weights of frequently occurring samples [26]

$$\omega_i = \left(\frac{1}{N} \cdot \frac{1}{P(i)}\right)^{\beta}, \tag{10.10}$$

where β controls how much the ISW affect learning, and N is the size of replay buffer.

10.3.3 Combining Level-K Reasoning with D3QN PER

In order to produce varied vehicles with distinct reasoning abilities, our simulator utilizes the D3QN PER RL algorithm. Our level-k learner is represented by the ego vehicle. Adhering to the level-k principle, we allocate our level-(k-1) trained policies (or predetermined level 0 behavior) to the remaining vehicles that make up the environment. The proposed method employs a pre-established level 0 policy that serves as the foundation for all subsequent policies generated through D3QN PER. The level-1 policy is produced by designing a traffic situation where only the ego car is trained to respond to the level 0 policy, while all other drivers operate at level 0. Five velocities (0.3 m/s, 0.6 m/s, 0.9 m/s, 1.2 m/s, and 1.5 m/s) are available for opponents to randomly choose from. At level $k = 0$ of reasoning, opponent vehicles travel at the chosen constant speed, without taking into account the movements of others. To prevent collisions of opponent vehicles from interfering with

Table 10.3 Training of level K agent

Algorithm 1 Obtaining the level K Policy
1: Load the trained level-(k-1) (or rule-based level 0) policy, π_{k-1};
2: Set the reasoning levels of all opponents in the environment, p_i, to level-(k-1): $\pi_{p_i} = \pi^{k-1}, i = 2, 3, \dots, n_d$;
3: Initialize the ego driver's policy to a uniform action probability distribution over all states: $\pi_{p_1} = \pi^{uniform}$;
4: Train the ego driver using D3QN PER algorithm;
5: At the end of the training, ego driver learns to best respond to π^{k-1},therefore the resulting policy is the level K policy, π^k.

Table 10.4 Training of adaptive agent.

Algorithm 2 Obtaining the Adaptive Policy

1: Load the previously obtained level-1 and level-2 policy randomly, π^1 and π^2 (see Algorithm 1);
2: Set the agents in the environment, p_i, as level-1 agents or level-2 agents: $\pi_{p_i} = \pi^1$ or $\pi_{p_i} = \pi^2, i = 2, 3, \ldots, n_d$;
3: Initialize the ego driver's policy to a uniform action probability distribution over all states: $\pi_{p_1} = \pi^{uniform}$;
4: Train the ego driver using D3QN PER algorithm;
5: At the end of the training, ego driver learns to best respond to π^1 and π^2. Thus, the resulting policy is the adaptive policy, π^{adap}.

---

the ego vehicle's training process, we prioritize the speed selection of each opponent. For instance, car 2 may choose a speed first, and then car 4 will randomly select a speed from our established "safe speed set" to prevent collisions with car 2 at collision area 2 (refer to Figure 10.1). Ultimately, car 3 will randomly choose a safe speed to avoid colliding with car 4. After finishing the training, the ego transitions to a driver at level 1. Table 10.3 outlines the steps to obtain the level-k policy using a combination of level-k reasoning and D3QN PER. It is worth noting that the number of drivers is denoted as n_d.

We trained level-1 and level-2 policies sequentially, but the issue is that these models assume all opponents play level(k-1). If their true strategies differ, conflict may occur, as Section 10.5 results reveal. To address this, opponents will choose a trained policy at random, and our ego car will explore

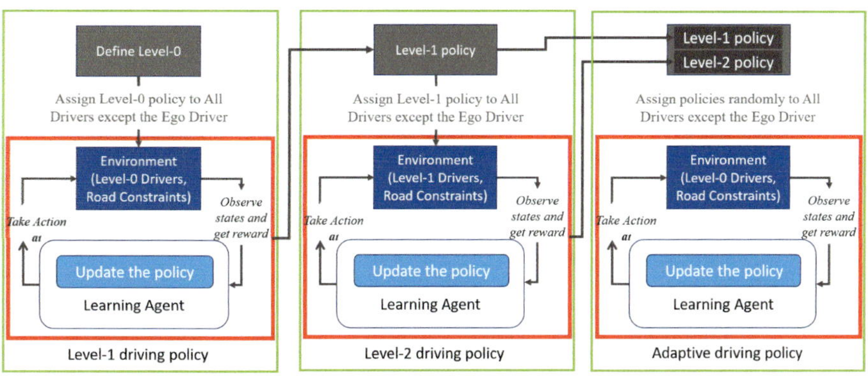

Figure 10.5 Combination of level K theory and RL.

adaptive strategies using D3QN PER in this mixed environment. Since each vehicle has limited access to traffic state via Lidar sensor, they cannot observe other vehicles' driving policies.

In order to tackle the partial observable Markov decision process, the D3QN PER algorithm is employed in tandem with an LSTM recurrent neural network. This enables the selection of an action that is based on not only the present observation but also a set number of the most recent observations. This black box approach helps in learning diverse driving policies, as opposed to estimating the belief of driver model of other vehicles through a Bayesian-based method. Refer to Table 10.4 for an overview of the adaptive policy training process.

The learning process described above reduces computational cost by utilizing previously trained policies for non-ego agents, thereby incorporating them into the environment. This allows for traffic scenarios with a range of levels, where all agents make strategic decisions simultaneously. This differs greatly from traditional decision-making methods in congested traffic, where only one driver makes strategic decisions while others follow predetermined policies to meet kinematic restrictions. Figure 10.5 visually illustrates the combination of level K reasoning and D3QN PER.

10.3.4 Setting of Algorithm Parameters

The D3QN PER algorithm's parameter settings are presented in Table 10.5. To accelerate training, subsequent training following the acquisition of the

Table 10.5 Parameter settings.

Parameters	Values
Discount factor (γ)	0.95
Learning rate	0.001
Starting (ending) value of ε greedy policy (ε)	1 (0.01)
Number of actions	5
Size of replay memory	6000
Maximum value of training steps	150
Number of steps to update the target network	500
Mini-batch size	32
Training steps (N_{training})	15,000
e	0.00001
Priority (α)	0.6
Adjusting the deviation (β)	0.4

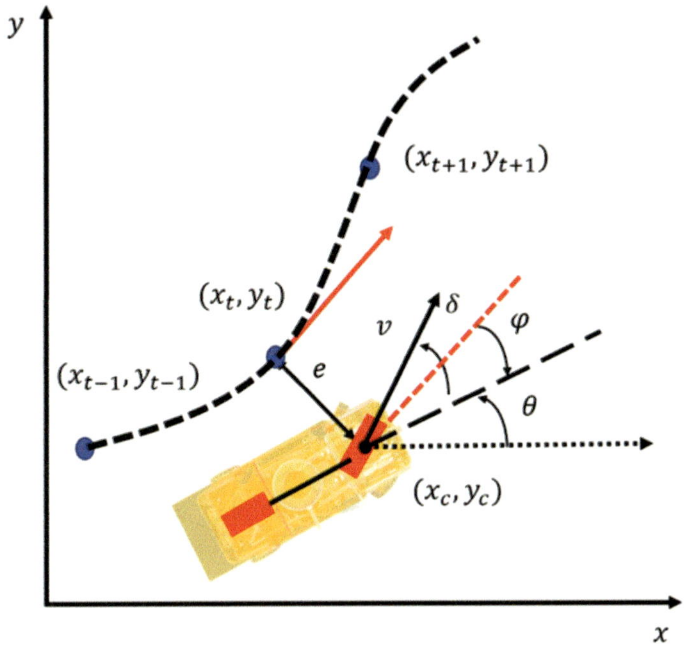

Figure 10.6 Stanley controller.

level-1 policy involved loading the prior model onto the ego vehicle and initializing ε to 0.5 instead of 1.

To prevent inadequate exploration and speed up convergence, the ε-greedy parameter decreases linearly from 1 based on training steps, as indicated in eqn (10.11), and remains constant until it reaches the end value

$$\varepsilon = \varepsilon - \frac{1.0}{N_{\text{training}}}. \tag{10.11}$$

The PER algorithm uses parameter e to avoid ignoring stored experience when the TD-error reaches 0. The degree of prioritization is controlled by the exponent α, where $\alpha = 0$ means uniform distribution. Parameter β fully accounts for the nonuniform probabilities $P(i)$ if $\beta = 1$. Its value follows a linear equation, as shown in eqn (10.12)

$$\beta = \beta + \frac{1.0}{N_{\text{training}}}. \tag{10.12}$$

10.4 Low Level Control

The preceding section presented the concept of car following and lane-changing tasks through end-to-end deep RL approach. Yet, the control performance needs to be guaranteed using traditional controllers once the decision-making module instructs acceleration and steering commands. Consequently, we implemented the PID controller at the low-level module for longitudinal control and the Stanley controller, illustrated in Figure 10.6, for transverse control.

10.4.1 Trajectory generation module

To achieve smoother driving, we assume zero acceleration at the start and end of motion with nonzero speed. Thus, we employ a fifth-order polynomial to create a trajectory for lane changing. This approach enables us to set six boundary conditions (position, velocity, and acceleration) at both $t_i = 0$ and $t_f = T$, which represents the time it takes to complete the lane change. The reference trajectory can be represented in eqn (10.14), which includes six coefficients for each equation

$$\mathrm{x(t)} = a_5t^5 + a_4t^4 + a_3t^3 + a_2t^2 + a_1t + a_0 \tag{10.13}$$

$$\mathrm{y(t)} = b_5t^5 + b_4t^4 + b_3t^3 + b_2t^2 + b_1t + b_0 \tag{10.14}$$

Then the time-dependent parameter matrix can be defined as (10.15):

$$M_{6\times6} = \begin{bmatrix} t_i^5 & t_i^4 & t_i^3 & t_i^2 & t_i & 1 \\ 5t_i^4 & 4t_i^3 & 3t_i^2 & 2t_i & 1 & 0 \\ 20t_i^3 & 12t_i^2 & 6t_i^1 & 2 & 0 & 0 \\ t_f^5 & t_f^4 & t_f^3 & t_f^2 & t_f^1 & 1 \\ 5t_f^4 & 4t_f^3 & 3t_f^2 & 2t_f^1 & 1 & 0 \\ 20t_f^3 & 12t_f^2 & 6t_f^1 & 2 & 0 & 0 \end{bmatrix} \tag{10.15}$$

10.4.2 Lateral control

The Stanley control method achieves lateral control by minimizing heading error $\psi(\mathrm{t})$ and cross-track error $e(t)$, as illustrated in Figure 10.6. $\psi(\mathrm{t})$ is the angle between the car heading and the trajectory heading, while $e(t)$ is the distance between the front wheels' center reference point (x_c, y_c) and the path (x_t, y_t) at the current time t. The linear speed of the front wheels is denoted as $v(t)$, and the steering angle as $\delta(t)$. The steering angle is limited

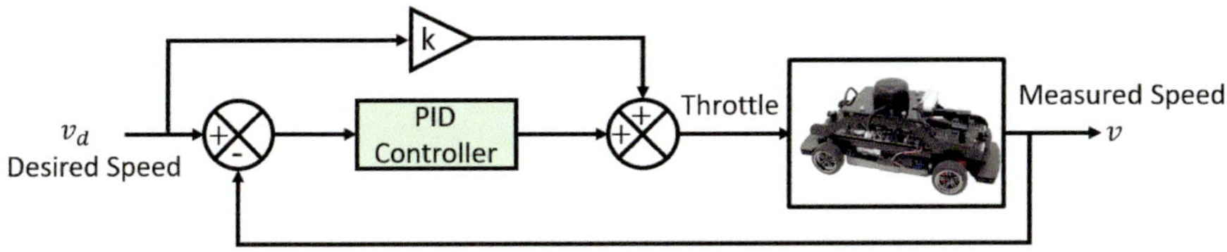

Figure 10.7 Longitudinal control.

by the vehicle's angle constraint, $\delta(t) \in [\delta_{\min}, \delta_{\max}]$. The controller equation is expressed as eqn (10.16).

$$\delta(t) = \psi(t) + \tan^{-1}\left(\frac{ke(t)}{k_s + v(t)}\right) \tag{10.16}$$

where softening constant, k_s is added to ensure the denominator be nonzero, and k is constant parameter.

10.4.3 Longitudinal control

To regulate speed, the controller examines the car's velocity and modifies the throttle to align with the predetermined speed determined by the decision-making module. The PID consists of three components: the proportional, integral, and derivative terms. The proportional term K_p is in direct proportion to the difference between the desired and actual speeds, while the integral term K_I is proportional to the error's integral, and the derivative term K_d is proportional to the error's derivative. The controller can be represented as eqn. (10.17)

$$u = K_P(v_d - v) + K_I \int_0^t (v_d - v)\, dt + K_D \frac{d(v_d - v)}{dt} \tag{10.17}$$

$$T = k * v_d + u \tag{10.18}$$

Hence, the control law (10.18) can depict the throttle command T.

10.5 Simulation and Experimental Results

10.5.1 Unsignalized intersection case

10.5.1.1 Comparison of different algorithms

In order to determine learning efficiency, each algorithm underwent 8000 episodes of training in an environment where opponents had level 0

Table 10.6 Comparison with existing work.

No.	Interaction type	GTDRL (%)	AV controller (%) [10]
1	L1 vs. L0	96.2	**99**
2	L2 vs. L1	**99.6**	95
3	L1 vs. L1	**94.9**	84
4	L2 vs. L2	**85.9**	57
5	L2 vs. L0	**83.1**	41
6	Adapt. vs. mixed env.	**96.2**	94–95

reasoning. To ensure consistency, all curves depicted in Figure 10.8 have been smoothed using a moving average over 300 episodes. Results indicate that the prioritized reply and dueling structure are critical components of the D3QN PER algorithm, as the removal of either resulted in a significant decrease in learning performance. It is noteworthy that the nature DQN and Dueling DQN exhibited poor performance during training, potentially due to the overestimation discussed in Section 10.3.2.

10.5.1.2 Simulation results

We have successfully trained different driving strategies, as per the scheme proposed in Section 10.3.3. Level-1, level-2, and adaptive driving strategies have been trained (see video[2]). We tested the trained policies in two scenarios, where ego vehicles are level K and opponents' vehicles are level(k-1), for 1000 episodes each (k=1, 2). The success rate was over 96%, meeting the

[2]Video for both training and testing is available at https://youtu.be/mPtoojXh2-s.

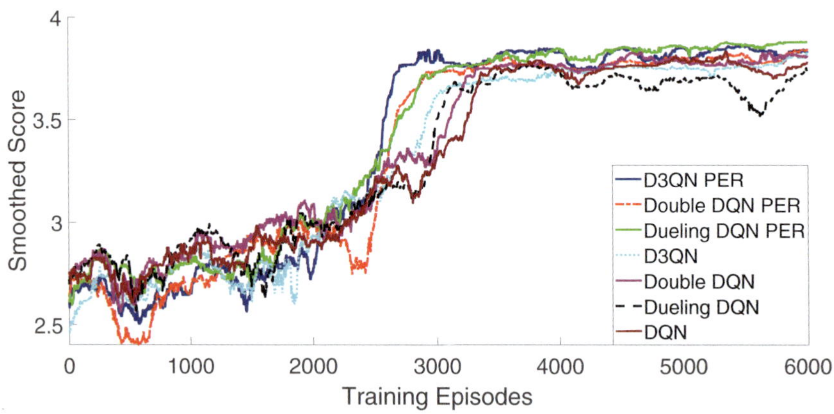

Figure 10.8 Comparison of various deep RL algorithms.

Table 10.7 Comparison of policies in mixed environment.

Interaction type	Deadlock (%)	Collision (%)	Success (%)
L1 vs. mixed	4.3	1.7	94.0
L2 vs. mixed	2.3	9.3	88.4
Adapt. vs. mixed	**3.5**	**0.3**	**96.2**

level K theory expectations. Then, we tested level-1 and level-2 policies on level K versus level K scenarios for 1000 episodes each (k=1, 2). These scenarios had a lower success rate as all cars had incorrect assumptions about others' driving policy. Level-1 policies resulted in deadlock due to conservative driving behaviors, and level-2 policies resulted in collisions due to aggressive driving behaviors (see Table 10.6).

Our algorithm, GTDRL, was compared to an AV controller studied in [10]. The AV controller employs driver interaction models and online model estimation. We assessed the performance of both algorithms and recorded the results in the Table 10.6. Our approach, which encompasses perception, decision-making, and control, resulted in a greater rate of success than the results reported in [10], in which only a less complex scenario involving two cars is studied without considering the noise in the observation data.

After running 1000 tests on the level-1, level-2, and adaptive policies in 81 scenarios (as described in Section 10.2.1), we found that the adaptive policy had the highest success rate for the ego vehicle to pass through intersections without collision or deadlock. Table 10.7 shows that the level-1 policy had the highest deadlock rate due to its cautious driving behavior of stopping abruptly. On the other hand, the collision rate was highest for the level-2 policy as it imitates aggressive driving patterns that frequently result in crashes with other vehicles. The adaptive policy efficiently reduces both deadlocks and collisions in mixed traffic scenes by integrating the benefits of level-1 and level-2 policies.

After conducting 1000 tests in 81 different scenarios outlined in Section 10.2.1, we evaluated the performance of level-1, level-2, and adaptive policies. The results, presented in Table 10.7, reveal that the adaptive policy achieved the highest success rate across all three policies. This success rate reflects the ego vehicle's capacity to navigate the intersection safely and avoid collisions and deadlocks. The conservative driving behavior associated with level-1 policy leads to the highest rate of deadlocks, while level-2 policy, which models aggressive driving behavior, has the highest collision rate. However, the adaptive policy effectively integrates the strengths of both level-1 and level-2 policies to minimize the occurrence of both deadlocks and collisions in mixed scenarios.

10.5.1.3 Hardware Implementation

Due to limited space, we have chosen four scenarios with four cars to demonstrate the trained model's performance. These scenarios depict interactions among level K vehicles at an unsignalized four-way intersection (refer to Figure 10.1 left). We have assigned different level K policies to four vehicles and analyzed how each traffic scenario evolves based on varying level K policy combinations. The results presented in Figure 10.11 indicate that conflicts between level-1 (l1) and level-2 (l2) vehicles can be resolved as expected. Level-1 vehicles (i.e., cautious drivers) yield the right of way while level-2 vehicles (i.e., aggressive drivers) proceed ahead.

Steps (a)–(d) illustrate five sequential stages of a hardware examination process. Our simulator has pretrained level-1 and level-2 policies that can regulate each vehicle. The lower panels exhibit the time records of the motion state of the four vehicles (refer to Figure 10.10). The routes are partitioned into 200 waypoints, and as the car reaches each point, the number of passed waypoints increments by one. For instance, the number of passed waypoints is one when the car is at its starting position and 200 when it arrives at the destination.

At time $t = 0$ s, all vehicles are located outside the intersection. In column (a), car 1 follows the level-1 policy and interacts with three cars following the

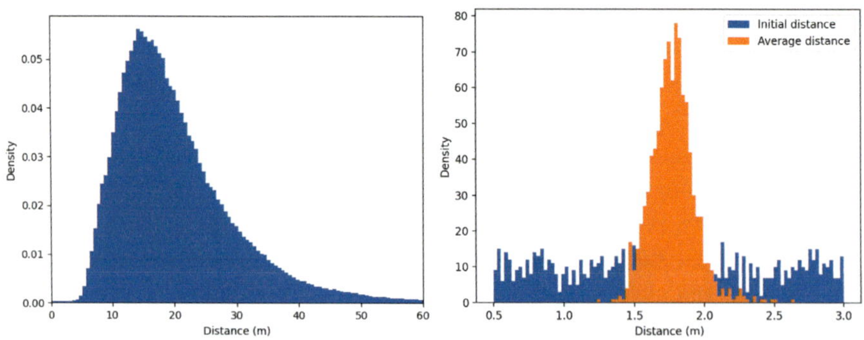

(a) (b)

Figure 10.9 Distribution of distance to car in front. (a) US-101 data set (b) Simulated environment

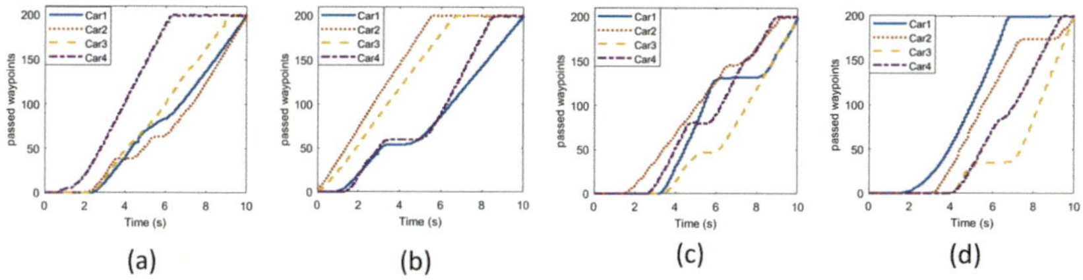

Figure 10.10 Travel efficiency under: (a) policy [l1, l2, l2, l2]; (b) policy [l1, l2, l2, l1]; policy [l1, l2, l1, l1]; (d) policy [l2, l1, l1, l1].

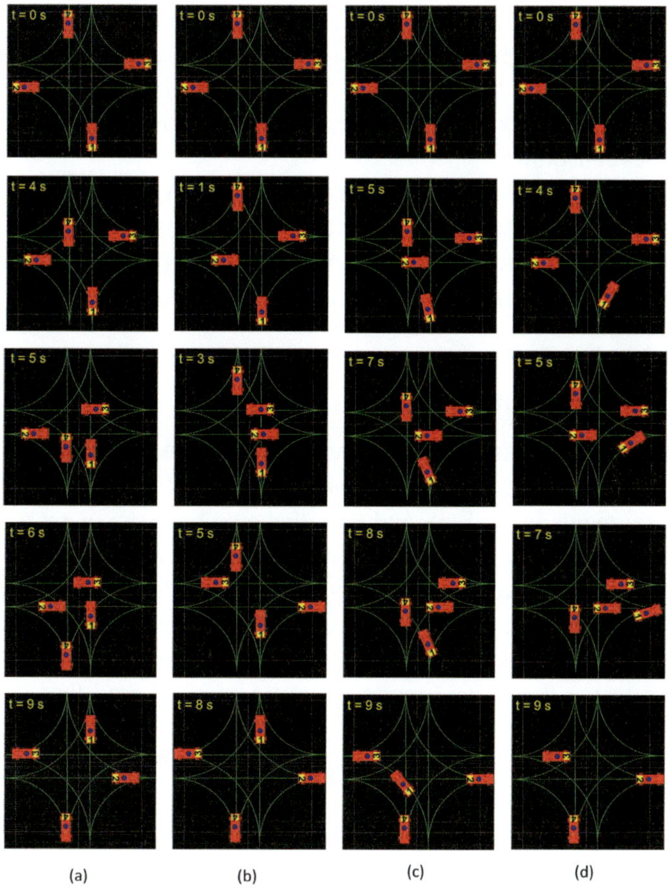

Figure 10.11 10 s motion states of car1 to car 4 shown in the column under (a) police [l1, l2, l2, l2] with car 1 going straight; (b) policy [l1, l2, l2, l1] with car 1 going straight; (c) policy [l1, l2, l1, l1] with car 1 turning left; (d) policy [l2, l1, l1, l1] with car 1 turning right. All surrounding vehicles are going straight. The numbers represent the car IDs.

level-2 policy. Cars 2, 3, and 4 prefer to pass the intersection quickly due to their level-2 policy. Car 4 passes the intersection first without decelerating, and cars 2 and 3 choose different degrees of deceleration based on their observation of car 4's motion state at around t = 4 s. Car 3 is ahead of car 1 at around t = 6 s, so car 1 decelerates at around t = 5 s. Car 2 waits for car 1 to pass the collision area before proceeding. Eventually, all four cars pass the intersection safely and in turn. In column (b), cars 1 and 4 follow the level-1 policy and interact with cars 2 and 3 following the level-2 policy. Cars 2 and 3 enter the collision area first, and cars 1 and 4 decelerate at around t = 3 s to wait for them to pass. In column (c), car 2 follows the level-2 policy and interacts with cars controlled by level-1, including car 1 turning left. In column (d), car 1 turns right and follows the level-2 policy while interacting with vehicles controlled by level-1. In both situations, the vehicle using the level-2 strategy passes through the intersection first, followed by the vehicles following the level-1 policy.

The aforementioned instances demonstrate the ability of models trained in the simulator to handle complex interaction scenarios even without knowledge of others' policies, thereby confirming the feasibility of our method for modeling varied driving behaviors outlined in this chapter.

10.5.2 Urban lane-change case

10.5.2.1 Compare the trained policy with human drivers

The training environment comprises six cars and three 100-m road segments. One of these segments is for Ego, and the other two have opponent vehicles with rule-based driving policies. To learn the car following policy, we set the initial intervehicle distance uniformly between 0.5 m and 3 m. In each episode, the maximum speed limit for opponent vehicles follows a uniform distribution between 0.5 m/s and 1.5 m/s, while Ego's maximum speed limit is 1.5 m/s. We tested the trained model 1000 times and found that Ego can maintain a safe driving distance from the car in front when following it. The distribution of the distance between Ego and the front vehicle follows the normal distribution with $\mu = 1.85$ and $\sigma = 0.8$ m, as shown in Figure 10.9b. Note that the car used in our experiment is only one-tenth the size of a real car, so the distance in the x-axes is divided by 10. The LiDAR's maximum detection distance is 2.5 m, so the distance distribution is limited to between 0.5 m and 3 m. The blue bar represents the initial distance between Ego and the car in front, while the orange bar represents the average intervehicle distance for each episode. This distance is similar to the distribution of US101 traffic, especially in the interval we analyzed, as shown in Figure 10.9a.

To assess our ACCs' performance, we compared them to human players in various traffic scenarios. Both groups utilized the same observation information, such as the real driving situation, to ensure a fair comparison. Human drivers relied on the point cloud data detected by the lidar to obtain position and speed information of the leading vehicle, while controlling their own car with a joystick. Each test group involved three cases where the leading vehicle was located in the danger zone (gap < 1.5 m), interaction zone ($1.5 <$ Gap < 2.7 m), and safety zone (Gap > 2.7 m) of the controlled vehicle, with a varying maximum speed limit of the leading vehicle within the range of [0.5, 1.5] m/s. The evaluation criteria included safety, comfort, and energy consumption, with the latter two reflected by the vehicle's acceleration. The action space consisted of five actions: maintaining the current speed, hard acceleration, hard deceleration, acceleration, and deceleration. To minimize fuel consumption, drivers typically reduce the frequency of deceleration actions. Comfort was determined by the following order of actions: maintain current speed $>$ acceleration (or deceleration) $>$ hard acceleration (or hard deceleration). Ultimately, to ensure safety, the car following policy must maintain a stable distance from the leading vehicle and maintain a relative speed of 0.

Based on the results of three comparison tests, it was observed that our ACCs exhibit comparable driving performance to humans, and even surpass them in certain scenarios. With regards to fuel consumption, our ACCs showed a significant decrease in the frequency of deceleration compared to human drivers, as evidenced by Figures 10.12–10.14. Moreover, our ACCs demonstrated a higher preference for maintainance action in terms of comfort, compared to human drivers. In the context of safety, human drivers were observed to occasionally enter the danger zone while adjusting the speed of the vehicle, as depicted in Figures 10.12f and 10.14f. Additionally, our trained policy exhibited superior performance in maintaining zero relative speed to the leading vehicle, compared to human drivers.

Afterward, we assess the ability to change lanes. As demonstrated in Figure 10.15, the outcomes for longitudinal and lateral control are presented. It is noticeable that the speed tracking error measures below 0.02 m/s, the trajectory tracking error measures below 0.05 m, and the path tracking error measures below 0.01 m. These values satisfy the criteria for self-driving vehicles.

10.6 Conclusions

An adaptive decision-making strategy with DRL was proposed for ADVs sharing the road in a multi-agent traffic scenario. The interactions between vehicles were modeled using a level K game-theoretic framework, and the ego adapted its behavior based on real sensor data. Simulation and hardware tests showed reasonable behavior expected in traffic. Performance was evaluated based on success rate, collisions, deadlock, and hardware testing snapshots, with the adaptive model demonstrating high rates of success in resolving traffic conflicts at each reasoning level.

Second, a hierarchical decision-making model is studied to imitate human driving behavior on highways via an end-to-end self-play training method. A high-fidelity simulator is developed to simulate vehicle dynamics and the LiDAR information acquisition, thus making the training environment closer to real-world scenarios. To design the reward function and action space, two traffic datasets are referred to. In order to make decisions, only a noisy 2D point cloud data and ego speed are used, instead of accurate position and speed information of surrounding vehicles. Experimental outcomes reveal that in this partially observable situation, the AV's policy learned to consistently mimic human driving behavior. Moreover, our ACCs exhibit equivalent or even superior driving performance to human drivers in some cases, verifying the efficacy of our approach.

The proposed framework in this study models multi-vehicle interactions and serves as a simulation tool for autonomous driving systems' calibration, validation, and verification. It can also support traffic management and high-level decision-making algorithms of ADVs. Additionally, the framework can model vehicle interactions in other traffic scenarios like highway merging and parking lot driving with modified road layouts and geometries.

10.6.1 Appendix

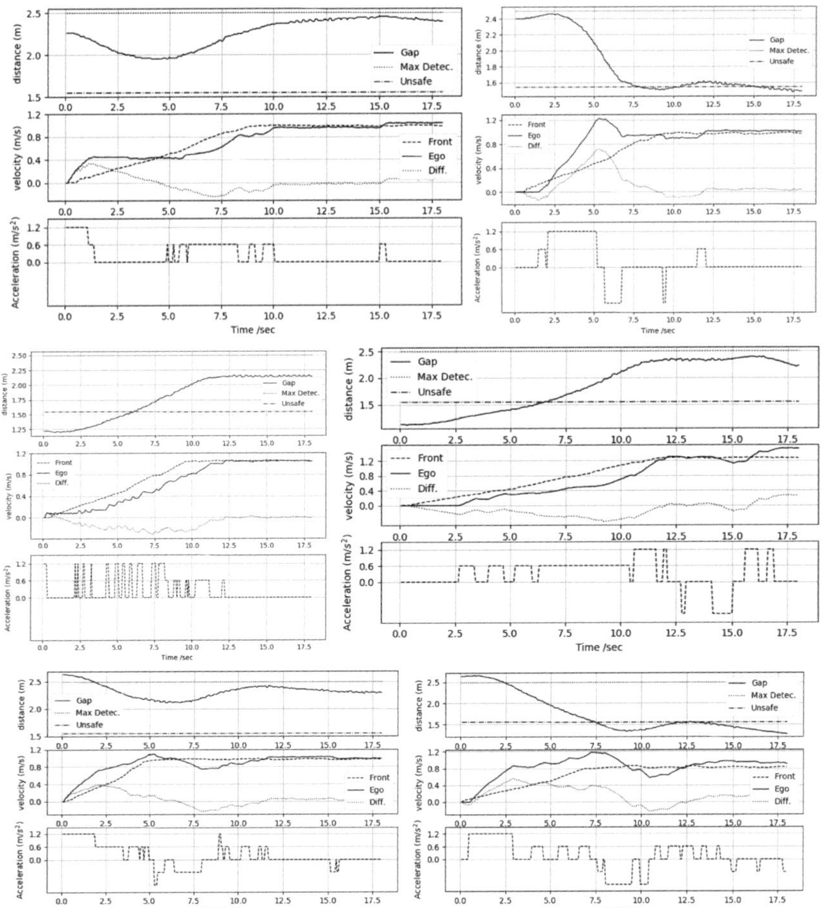

Figure 10.12 Gap between leader and follower, speed difference, and actions history: different initial gaps, (a) and (b) are examples when the leader of ego is located at the interaction zone; (c) and (d) are examples when the leader of ego is located at the unsafe zone; (e) and (f) are examples when the leader of ego is located at the safe zone. (a) ACCs (interaction zone), (b) player 1 (interaction zone), (c) ACCs (unsafe zone), (d) player 1 (unsafe zone), and (e) ACCs (safe zone), and (f) player 1 (safe zone).

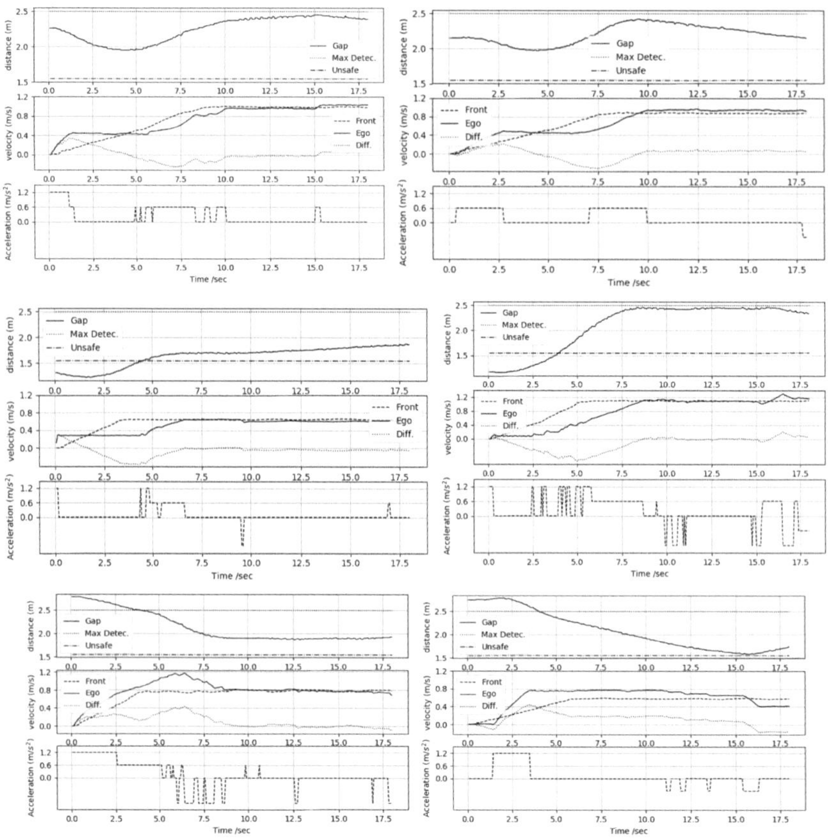

Figure 10.13 Gap between leader and follower, speed difference, and actions history: different initial gaps, (a) and (b) are examples when the leader of ego is located at the interaction zone; (c) and (d) are examples when the leader of ego is located at the unsafe zone; (e) and (f) are examples when the leader of ego is located at the safe zone. (a) Trained policy (interaction zone), (b) player 2 (interaction zone), (c) trained policy (unsafe zone), (d) player 2 (unsafe zone), (e) trained policy (safe zone), and (f) player 2 (safe zone).

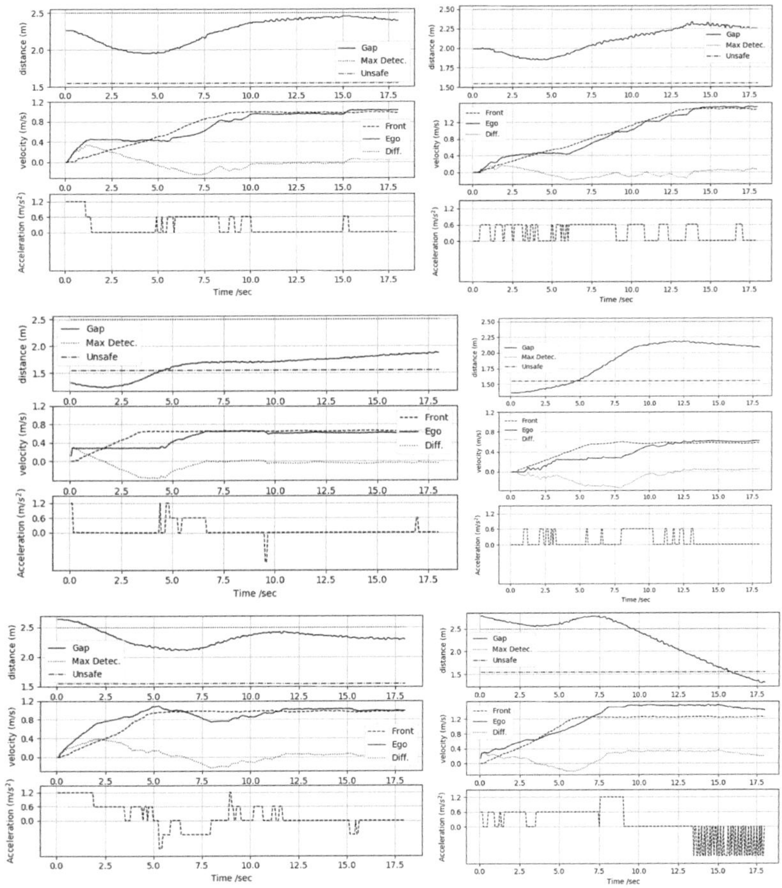

Figure 10.14 Gap between leader and follower, speed difference, and actions history: different initial gaps, (a) and (b) are examples when the leader of ego is located at the interaction zone; (c) and (d) are examples when the leader of ego is located at the unsafe zone; (e) and (f) are examples when the leader of ego is located at the safe zone. (a) Trained policy (interaction zone), (b) player 3 (interaction zone), (c) trained policy (unsafe zone), (d) player 3 (unsafe zone), (e) trained policy (safe zone), and (f) player 3 (safe zone). (a) Trained policy (interaction zone), (b) player 3 (interaction zone), (c) trained policy (unsafe zone), (d) player 3 (unsafe zone), (e) trained policy (safe zone), and (f) player 3 (safe zone).

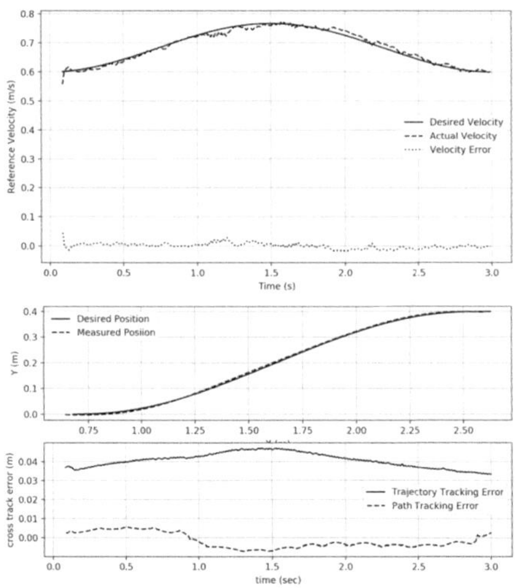

Figure 10.15 Speed and trajectory tracking performance. (a) Speed tracking and (b) trajectory tracking.

References

[1] Amir Efrati. Waymo riders describe experiences on the road. *URL https://www.theinformation.com/articles/waymo-riders-describe-experiences-on-the-road.(Accessed on 09/22/2020)*, 2020.

[2] N Kalra. With driverless cars, how safe is safe enough? *The RAND Center Decision Making Under Uncertain., Santa Monica, CA, USA, Tech. Rep*, 2016.

[3] Ujjwal Das Gupta, Erik Talvitie, and Michael Bowling. Policy tree: Adaptive representation for policy gradient. In *Proceedings of the AAAI Conference on Artificial Intelligence*, volume 29, 2015.

[4] Laurène Claussmann, Ashwin Carvalho, and Georg Schildbach. A path planner for autonomous driving on highways using a human mimicry approach with binary decision diagrams. In *2015 European Control Conference (ECC)*, pages 2976–2982. IEEE, 2015.

[5] Trong Nghia Hoang and Kian Hsiang Low. Interactive pomdp lite: Towards practical planning to predict and exploit intentions for interacting with self-interested agents. In *Twenty-Third International Joint Conference on Artificial Intelligence*, 2013.

[6] Julian FP Kooij, Fabian Flohr, Ewoud AI Pool, and Dariu M Gavrila. Context-based path prediction for targets with switching dynamics. *International Journal of Computer Vision*, 127(3):239–262, 2019.

[7] Mithun Babu, Raghu Ram Theerthala, Arun Kumar Singh, BP Baladhurgesh, Bharath Gopalakrishnan, K Madhava Krishna, and Shanti Medasani. Model predictive control for autonomous driving considering actuator dynamics. In *2019 American Control Conference (ACC)*, pages 1983–1989. IEEE, 2019.

[8] Charlott Vallon, Ziya Ercan, Ashwin Carvalho, and Francesco Borrelli. A machine learning approach for personalized autonomous lane change initiation and control. In *2017 IEEE Intelligent Vehicles Symposium (IV)*, pages 1590–1595. IEEE, 2017.

[9] Mario Garzón and Anne Spalanzani. Game theoretic decision making based on real sensor data for autonomous vehicles' maneuvers in high traffic. In *2020 IEEE International Conference on Robotics and Automation (ICRA)*, pages 5378–5384. IEEE, 2020.

[10] Nan Li, Ilya Kolmanovsky, Anouck Girard, and Yildiray Yildiz. Game theoretic modeling of vehicle interactions at unsignalized intersections and application to autonomous vehicle control. In *2018 Annual American Control Conference (ACC)*, pages 3215–3220. IEEE, 2018.

[11] Nan Li, Yu Yao, Ilya Kolmanovsky, Ella Atkins, and Anouck R Girard. Game-theoretic modeling of multi-vehicle interactions at uncontrolled intersections. *IEEE Transactions on Intelligent Transportation Systems*, 2020.

[12] Ran Tian, Sisi Li, Nan Li, Ilya Kolmanovsky, Anouck Girard, and Yildiray Yildiz. Adaptive game-theoretic decision making for autonomous vehicle control at roundabouts. In *2018 IEEE Conference on Decision and Control (CDC)*, pages 321–326. IEEE, 2018.

[13] Gokul S Sankar and Kyoungseok Han. Adaptive robust game-theoretic decision making strategy for autonomous vehicles in highway. *IEEE Transactions on Vehicular Technology*, 69(12):14484–14493, 2020.

[14] Guannan Qu, Adam Wierman, and Na Li. Scalable reinforcement learning of localized policies for multi-agent networked systems. In *Learning for Dynamics and Control*, pages 256–266. PMLR, 2020.

[15] Arbaaz Khan, Chi Zhang, Shuo Li, Jiayue Wu, Brent Schlotfeldt, Sarah Y Tang, Alejandro Ribeiro, Osbert Bastani, and Vijay Kumar. Learning safe unlabeled multi-robot planning with motion constraints. In *2019 IEEE/RSJ International Conference on Intelligent Robots and Systems (IROS)*, pages 7558–7565. IEEE, 2019.

[16] Markus Kuderer, Shilpa Gulati, and Wolfram Burgard. Learning driving styles for autonomous vehicles from demonstration. In *2015 IEEE International Conference on Robotics and Automation (ICRA)*, pages 2641–2646. IEEE, 2015.

[17] Shital Shah, Debadeepta Dey, Chris Lovett, and Ashish Kapoor. Airsim: High-fidelity visual and physical simulation for autonomous vehicles. In *Field and Service Robotics*, pages 621–635. Springer, 2018.

[18] Yevgen Chebotar, Ankur Handa, Viktor Makoviychuk, Miles Macklin, Jan Issac, Nathan Ratliff, and Dieter Fox. Closing the sim-to-real loop: Adapting simulation randomization with real world experience. In *2019 International Conference on Robotics and Automation (ICRA)*, pages 8973–8979. IEEE, 2019.

[19] Mingfeng Yuan, Jinjun Shan, and Kevin Mi. Deep reinforcement learning based game-theoretic decision-making for autonomous vehicles. *IEEE Robotics and Automation Letters*, 7(2):818–825, 2021.

[20] Zhongxiang Dai, Yizhou Chen, Bryan Kian Hsiang Low, Patrick Jaillet, and Teck-Hua Ho. R2-b2: Recursive reasoning-based bayesian optimization for no-regret learning in games. In *International Conference on Machine Learning*, pages 2291–2301. PMLR, 2020.

[21] Berat Mert Albaba and Yildiray Yildiz. Modeling cyber-physical human systems via an interplay between reinforcement learning and game theory. *Annual Reviews in Control*, 48:1–21, 2019.

[22] Nan Li, Dave W Oyler, Mengxuan Zhang, Yildiray Yildiz, Ilya Kolmanovsky, and Anouck R Girard. Game theoretic modeling of driver and vehicle interactions for verification and validation of autonomous vehicle control systems. *IEEE Transactions on Control Systems Technology*, 26(5):1782–1797, 2017.

[23] Gerd Gigerenzer and Reinhard Selten. *Bounded rationality: The adaptive toolbox*. MIT press, 2002.

[24] Ying Wen, Yaodong Yang, Rui Luo, and Jun Wang. Modelling bounded rationality in multi-agent interactions by generalized recursive reasoning. *arXiv preprint arXiv:1901.09216*, 2019.

[25] Volodymyr Mnih, Koray Kavukcuoglu, David Silver, Andrei A Rusu, Joel Veness, Marc G Bellemare, Alex Graves, Martin Riedmiller, Andreas K Fidjeland, Georg Ostrovski, et al. Human-level control through deep reinforcement learning. *nature*, 518(7540):529–533, 2015.

[26] Tom Schaul, John Quan, Ioannis Antonoglou, and David Silver. Prioritized experience replay. *arXiv preprint arXiv:1511.05952*, 2015.

Authors' Biography

Mingfeng Yuan received his M.S. (2019) in control engineering from Tianjin University of Technology, China. He is currently pursuing his Ph.D. degree with the Department of Earth and Space Science and Engineering, York University, Toronto, ON, Canada. His research interests include machine learning for prediction and decision-making in automated driving, game theory, deep RL, and nonlinear system modeling.

Jinjun Shan (SM'08) received Ph.D. degree in spacecraft design from the Harbin Institute of Technology, Harbin, China, in 2002. He is currently a Full Professor of Space Engineering and Chair of the Department of Earth and Space Science and Engineering, York University, Toronto, ON, Canada. His research interests include dynamics, control, and navigation. Dr. Shan is a Fellow of Engineering Institute of Canada (EIC) and a Fellow of American Astronautical Society (AAS). He was the recipient of the Alexander von Humboldt Research Fellowship and JSPS Invitation Fellowship in 2012. Since 2007, he has been a Professional Engineer in Ontario.

11

State of Modeling and Simulation in Autonomous Vehicles

Arpan Kusari

University of Michigan Transportation Research Institute, USA
Email: kusari@umich.edu

Abstract

This chapter provides an overview of the critical steps in developing general purpose simulation tools. While a variety of open-source and proprietary simulation tools already exist, it can be difficult to understand the steps required in constructing an effective world model and also how to input a realistic view of the world into the AV pipeline. An additional goal of this chapter is to provide the reader with an in-depth understanding of where the benefits of simulation could be maximized in terms of fidelity and control, the challenges in developing realistic sensor simulations and complex driving scenarios and finally discuss some of the current open problems existing in this space.

Keywords: Simulation, operational design domain, scene, situation, scenario, sensor simulation.

11.1 Introduction

The paramount goal of autonomous vehicles (AV) is to navigate through complex and dynamic traffic environments under variable weather and lighting conditions without any human assistance. However, this seemingly simple goal turns out to be a lot more complicated to formulate concretely given the inherent complexity in static road networks and traffic rules, not to mention the large, almost unbounded space of dynamic traffic conditions. According to some estimates [1] a fleet of fully autonomous vehicles would need to drive billions of miles or conversely hundreds of years to prove their safety and reliability in handling complex driving scenarios. It has become quite clear to the AV companies and the regulators alike that it is impossible to achieve anything remotely close to these numbers through current traditional testing methods such as closed-track and open-road testing. Thus, there has been a clear impetus towards providing multifaceted modeling and simulation tools with varying degrees of fidelity and control. Simulation tools can provide a very cheap, almost infinite source of variability, which can test the artifacts and behavior of the AV algorithms under hard-to-predict situations. These tools have become a critical component in scaling the capabilities of AV with an eye towards providing safety verification against the complexities encountered in the real world. A perfunctory analysis of simulation tools in industry shows the potential to accelerate throughout the development process in a manner that makes simulation a lucrative proposition for businesses.

The biggest advantage of using simulation lies in the tight control of the environment and conditions including behavior of the dynamic agents. The analogous benefit is the absence of uncertainty which helps in making the tests repeatable. Another advantage is using simulation for generating a potentially infinite number of diverse scenarios to test the system. While simulation tools help advance AV technology and safety in a myriad of ways, on the flip side, it is difficult to model dynamic agents to have full fidelity and randomness as real world, commonly referred to as sim-to-real domain gap. Our purpose in this chapter is to provide an overview of the critical steps in developing general purpose simulation tools. We would like to note that our goal is not to survey existing open source and/or commercial simulation tools but rather to provide an in-depth understanding of where the benefits of simulation could be maximized, the challenges in developing realistic sensor simulations and complex driving scenarios and the current open problems existing in this space.

In general, simulation tools enable developers and other stakeholders to understand the constraints of the system under test (SUT) given the caveat that there exists a gap between the simulation and reality. The SUT can

[1] Kalra, N., & Paddock, S. M. (2016). Driving to safety: How many miles of driving would it take to demonstrate autonomous vehicle reliability?. Transportation Research Part A: Policy and Practice, 94, 182-193.

then be subjected to rapid and fail-safe testing by modeling with sufficient fidelity to mimic its behavior to that of the real system. Each module of the AV software stack needs to be thoroughly tested and its functionality assured under a host of environmental and traffic conditions, referred to as Operational Design Domain (ODD). The testing procedure needs to be of the same caliber as the real environment that the AV would encounter, therefore needing to simulate all the operational conditions. Given a virtual environment, the developers can modify key parameters randomly or from a specific probability distribution, gathering data including failure modes for post processing and analysis. Recently, using previously collected driving data by decomposing it into spatio-temporal sequences known as scenarios is gaining a lot of traction. Scenario represents an evolution of the states of an operational world model in the time domain. In the first part of this chapter, we will delve into the creation of this world model, specifically the question of how to decompose driving data into scenarios. We first define the various terms associated with scenarios in an AV context and then talk of the various attempts to break down the world model into scenario layers. Our aim through this section is to make the reader aware of the quality that would dictate the capability of the simulator itself in building a test that can match real world variability.

Given the world model, we need a way to send the relevant information into the AV software stack. This can be accomplished by sensor simulation which can then feed realistic sensor output data into perception algorithms. Simulating realistic sensor data is a complex task in itself. Simple ray-casting or ray-tracing techniques lack the richness required in creating sensor data and do not take into account the physical nature of the objects along with the physical laws of occlusions, distortions etc. In the second part of the chapter, we discuss various efforts taken towards simulating sensor data with a goal towards emulating the sensor properties such that the decisions made by the controls algorithms match that made in reality. The goal of this section is to provide an understanding of the steps to be taken in sensor simulation and the current state-of-the-art methods.

Finally, we try to understand what the open problems are in the simulation space. While there is a lot of research being conducted in the AV industry, there are a few big disagreements on fundamental concepts which are hindering collective progress in this sphere. We specifically shed some light on some of the work going on towards building a consensus in usability and interoperability of scenarios.

11.2 Building a world model in simulation

The fundamental challenge in creating a virtual testing environment is to create a sufficiently rich and expressive operational world model that can be used to specify driving tasks and requirements. These tasks and requirements in turn drive the verification and validation of the intended AV behavior. For advanced driver assistance systems (ADAS), testability of the driving functions have to be demonstrated via standardized approvals such as ISO 26262. However, with automated driving functions, such standardized approvals for providing safety certification is not only currently absent but is rather difficult to arrive at since the context under which the testing is performed becomes a central component. This driving context is then referred to as ODD and is defined by SAE J3016 as "operating conditions under which a given driving automation system or feature thereof is specifically designed to function, including but not limited to, environmental, geographical, and time-of-day restrictions, and/or the requisite presence or absence of certain traffic or roadway characteristics" [23]. The ODD depends on levels of automation where the last automation level, level 5, corresponds to an unbounded ODD, similar to an human driver. In order to make the problem tractable, certain constraints are applied to create a structured testing framework from:

1. static road network - may be limited to broad types such as highways, urban, semi-urban and rural but often needs to incorporate specific elements such as number of lanes, roundabouts, entry and exit points and state/region specific road structures such as Michigan left. Another area which is crucial for laying out a realistic road network is temporary modifications to road networks such as construction zones, which can lead to a change in traffic behavior as well.
2. behavior of dynamic objects - dynamic objects in the simulation environment can be the subject vehicle (SV), other background vehicles (BV) including motorized 2-wheeled vehicles, pedestrians and bicyclists and other smaller rare occurrences of animals which interact with the traffic. The behavior of each dynamic object in the simulation environment can then be defined as fulfilling its own goal subject to the constraints of the external environment. Although this behavior definition is intuitive and can be succinctly described, defining the goal and the constraints for each dynamic object is a highly complex problem and one that is still a very active area of research.
3. external environmental conditions - simulating realistic external environmental conditions is a very complex problem as well. Some

simulation environments provide preset weather and lighting conditions (for e.g. CARLA which provide clear, cloudy, wet weather and noon, sunset lighting) but it's not a continuous change and it is not immediately clear how the continuous transition could be made.

The training and testing of AV then requires it to be validated in every aspect of the ODD and similarly recognize when an exclusion occurs out of the ODD. It is immediately apparent to almost any practitioner that application of every constraint still leads to a very high dimensional space capturing the granularity of which coupled with the extent is still elusive for AV simulation experts.

11.2.1 Components of a scenario

Defining granularity for such a high-dimensional spatio-temporal problem is a challenging task in itself. In recent years, researchers in Europe have grappled with the definitions and discussions of granularity [1, 2, 8, 28]. They have specifically noted that simple and ubiquitous terms such as scene, situation and scenario, which are used to define the aforementioned ODD constraints, have been vague or in some cases, contradictory, which has led to inconsistencies in results. Our definitions and discussions of these terms are informed by these papers.

11.2.1.1 Defining scene

We start with defining a scene which has the smallest granularity. [27] in 1985 talked about scene as part of digital scene analysis where the physical objects were captured in a digital image from which they are extracted and labeled. [8] utilized the analogy of theater to define a scene as a collection of physical objects in a scenery, dynamic elements and optional driving instructions. The authors also suggested that a scene starts with the end of the previous scene or in the case of the first scene, with a predefined starting scene. Thus, the scene is then not a single time instance but a certain time bounded phenomenon. This can lead to various technical challenges, where the exact start or the amount of time allotted for a given scene are not clearly defined. Ulbrich et al. [28] provide a differing definition which deviates from the previous definition as:

Definition 11.1 (Scene) *A scene describes a snapshot of the environment including the scenery and dynamic elements, as well as all actors' and observers' self-representations, and the relationships among those entities. Only a scene representation in a simulated world can be all-encompassing* *(objective scene, ground truth). In the real world it is incomplete, incorrect, uncertain, and from one or several observers' points of view (subjective scene).*

An actor represents an element in the scene which is acting on its own behalf while an observer is an element of the scene which is perceiving the scene. Self-representation can then be states and attributes of the actor and/or observer, along with other properties such as field of view or occlusions. Using these verbal definitions, Andreotti et al. [1] in 2020 provided a mathematical definition of the scene. A set of preliminary concepts need to be defined starting with a portion of space, $\mathscr{S}$, in the real world. The static objects including the road network and the infrastructure can then be defined as the static object set $\zeta^s(\mathscr{S})$ and analogously, the dynamic object set be defined as $\zeta^d(\mathscr{S},t)$ (representing all vehicles, pedestrians, bicyclists etc) where each dynamic object occupies space $\mathscr{S}$ in real world t. Each element within the dynamic set $\zeta^d(\mathscr{S},t)$ can then be referred to as an actor. The union of the static and dynamic object set is the set of all objects and is denoted as $\zeta(\mathscr{S},t) = \zeta^s(\mathscr{S}) \bigcup \zeta^d(\mathscr{S},t)$, with each element being referred to as an object. While we have defined the objects, we have not provided a self-representation or established a relationship between them. We utilize the position function for this purpose defining it as $\mathscr{C} : \zeta(\mathscr{S},t) \to \mathscr{S}$, such that each object is then defined using a geographical coordinate with respect to an origin of choice. The set of all object positions is then referred to as the position set $\mathscr{C}(\zeta)$. The couple $(\zeta,\mathscr{C})$ provides the ground-truth snapshot of all objects of the portion of space $\mathscr{S}$ at time t. However, for each actor acting as an observer, the subset of the object set that it perceives can be represented as ζ_x where x is the given observer. The position function can then also be defined from x's point of view, given as $\mathscr{C}_x$, where

$$\mathscr{C}_x : \zeta_x \to \mathscr{S}, y \mapsto \mathscr{C}_x(y) \tag{11.1}$$

The object y can be time-dependent, in which case, it is written as $y(t)$, and through $\mathscr{C}_x$ is associated to its position at time t. It follows that position from x's point of view set (or x-position set) can be given as the image set ζ_x under the x-position function $\mathscr{C}_x(\zeta_x)$. Given that there are measurement errors (among other sources of errors), even if the object y can be perceived from the ego-vehicle, the position of y from x, $\mathscr{C}_x(y)$ in general will not be the true $\mathscr{C}(y)$. In mathematical terms, $y \in \zeta_x \subset \zeta \Rightarrow \mathscr{C}_x(y) \neq \mathscr{C}(y)$.

With the discussion of individual objects and their positions (both absolute and relative to a particular object), now we intend to categorize them based on their attributes into classes. This is done using the equivalence relation $\sim$ where $\sim$ is defined as the "is the same type of object". The quotient set V represents all categories in ζ where for objects x and y belonging to the same class, the quotient set can be defined as $V_x \varepsilon \{x, y \in V | x \sim y\}$. The equivalence relation can be relaxed or stringent depending on the model used, for e.g. in one context, it is enough to categorize vehicles, pedestrians and bicyclists but in other contexts, cars need to be separated from trucks, motorcycles and so on. The states and attributes of equivalence class set V_x is defined as A^{V_x}, as the long-term dynamic motion (lane keeping, lane change, turning etc.) and the current action being taken during this snapshot (turn indicator on, waiting etc.). With $P(A^{V_x})$being defined as the power set of A^{V_x} and $A \varepsilon \bigcup_{V_x \in V} P(A^{V_x})$ denoting the states & attributes set, the states & attributes function is defined as follows:

$$f_{sa} : \zeta \rightarrow A, x \mapsto f_{sa}(x) \tag{11.2}$$

where each object $x \in \zeta$ is associated to at least one element of A. The states & attributes from x's point of view function, or x-states & attributes function is defined as follows:

$$f_{sa}^{x} : \zeta_x \rightarrow A_x, x \mapsto f_{sa}(x) \tag{11.3}$$

where $A_x \varepsilon \bigcup_{V_y \in V_x} P(A_x^{V_y})$, $A_x^{V_y}$ is the states & attributes of the equivalence class of y from x's point of view set and f_{sa}^{x} is the states & attributes from x's point of view set. The scene can then be defined from the amalgamation of the previous definitions as:

Definition 11.2 (Scene) *A scene $\xi_x(\mathscr{S}, t)$ of the portion of space $\mathscr{S}$ at time t from x's point of view is defined as the 3-tuples $\xi_x(\mathscr{S}, t) \varepsilon (\zeta_x, \mathscr{C}_x(\zeta_x), f_{sa}^{x}(\zeta_x))$, where ζ_x is the object set from x's point of view at time t, $\mathscr{C}_x(\zeta_x)$ and $f_{sa}^{x}(\zeta_x)$ are the x-position set and the x-states & attributes set respectively.*

A specific case of the scene is scenery which is restricted to only static objects. If ζ_x^s is defined as the static object set from x's point of view i.e. $\zeta_x^s \varepsilon \zeta^s \cap \zeta_x$, then the scenery is defined as follows:

Definition 11.3 (Scenery) *A scenery $\mathscr{Y}_x(\mathscr{S})$ of the portion of space $\mathscr{S}$ is defined as the 3-tuples $\mathscr{Y}_x(\mathscr{S}) \varepsilon (\zeta_x^s, \mathscr{C}_x(\zeta_x^s), f_{sa}^{x}(\zeta_x^s))$ where ζ_x^s is the x-static object set, $\mathscr{C}_x(\zeta_x^s)$ is the x-static position set and $f_{sa}^{x}(\zeta_x^s))$ is the x-static states & attributes set respectively.*

Note here that the scenery is defined, similar to the scene, from the viewpoint of a specific observer x. This can be replaced with the omniscient observer, in which case, the complete and uncertainty-free scenery is defined as: $\mathscr{Y}(\mathscr{S})\varepsilon(\zeta^s,\mathscr{C}(\zeta^s),f_{sa}(\zeta^s))$.

A concrete example of elements of scene is provided in Figure 1 below. The area of interest or the portion of space $\mathscr{S}$ is delimited by a black perimeter (Fig. 11.1(a)). The static object set $\zeta^s(\mathscr{S})$ represents all static objects within the portion of space $\mathscr{S}$, which in this case, is the road network and the trees, while the vehicle and bicyclist represent the dynamic set, $\zeta^d(\mathscr{S},t)$. Each object can then be represented as a center point and size or a collection of four points representing the four vertices of the enclosing rectangle (Fig. 11.1(b)). Fig. 11.1(c) provides the field of view of the dynamic objects and the subsequent object set from x's point of view, ζ_x. As is apparent from the figure, the vehicle driver cannot view the cyclist due to occlusion. So the x-position set of the vehicle will not contain the bicycle. The states & attributes of the equivalence class of x-set, A^{V_x}, can be made up of elements $\{i,f,s,lk,llc,rlc\}$ (where i= indicator activated, f = moving

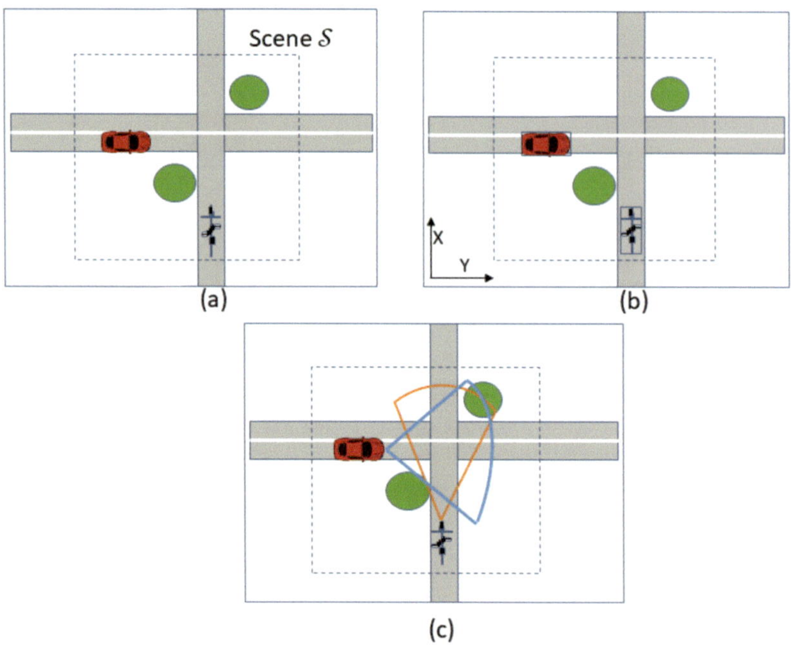

Figure 11.1 Explanation of scene in three steps - (a) provides an example of a vehicle and a bicyclist at an intersection; (b) shows the dimensions of each dynamic object with the reference coordinate system and (c) shows the fields of view of each dynamic object where blue represents the FOV of the vehicle and orange represents the FOV of the bicyclist. As can be seen, neither objects are visible to the other person due to occlusion.

forward, s = stopped, lk = lane keeping, llc = left lane change, rlc = right lane change). The states & attributes set for object x can then be defined as $f^{x}_{sa} = \{f, lk\}$.

11.2.1.2 Defining situation

A situation is different from a scene in the aspect of "possible actions" [11]. [15] defines a situation as an extended state, in which an element is not only seen as a physical object but also its action and action alternatives. [21] defines a situation to be the sum of all behavior decision relevant aspects. The current scene, the intentions and actions of all subjects in a scene (including the ego-vehicle), and the abilities of the ego-vehicle are all parts of the behavior decision making. Mock-Hecker considers the traffic situation to be an extract of the traffic (world) at a certain point of time. Geyer et al. [8] clarifies that a "situation is defined by the set of criteria that need to be true to conduct an associated action". The authors surmise that the choice of action can lead to different situational outcomes. Ulbrich et al. [28] argue that [8]'s definition of scene should realistically apply for situation rather than scene. They also provide examples of other authors to point to the fact that the situation arises from the information selection and augmentation of the scene information based on goals (permanent or transitory) of the ego vehicle. They provide the following definition of a situation:

Definition 11.4 *A situation is the entirety of circumstances, which are to be considered for the selection of an appropriate behavior pattern at a particular point of time. It entails all relevant conditions, options and determinants for behavior. A situation is derived from the scene by an information selection and augmentation process based on transient (e.g. mission-specific) as well as permanent goals and values. Hence, a situation is always subjective by representing an element's point of view.*

According to the authors' definition of a situation, it can be fully derived from a scene and the system's goals and values. Situation therefore, contains all parts of the scene including all relevant parts of the scenery, dynamic elements and self-representation. Information selection simplifies the situation representation and goals and values add the temporal aspect.

Andreotti et al. [1] defines the goals & values position from x's point of view function or x-goals & values position function as

$$\bar{\mathscr{C}}_x : \zeta_x \rightarrow \mathscr{S}, y \mapsto \bar{\mathscr{C}}_x(y) \tag{11.4}$$

which associate each object of $y \in \zeta_x$ to its future position based on x's expectation. y's future position is based on two disparate identities: y's near term goal and x's belief of y's current state and its propagation into the future. The set $\bar{\mathscr{C}}_x(\zeta_x)$ is known as the corresponding x-goals & values set. However, this set does not take into account the real goals of the objects in $\zeta(\mathscr{S},t)$ or that the number of objects in the set could change with time. Therefore, the general goals & values position function is defined as:

$$\bar{\mathscr{C}} : \zeta \to \mathscr{S}, y \mapsto \bar{\mathscr{C}}(y) \tag{11.5}$$

where the goals & values position function associates each object y in ζ to the dimensional coordinate of y in the future, based on y's expectation. The goals & values position function knows the goals of each object at time t i.e. what the future action of each object in ζ is going to be. Another mathematical property of interest to us is the relevant function which filters relevant information given all information. For this, we define the enlarged situation set, $\mathscr{N}_x^e(\mathscr{S},t)$ as the 4-tuples made up of the x-object, x-position, x-states & attributions and x-goals & values position sets $\mathscr{N}_x^e(\mathscr{S},t)\varepsilon(\zeta_x,\mathscr{C}_x(\zeta_x),f_{sa}^x(\zeta_x),\bar{\mathscr{C}}_{\S}(\zeta_x))$. The relevant function operates on the relevant objects and provides the rest of the states and attributes of those objects. It is defined as:

$$f_r^x : \mathscr{N}_x^e \xrightarrow{f_{ro}^x} \mathscr{N}_{xo}^e \xrightarrow{f_{ri}^x} \mathscr{N}_{xi}^e \tag{11.6}$$

where $\mathscr{N}_{xo}^e\varepsilon(\zeta_x^r,\mathscr{C}_x(\zeta_x^r),f_{sa}^x(\zeta_x^r),\bar{\mathscr{C}}_{\S}(\zeta_x^r))$, with ζ_x^r being any subset of ζ_x; and $\mathscr{N}_x\varepsilon(\zeta_x^r,\mathscr{C}_x^r(\zeta_x^r),f_{sa}^{xr}(\zeta_x^r),\bar{\mathscr{C}}_x^r(\zeta_x^r))$ with all r-superscript terms being subsets of the original terms.

The set ζ_x^r is referred to as x-relevant object set and the sets $\mathscr{C}_x^r(\zeta_x^r), f_{sa}^{xr}(\zeta_x^r)$ and $\bar{\mathscr{C}}_x^r(\zeta_x^r)$ are the x-relevant position, x-relevant states & attributions and x-relevant goals & values position sets respectively. Given all this background information, we can now define a situation as:

Definition 11.5 (Situation) *A situation $\mathscr{N}_x(\mathscr{S},t)$ of the portion of the space $\mathscr{S}$ at time t from x's point of view is defined as any 4-tuples $\mathscr{N}_x(\mathscr{S},t)\varepsilon(\zeta_x^r,\mathscr{C}_x^r(\zeta_x^r),f_{sa}^{xr}(\zeta_x^r),\bar{\mathscr{C}}_x^r(\zeta_x^r))$ where ζ_x^r is a x-relevant object set, $\mathscr{C}_x^r(\zeta_x^r)$ is the x-relevant position set, $f_{sa}^{xr}(\zeta_x^r)$ is the x-relevant states & attributions set and $\bar{\mathscr{C}}_x^r(\zeta_x^r))$ is the x-relevant goals & values position set respectively.*

An example of a situation is given in Figure 11.2 below. Continuing with the example in Figure 11.1, we assert that due to occlusion, the vehicle driver of vehicle x cannot observe the bicycle y, which is why, in the future, the driver sees oneself crossing the intersection without braking (Fig. 11.2(a)).

However, due to bicycle being present and their respective occlusions, the vehicle and the bicycle will have a collision (Fig. 11.2(b)). Even though this is a case where physical barrier, in another case where the vehicle x is taking an unprotected left turn at an intersection while bicyclist y is going straight (Fig. 11.2(c)), if x does not have the turn indicator on, observer y will expect x to drive straight in the near future. Given that x decelerates at the intersection

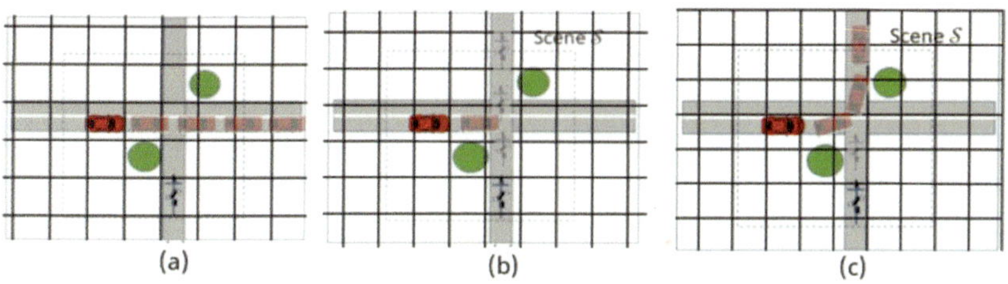

Figure 11.2 Explanation of situations from point of view of the dynamic objects - (a) provides the example from the vehicle's point of view without recognition of bicyclist (b) possible collision due to bicycle present (c) left turn at intersection

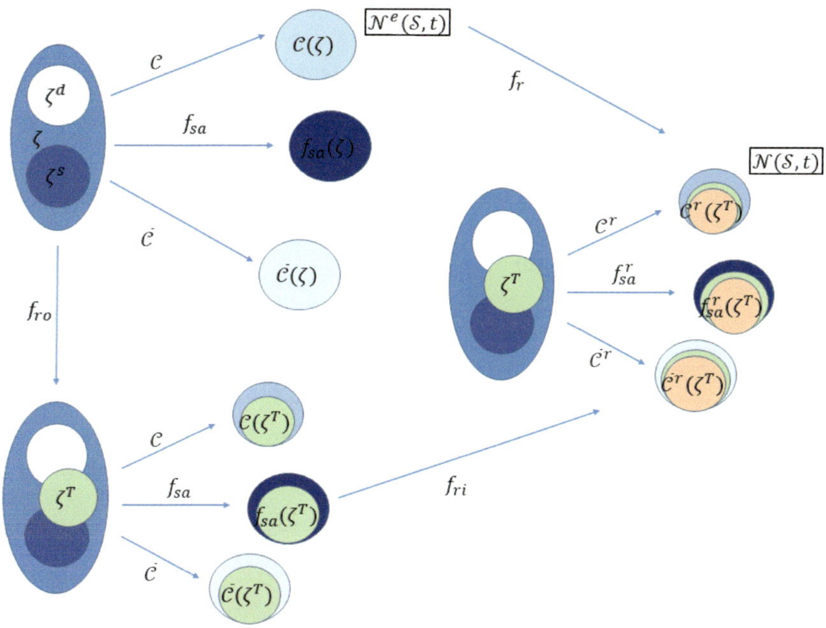

Figure 11.3 Relevant function set diagram

as observed by y, the image set of ζ_x under the x-goals & values position function change over time based on the observed deceleration.

We also visualize the object set and the three functions: position function $\mathscr{C}$, states & attributes function f_{sa} and goals & values function $\bar{\mathscr{C}}$ in Figure 11.3(a) along with the relevant function definition in Figure 11.3(b).

11.2.1.3 Defining scenario

Scenario is used often in the context of simulation and testing. [9] essentially categorizes a scenario as having four distinct elements: actors, background information on the actors and assumptions about their environment, goals or objectives and sequence of actions or events. [8] define that "a scenario includes at least one situation within a scene including the scenery and dynamic elements. However, scenario further includes the ongoing activity of one or both actors. According to the movie and theater metaphor previously introduced, the term scenario can be understood as some kind of storyline - including the expected action of the driver - but does not specify every action in detail." While, this definition is close to the accepted definition, Ulbrich et al. [28] modifies it as:

Definition 11.6 (Scenario) *A scenario describes the temporal development between several scenes in a sequence of scenes. Every scenario starts with an initial scene. Actions & events as well as goals & values may be specified to characterize this temporal development in a scenario. Other than a scene, a scenario spans a certain amount of time.*

[1] begin by defining goals & values function as a function that describes the path taken by object y in order to reach its goal, i.e. position $\bar{\mathscr{C}}(y)$. Mathematically, it can be written as:

$$f_{gv} : \mathscr{C}(\zeta) \to \bar{\mathscr{C}}(\zeta) \times T, \mathscr{C}(y) \mapsto (\bar{\mathscr{C}}(y), \bar{t}) \tag{11.7}$$

where $\bar{t}$ is the time to goal $\bar{\mathscr{C}}(y)$ for the object y. The goals & values function is updated at each time step depending on the interactions of the object with the environment. An actions & events function is described as the path that object y, located in $\mathscr{C}(y(t))$ at time t, takes in order to reach its position at time $t+\Delta t$, i.e. to reach the position $\mathscr{C}(y(t+\Delta t))$

$$\begin{aligned} f_{ae} : \mathscr{C}(\zeta(\mathscr{S},t) \times T \quad & \to \mathscr{C}(\zeta(\mathscr{S},t+\Delta t)) \\ (\mathscr{C}(y(t)),t+\Delta t) \quad & \mapsto \mathscr{C}(y(t+\Delta t)) \end{aligned} \tag{11.8}$$

A very important contrast is between the perceived goal $\bar{\mathscr{C}}(y(t))$ at time t and the true goal $\mathscr{C}(y(\bar{t}))$. Figure 11.4 shows the set diagrams for the functions

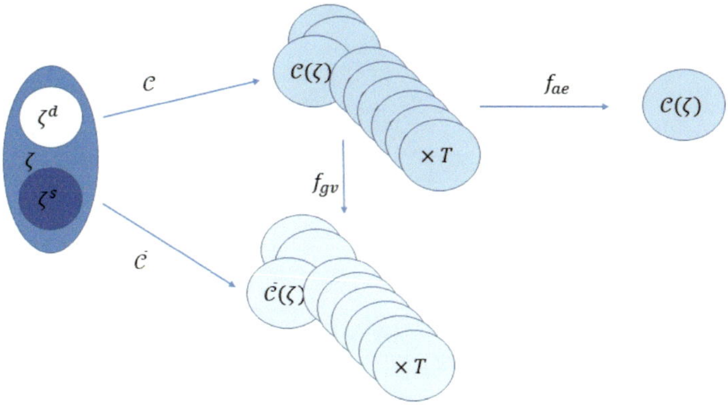

Figure 11.4 Goals & values and Actions & events functions

described above. For each function defined above, we can get a counterpart function from object x's point of view, x-goals & values function f^x_{gv} and x-actions & events function f^x_{ae}. Given the definition of the functions above, the scenario is mathematically defined as:

Definition 11.7 (Scenario) *A scenario $\mathcal{O}^x_{\mathcal{T}}(\mathcal{S})$ on the portion of space $\mathcal{S}$ in the time set $\mathcal{T}$ from x's point of view is defined as the $\mathcal{O}^x_{\mathcal{T}}(\mathcal{S})\varepsilon(\xi^x_t(\mathcal{S}), f^x_{gv}, f^x_{ae})_{t\in T}$, where the time set is a set of ordered sequence of times $\mathcal{T} = \{t_i : t_i < t_{i+1}, i = 0, \ldots, n-1\}$.*

Figure 11.5 provides the set diagram for a scenario.

11.2.2 Scenario descriptions in frameworks

With the scenario definition provided above, there is still the unstructured ODD that needs to be decomposed systematically for the purpose of decreasing the complexity and providing a structured process towards building the scenario. Different organizations in countries around the world have been working on developing standards for organizing scenarios. We provide a few main examples of scenario categorization below:

11.2.2.1 PEGASUS and related projects - Germany

The research project PEGASUS (**P**roject for the **E**stablishment of **G**enerally **A**ccepted quality criteria, tools and methods as well as **S**cenarios and **S**ituations) was promoted by the German Federal Ministry for Economic

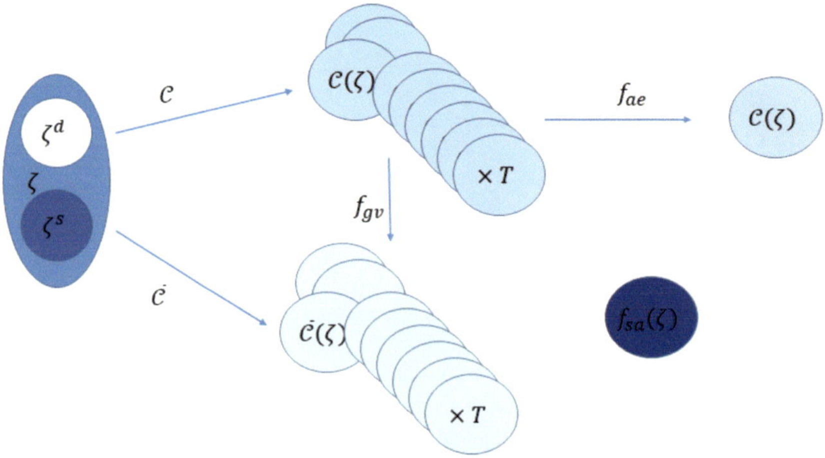

Figure 11.5 Scenario set diagram

Affairs and Energy (BMWi) as a means to provide a systematic and structured approach to verification and validation of AV instead of naive randomized testing [19]. Some of the high level research questions sought by the project were *"What level of performance is expected of an automated driving system?"* and *"How can we verify that it achieves the desired performance consistently?"* As part of this project, there has been a focus on utilizing a layered model to structure an environment description for scenes and scenarios. Schuldt et al. [26] introduced the layered model featuring four layers: first layer for road network, second layer for situation-specific adaptations of the road network, third layer to describe the actors and their control and the last layer for environment conditions. While lane marking were included in the first layer, signs, guardrails and other roadside structures were located in the second layer. This work left some ambiguity in terms of road level and traffic level infrastructure being in the same layer, which was addressed by the subsequent work [2] which separated the road description and traffic rules into two separate layers. The road-level only contains the layout of the road (geometry) and its topology while the layer for traffic infrastructure contains structural boundaries. The third layer was also modified to contain temporary manipulation of road and traffic layers such as in the event of construction. The fourth layer was modified to contain only objects (both stationary and non-stationary) along with their interactions and maneuvers. [3] finally introduced a sixth layer with the digital information for V2X communication. The authors also presented some important changes in Layer

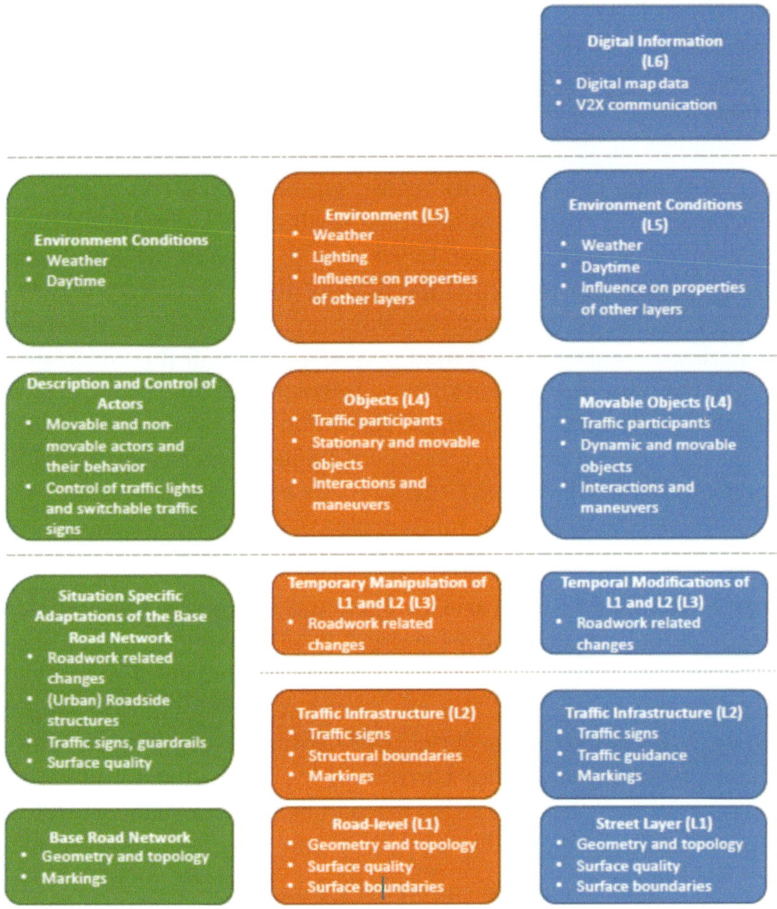

Figure 11.6 Development of scenario layered model. Four layers of Schuldt (left), compared to five layers of Bagschik (middle) and six layers of Bock (right).

2 with the traffic guidance part of the layer while Layer 4 was exclusively reserved for dynamic objects (vehicles, bicyclists, pedestrians) along with their interactions and maneuvers.

In the current research project 'VVM-Verification & Validation Methods for Automated Vehicles Level 4 & 5', [25] provided a refinement of the six-layer model (referred to as 6LM) focusing on simple and unambiguous design towards an end goal of comparison of scenarios and subsequent reduction of given scenario sets. Specifically, in order to test AV functions independently of one another, the testing and debugging could be focused on single layers instead of the entire scenario description. Thus, a requirement of independence of layers exist in order to simplify testing and validating specific AV functionalities. For example, testing the motion planning module can be undertaken with only the road network, traffic infrastructure and dynamic objects, but without consideration for environmental conditions. We provide a definition of the layers below:

- **Layer 1 (Road network and traffic guidance objects)** - The first layer contains the road network structure along with permanent traffic infrastructure. The purpose of Layer 1 is to provide the area where dynamic objects can move and the rules of their movement. The road network contains the geometry, topology and topography of the roads. The surface material of each road is also assigned in this layer along with any irregularities present such as potholes. The lane semantics can be derived from the lane markings. The traffic infrastructure consists of signage that regulate traffic such as speed limits, stop signs, traffic lights, roundabout signs. All the road infrastructure helps in planning the route for a given vehicle - known as mission planning; on a macroscopic scale, it can also provide complete traffic flows.
- **Layer 2 (Roadside Structures)** - There are a large number and type of roadside objects that can be present in urban setting, which justifies a separate layer to aggregate these objects. Examples of roadside structures are static objects that are placed alongside the road such as buildings, vegetation, street lamps, fire hydrants etc. Safety structures such as guardrails and other barriers which are meant to lessen the crash severity for vehicles are also listed in this layer. Finally, pedestrian bridges are also present in this layer.
- **Layer 3 (Temporary modifications of L1 and L2)** - This layer is constructed on top of Layers 1 and 2 and is concerned with changes which come up due to construction or sudden change in roadway structure. Layer 3 does not introduce new objects but deals with modifications of the two previous layers. For example, a construction site has corresponding traffic signs and temporary lane markings along with temporary barriers. Similarly, beacons or traffic cones are classified into traffic guidance objects which are instantiated in this layer. However, dynamic objects related to construction are assigned to Layer 4 even though they are immobile during the duration of the scenario. A thing to note is that even though the modifications are temporary, they are constant during the duration of the scenario.

- **Layer 4 (Dynamic objects)** - This layer contains all objects which either move or can possibly move. The latter distinction is for objects which can move but don't during the duration of the scenario. All traffic participants are examples of this layer including all kinds of vehicles, motorcycles, bicyclists, pedestrians, animals and other miscellaneous objects that are movable. Although scenario definition contains goals and values, it is not explicitly mentioned in this layer.
- **Layer 5 (Environmental conditions)** - This layer contains weather, atmospheric and lighting conditions. Weather and atmospheric conditions include type and amount of precipitation, visibility, winds and cloud cover. For daytime lighting conditions, the positioning of the sun could be listed. For nighttime, artificial lighting sources and their intensities can be listed. This layer also specifies the weather related conditions of the road surfaces such as dry, wet or icy roads including the absence of roadway features due to weather conditions.
- **Layer 6 (Digital information)** - The final layer focuses on all kinds of digital information exchange, communication and cooperation based on data. This layer focuses on all information sources between vehicles (V2V), infrastructure (V2I) or both (V2X). States of traffic signals are also part of this layer.

11.2.2.2 SAKURA - Japan

The SAKURA project (**S**afety **A**ssurance **KU**dos for **R**eliable **A**utonomous vehicles) is one of the coordinated initiatives funded by the Ministry of Economy, Trade and Industry (METI) of Japan aimed at creating a roadmap for deployment of AVs, demonstration tests and coordination efforts [24]. Two issues have been identified by them - 'evaluation scope efficiency' or coverage metrics and 'explainability in emergencies' or responsibility assignment during emergency interventions. The researchers utilize a 'Physical Principles Approach Process' for safety evaluations. Their argument is that the number of potential hazardous situations that an AV may encounter in real traffic is infinite but the number of physics principles that the system can apply for safely handling such situations is limited. Therefore, breaking down each AV subtask - perception, decision-making (referred to here as judgement) and controls (referred to here as operations), we find that each is associated with specific physics principles which would be unchanged even with change in scenarios. Thus, they decompose scenarios logically in consideration of the physics of an AV to provide complete coverage of the

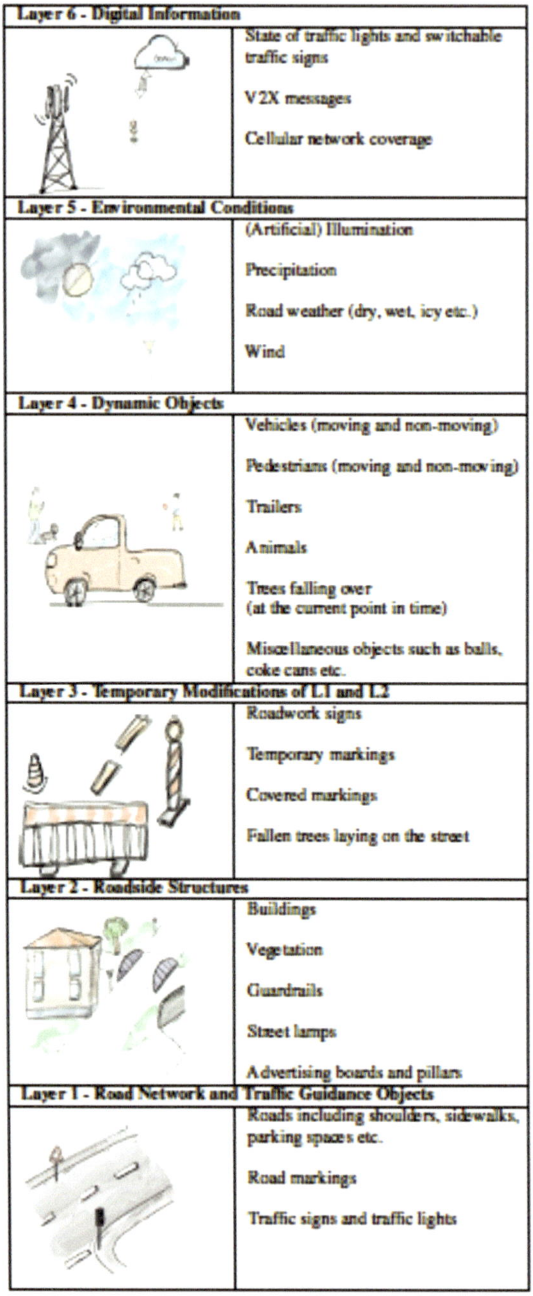

Figure 11.7 6-Layer model as given in [25]

scenarios. In each subtask, the possible disturbances are incorporated into the analysis and the corresponding physics principles are provided (example given in Figure 11.8).

Perception disturbance can be any condition in which sensor system fails to correctly judge a hazard or a non-hazard for sensor or vehicle intrinsic or extrinsic reasons. Intrinsic factors might be faulty sensors which can lead to partial or complete data loss, miscalibration or incorrect sensor mounting. Extrinsic reasons refer to environmental conditions which can lead to occlusion such as fog, rain, snow, dirt etc. Traffic disturbance refers to traffic conditions that may lead to hazard based on a combination of the following factors: road geometry (e.g. on-ramps on highways), ego-vehicle behavior (e.g. lane change maneuver), surrounding vehicle location and motion (e.g. sudden aggressive cut-in from vehicle in neighboring lane). Vehicle disturbance refers to situations where the upstream subtasks (perception and decision-making) function correctly but the subject vehicle fails due to incorrect dynamics. Intrinsic factors include total weight, weight distribution etc. while extrinsic factors include road surface irregularities and weather related conditions of the road surface. These three kinds of disturbances form the basis of scenarios that provides systematic analysis and classification.

Since the main safety risk is in collision with surrounding traffic participants or static obstacles, it can be systematically solved using the traffic disturbance scenarios. Quantified ranges of reasonable foreseeability and preventability are defined to provide quantitative criteria for the tests. Given these traffic related hazardness scenarios, the evaluation can be expanded to include perception and control disturbances into the assessment to create a comprehensive safety evaluation.

Task	Processing results	Disturbance	Governing physics principles
Perception	Own position, surrounding traffic environment positional information and other traffic information	Perception disturbance	Light, radio wave, infrared light propagation principles that affect camera, mili-wave radar and LiDAR sensors, respectively
Judgement	Path, speed plan instructions	Traffic disturbance	Kinematics describing the motion of traffic participants, objects and systems of groups of objects, without reference to the causes of motion.
Operation	Movement instruction allocation for each ACT for achieving path and speed plan instructions	Vehicle control disturbance	Dynamics, concerned with forces applied on the vehicle's body and tires, and their effects on motion.

Figure 11.8 Subtask breakdown with the corresponding physics principles [24]

11.2.2.3 Other relevant frameworks

As stated previously, there have been other approaches towards scenario frameworks in various regions of the world. Below, we provide a summary of some scenario description approaches -

- **MOOVE - France** - The MOOVE project (**MO**nitoring **O**utillé pour le Véhicule dans son Environnement) was created by VEDECOM, an Institute for Energy Transition (ITE) in partnership with car manufacturers and suppliers - Renault, PSA, Valeo to analyze how cars move in roads with other users, how to control dynamic of vehicle in real world traffic etc. The project aimed to build a model of real world driving by collecting real world data and identify relevant safety critical scenarios [17].
- **HEADSTART - European Union** - The HEADSTART project (Harmonised European Solutions for Testing Automated Road Transport) is a research project funded by the European Union that aims to define testing and validation procedures for Connected and Automated driving functions [30]. The project has 17 European partner countries involved to find a coordinated solution. They utilize a database of relevant scenarios as surrogate for real world driving for testing AV functionalities.

11.3 Sensor simulation - Inputting data into AV pipeline

With the development of the world model, our focus now shifts towards simulation of sensors for AVs. Sensor simulation has become a critical component in virtual testing of software stack especially, models and methods to create realistic sensor information. Realistic sensor data can be used for accurate testing of downstream subtasks - perception, decision making and control. The goal of emulating the physical sensors is not to create visually appealing outputs but rather to achieve a level of data fidelity so as to match the control outputs with that of reality. A given sensor in an AV acquires knowledge of the surrounding by sensing the environment in order to detect the drivable area (i.e. the portion of space that can be driven on) and other traffic guidance structures (traffic lights and signs, pedestrian crossing and barriers among others) first. Given the drivable area detection, it then detects the dynamic objects in the environment (other vehicles, bicyclists, pedestrians). The sensor simulation data should take into account the physical nature of the objects along with the physical laws of occlusions, distortions etc. The analysis of sensor simulation presented below is informed by the review of [6] and

[22]. Specifically, the sensors on AV utilize different areas of the Electro-Magnetic (EM) Spectrum and are shown in the Figure 11.9. Figure 11.10 shows a comparison of sensor features in terms of field-of-view (FOV), range, accuracy, frame-rate, resolution, color perception where it needs to attain maximum value (3 points) while minimizing cost, visibility, maintenance, weather fluctuations. We now delve into specific sensor simulation below:

11.3.1 Camera

Cameras operate either in the visible range of the EM spectrum or the infrared range. The visible range cameras operate in the 400 nm to 780 nm wavelength range and is divided into three channels: red, green and blue. AV perception systems use these sensors frequently due to their low cost, much higher resolution than other sensors and high quality color information. The flip side of higher resolution is the huge amount of data generated per sensor which leads to difficulties for efficiently processing the data. Cameras are used in different safety related applications in AVs, namely localization, object detection and classification etc which make it invaluable in terms of sensing the surrounding environment. Generally, two different kinds of

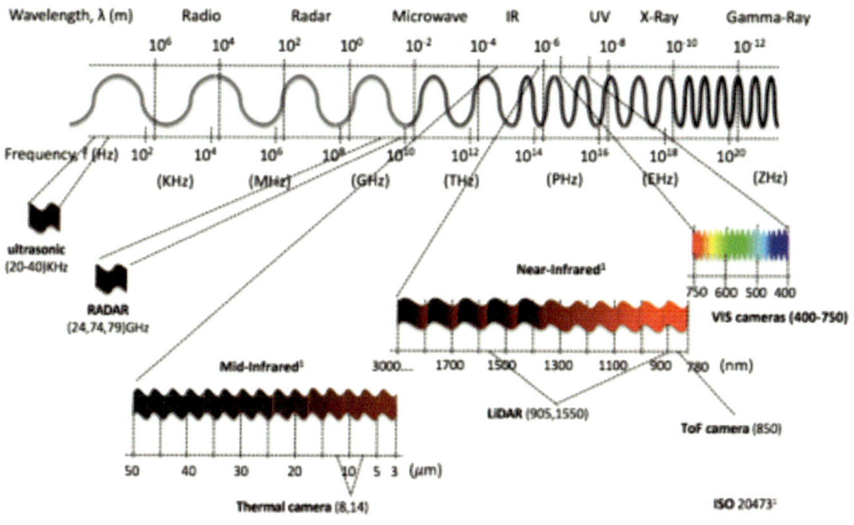

Figure 2. An overview of different spectra used for perception systems in autonomous vehicles.

Figure 11.9 The physical wavelengths and frequencies of the representative AV sensors [22]

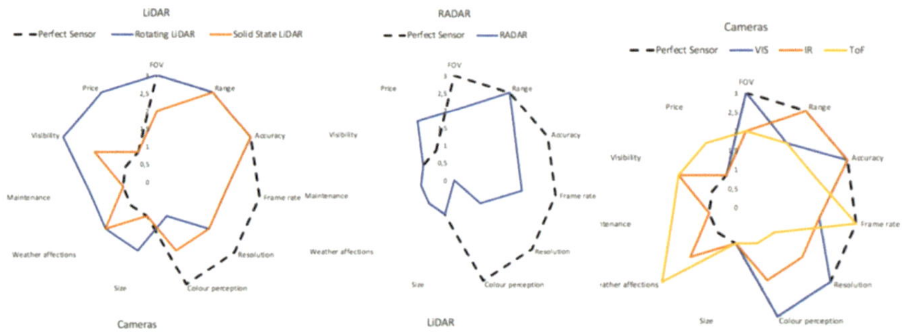

Figure 11.10 Sensor features in a spider diagram (adapted from Rosique et al. [22])

camera sensors are used: Charge-coupled device (CCD) and complementary metal oxide semiconductor (CMOS). CCD differs from CMOS in terms of expense of the process (and therefore, reduction in overall performance). While CMOS has lower energy consumption which leads it to be used in different mass-production cameras, CCD has high dynamic range and also, higher image quality in low lighting conditions.

In order to create a physics-based camera sensor model, we first need to understand the process of image capture in camera sensor [16]. Camera is a passive sensor, which means that the light source is usually the sun during daytime or headlights at night, from where the light travels through atmosphere (air, rain, fog etc.) before being reflected from surface materials (such as vehicle bodies, vegetation, road surfaces etc.) and being captured in the imaging sensor after traveling through the media again. The reflected area which enters the camera is called the scene. The spectral radiance from the scene first enters the lens system. The lens system distorts the scene representation at the camera lens and generates the spectral radiance image on the sensor plane. The shutter determines the exposure to the spectral radiance. The imaging sensor (both CCD and CMOS) contains 2D array of pixels, each pixel containing photodetectors which convert light into photocurrent. Temperature, exposure time and other sensor settings such as sensitivity impacts the image. Typically even the visible range cameras used in AVs are sensitive beyond the visible range and also use up to 20 bits precision for their internal signal.

The cameras used in AV industry are not very high-end which leads them to have a number of artifacts in the lens system and the sensor system. A robust camera simulation model must take all these factors into consideration which generating an image of the scene. [16] describes the following requirements for modeling an accurate physics-based sensor:

- Spectral-based
- High dynamic range
- High color depth
- Real time performance
- Physically plausible reflection model
- Reproduce limitations and artifacts of real cameras

Additionally, current computational cameras have to consider both the physical and the digital domain and due to the large amount of possibilities to optically transform the scene representation using lenses, spectral filters etc., there can be a large number of possible camera designs for a given task [32]. There has been a number of recent research works focusing on testing image processing algorithms on simulated images and comparing them to real data. Butler et al. [4] utilized an open-source animated short film for optic flow evaluation by varying different simulation parameters such as depth-of-field. Irgenfried et al. [13] used raytracing to compare different image segmentation algorithms using simulated and real data, stating the need for physics based simulation in order to determine correspondence between simulated and real images. Nürnberg et al. [18] created a raytracing based simulation framework towards design and evaluation of computational camera systems.

We conclude the discussion on camera simulation on the topic of applying post-processing effects to augment the simulated images. In a lot of AV camera-based applications, it is of paramount importance to understand the effect of sensor artifacts and atmospheric conditions. Recently, there has been some work in augmenting camera images collected in normal weather conditions to provide corruptions due to atmospheric conditions [29]. In the example given in Figure 11.11, we showcase a real image from a front facing camera (Fig. 11.11(a)) and then apply the following effects:

- **Black grids**: Simulates corruption of imaging sensor by partial block of camera image (Fig. 11.11(b)).
- **Fog**: Simulates corruption of camera image by fog (Fig. 11.11(c)). Light traveling through fog is absorbed or scattered according to an exponential equation: $I_t = I_i e^{-\alpha_{ext} d}$ where I_i is the incident light intensity, α_{ext} is the attenuation factor and I_t is the final intensity reaching the observer [29]. Each channel is affected differently, i.e. have their own attenuation factor.

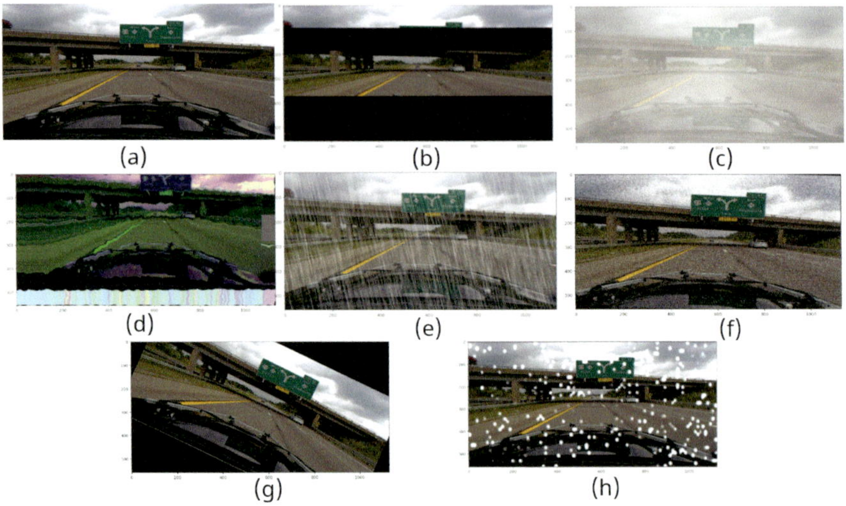

Figure 11.11 Various realistic image augmentation options - (a) Original image (b) Black grids added (c) Fog added (d) Hue, saturation, water effect and padding added (e) Rain added (f) Gaussian noise added (g) Image rotation applied (h) Snow added

- **Hue, Saturation, Water effect and Padding**: Simulates corruption due to change in hue (random alterations of color channels), change in saturation (changes towards primary colors), water effect left on camera lens (given by the distortion of lines) and water padding (at the bottom of the image) (Fig. 11.11(d)).
- **Rain**: Simulates corruption of camera image by rain (Fig. 11.11(e)). Rain is generated as a series of small lines in a specific angle with some blur added to the image.
- **Gaussian noise**: Adding a simple Gaussian noise on to the image (Fig. 11.11(f)).
- **Image rotation**: A very common cause of image corruption where the camera is tilted on its axis leading to rotated images (Fig. 11.11(g)).
- **Snow**: A simple image augmentation is done to add circular blurs which lead to the effect of snowfall (Fig. 11.11(h)).

11.3.2 LiDAR

Light Detection and Ranging (LiDAR) systems are active sensing systems, which utilize measurement of time of flight (ToF) from a pulsed light emitting from a laser diode to when it returns to a sensor. Usually, LiDAR uses

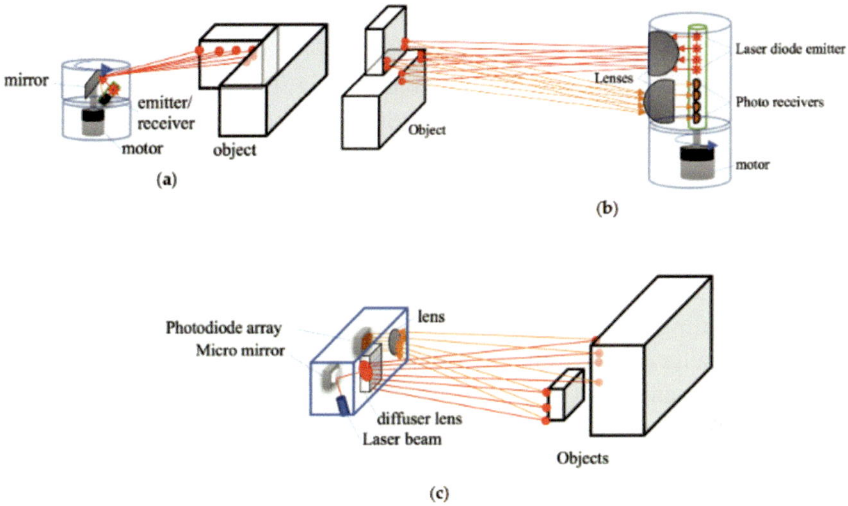

Figure 11.12 Various realistic image augmentation options - (a) Original image (b) Black grids added (c) Fog added (d) Hue, saturation, water effect and padding added (e) Rain added (f) Gaussian noise added (g) Image rotation applied (h) Snow added.

the near-infrared (905 nm) or the infrared (1550 nm) wavelength range. The ToF is then used to calculate the distance between the laser and the reflected object. LiDAR systems operate in two different kinds of sensors: 2D or 3D. In terms of 2D LiDARs, High end survey grade sensors have a single high-powered laser emitter and a rotating mirror. On the other hand, lower end automotive sensors have 3D LiDAR consisting of linear array of lasers or a 2D flash array of lasers (on a micromirror MEMS circuit) to capture the scene. Each of these different kinds have different error sources: for 2D LiDARs, there can be laser beam misalignment with respect to scanning mirror, clock error, scanner error and scanner torsion [31]; for 3D LiDARs, there can be misalignment between individual lasers, clock error, scanner error and rotation error. A visual explanation from [22] is provided in Fig. 11.12 for different kinds of LiDAR sensors and their operating principles.

LiDAR simulation primarily have simulation based on ray casting, ray tracing, rasterization and deep learning based approaches which have different levels of complexity and simulation performance.

The earliest models for LiDAR simulation used raycasting by leveraging physics-based collision detecting algorithms implemented in simulation engines. Collision measurements are precise but since the objects represented within the virtual environments are based on simple parametric surfaces, the resulting LiDAR simulations are simplistic. The open-source simulation software, Gazebo, implements this approach for LiDAR simulation [14]. Typically with the increase in complexity of the scene, the performance of this approach degrades significantly.

Ray tracing is a modification of raycasting where each laser beam is treated as a collection of rays sent from the sensor along a specific direction into the scene given specific beam divergence. In ray tracing, specific surface properties are modeled along with other graphics related approaches such as lighting, shading etc. Since, it has a multitude of collision detection happening, it results in high computational cost which is decreasing with the increasing in efficiency of GPUs. Raytracing is used frequently in applications where complex transmission and reflection characteristics are desired. [10] utilized CPU-based raytracking to model beam divergence, beam shape, surface reflectance and scattering effects.

Raster approach, unlike raycasting or ray tracing, leveraged raster graphics to accelerate LiDAR sensing simulation process. Raster graphics projects a 3D scene into a 2D image plane and performs per-vertex shading operations. The shading algorithms were modified to replace camera lighting conditions with LiDAR depth and intensity calculations [20]. Thus, by replacing vertex shader with a depth shader, this approach performs the fastest with low computational complexity. However, the approach has serious limitations since the entire 3D scene is projected onto the 2D imaging plane which leads to artifacts.

Finally, in recent years, deep learning based approaches have been proposed to automatically generate 3D point cloud, given real world LiDAR data and real traffic flow [7]. This approach takes in static real world data as training data to train a LiDAR renderer. The trained renderer generate realistic LiDAR points, annotated for synthetic objects. While this approach can be quicker than the previous approaches, it requires massive amount of training data which is difficult to collect.

11.3.3 Radar

Radio Detection and Ranging (Radar) systems operate on mm-level wavelengths which lead to them being robust to external atmospheric conditions such as rain and snow which affect other sensors. Radar operates in a ToF manner where it measures the distance between the emitter and the object by the time between the emitted signal and the received echo. In addition to the distance, the system accurately detects the speed of the object and the angle of the detection both of which are invaluable information for tracking neighboring objects. There has been a lot of applications utilizing radars in driver assistance technologies to enhance safety of the vehicle and its occupants and it has been used extensively in AV to increase the effective horizon of the sensing under adverse conditions providing complimentary benefits to other sensors. However, radars are plagued by lack of precision and the increased number of false positives due to multipath. Current automotive radar systems are designed at 24 Ghz and 77 GHz [5]. However, compared to the former, 77 GHz is a far more attractive solution due to superior range accuracy, range resolution, velocity accuracy and velocity resolution.

Typically, radar simulations are difficult to do at scale where at 77 GHz, the electrical size of the radar scenes run into hundreds of billions of cubic wavelengths making it impractical for simulation with finite element method (FEM)-based solvers. [5] provided a simulation solution for automotive radar antennas based on ANSYS's FEM-based High Frequency Structure Simulator (HFSS-FEM). Radar antennas are used to send and receive electromagnetic pulses and are therefore, integral components of the radar systems. The authors chose microstrip series-fed patch antenna array as the antenna design due to its simple structure, low profile and low manufacturing cost. There are alternative analytical models which can be applied to provide approximate solutions but they cannot provide accurate solutions for complex problems. Thus, FEM is used as an alternative since it can provide easier modeling of complex and irregular shapes and is adaptable and accurate than the analytical approaches. The performance of the antennas were investigated using ANSYS's shooting and bouncing (SBR+) solver (HFSS-SBR+). HFSS-SBR+ is a high-frequency asymptotic ray-tracing electromagnetic solver that uses geometrical optics (GO) ray tracing to efficiently solve for electrically large geometries. CAD models of packaged antenna, bumper and facia were imported and setup using HFSS-SBR+. The effects of the antenna packaging and integration were shown by comparing it to real radar sensing data.

11.4 Challenges and opportunities in AV simulation

Within the last few years, there has been a growing awareness among the AV industry leaders and academicians alike that the promises made by the industry were too lofty and it would take a lot longer to bring AVs to a production scale [2]. While every step of the pipeline comes with its own challenges, the current realization is that the process of verification and validation (V&V) is something that AV companies are struggling to achieve moderately. AV fatalities can be divided into "unavoidable accidents" and "AV bugs" [12], where the AV bugs could be simple but crucial errors that can be solved and the unavoidable accidents are those that cannot be avoided even when each subtask works correctly. The idea is that with the increase in testing, the AV safety will increase (i.e. the number of bugs would go lower) while the human safety stays the same. However, analyzing past AV-related accidents reveals a pattern of specific module failure which point towards lack of coverage in V&V. Specifically, even though the industry mindset has shifted towards catching software bugs in simulation, there has not been agreement in how to break down scenarios and perform scenario based testing. Secondly, even though sensor simulation models exist, they are rudimentary and do not compare to real sensor outputs.

Firstly, even though the scenario-based frameworks exist, they have not been used extensively in crafting industrial-scale usage of scenarios. The scenario library needs to have adequate coverage in order to have some confidence in testing on these scenarios. Some of the criteria of coverage is laid out by Hollander [12] in their blog and is provided here as an example:

- The library needs to cover "normal" driving scenarios as well as known edge cases
- The library needs to pay attention towards interactions between AVs and other dynamic objects such as human driven vehicles, bicyclists and pedestrians
- The library needs to have real world road layouts as well as "extreme" artificial topologies to test the software modules
- The traffic scenarios should be scalable - they should be adaptable to a new location/domain etc.

This list already establishes the challenge that V&V engineers face in balancing the high-dimensional space of scenarios versus making them tractable and solvable. In particular, the trade-off would be realistically mixing various dimensions of the scenarios in terms of static objects, dynamic objects and weather conditions. Real world driving as a way to seed scenario databases has been proposed in various frameworks. But even with a fleet of vehicles, finding and extracting scenarios is not an easy task. A solution would be

[2]https://www.theverge.com/22423489/autonomous-vehicle-consolidation-acquisition-lyft-uber

creating a industry wide AV agnostic database of scenarios which could leverage sharing anonymized trip data in an agnostic approach which can help in bringing the specific edge-cases to the forefront and solve it in an industry-wide fashion.

Secondly, a major challenge in creating realistic simulations is sensor simulation. Simulating sensors realistically is a very complex high-dimensional research topic which is far from being solved. Synthetic inputs such as those present in CARLA simulator provide sensor outputs which look pretty good visually but it is not very clear how closely it can replicate realistic sensor data and whether it can achieve similar controls output as in real data. Another approach is utilizing recorded data to seed the sensor generating pipeline. In this case, full 3D models of entire areas/cities could be imported into simulation. 3D models of cars and people are then introduced into the simulation and "reasonable" sensor streams can be generated. While this idea has become popular recently, it has high computational cost and the sensor outputs are not perfect.

As is evident from the previous discussion, there is a lot of opportunities in developing scalable and accurate simulation models that can be applied towards AV testing. Most of the AV simulation companies are working towards tackling these challenges but a lot of work needs to be done before the mass-produced AVs can hit the market.

References

[1] E. Andreotti, P. Boyraz, and S. Selpi. Mathematical definitions of scene and scenario for analysis of automated driving systems in mixed-traffic simulations. *IEEE Transactions on Intelligent Vehicles*, 6(2):366–375, 2020.

[2] G. Bagschik, T. Menzel, and M. Maurer. Ontology based scene creation for the development of automated vehicles. In *2018 IEEE Intelligent Vehicles Symposium (IV)*, pages 1813–1820. IEEE, 2018.

[3] J. Bock, R. Krajewski, L. Eckstein, J. Klimke, J. Sauerbier, and A. Zlocki. Data basis for scenario-based validation of had on highways. In *27th Aachen colloquium automobile and engine technology*, pages 8–10, 2018.

[4] D. J. Butler, J. Wulff, G. B. Stanley, and M. J. Black. A naturalistic open source movie for optical flow evaluation. In *European conference on computer vision*, pages 611–625. Springer, 2012.

[5] U. Chipengo, P. M. Krenz, and S. Carpenter. From antenna design to high fidelity, full physics automotive radar sensor corner case simulation. *Modelling and Simulation in Engineering*, 2018, 2018.

[6] A. Elmquist and D. Negrut. Methods and models for simulating autonomous vehicle sensors. *IEEE Transactions on Intelligent Vehicles*, 5(4):684–692, 2020.

[7] J. Fang, F. Yan, T. Zhao, F. Zhang, D. Zhou, R. Yang, Y. Ma, and L. Wang. Simulating lidar point cloud for autonomous driving using real-world scenes and traffic flows. *arXiv preprint arXiv:1811.07112*, 1, 2018.

[8] S. Geyer, M. Baltzer, B. Franz, S. Hakuli, M. Kauer, M. Kienle, S. Meier, T. Weißgerber, K. Bengler, R. Bruder, et al. Concept and development of a unified ontology for generating test and use-case catalogues for assisted and automated vehicle guidance. *IET Intelligent Transport Systems*, 8(3):183–189, 2014.

[9] K. Go and J. M. Carroll. The blind men and the elephant: Views of scenario-based system design. *interactions*, 11(6):44–53, 2004.

[10] C. Goodin, M. Doude, C. R. Hudson, and D. W. Carruth. Enabling off-road autonomous navigation-simulation of lidar in dense vegetation. *Electronics*, 7(9):154, 2018.

[11] M. Haag. *Bildfolgenauswertung zur Erkennung der Absichten von Strassenverkehrsteilnehmern (English title: Image sequence analysis for detecting intentions of traffic participants)*. PhD thesis, 1998.

[12] Y. Hollander. The uber accident and the bigger picture, 2018. Accessed July 30, 2022. https://blog.foretellix.com/2018/03/27/the-uber-accident-and-the-bigger-picture/.

[13] S. Irgenfried, F. Dittrich, and H. Wörn. Realization and evaluation of image processing tasks based on synthetic sensor data: 2 use cases. In *Forum Bildverarbeitung*, volume 82, pages 35–46. KIT Scientific Publishing, 2014.

[14] N. Koenig and A. Howard. Design and use paradigms for gazebo, an open-source multi-robot simulator. In *2004 IEEE/RSJ International Conference on Intelligent Robots and Systems (IROS)(IEEE Cat. No. 04CH37566)*, volume 3, pages 2149–2154. IEEE, 2004.

[15] W. Kruger. Situationsmodellierung in der bildfolgenauswertung (english title: Situation modeling in image sequence analysis),. In *KI -Informatik Fachbachrichte*. Springer, 1991.

[16] R. Molenaar, A. van Bilsen, R. van der Made, and R. de Vries. Full spectrum camera simulation for reliable virtual development and validation

of adas and automated driving applications. In *2015 IEEE Intelligent Vehicles Symposium (IV)*, pages 47–52. IEEE, 2015.

[17] Moove. Scenario database, 2018. Accessed July 30, 2022. https://www.vedecom.fr/presentation-du-projet-moove-a-lautonomous-vehicle-test-development-symposium-de-stutggart/?lang=en.

[18] T. Nürnberg, M. Schambach, D. Uhlig, M. Heizmann, and F. P. León. A simulation framework for the design and evaluation of computational cameras. In *Automated Visual Inspection and Machine Vision III*, volume 11061, page 1106102. SPIE, 2019.

[19] PEGASUS. Scenario database, 2018. Accessed July 30, 2022. https://www.pegasusprojekt.de/files/tmpl/Pegasus-Abschlussveranstaltung/15_Scenario-Database.pdf.

[20] N. Peinecke, T. Lueken, and B. R. Korn. Lidar simulation using graphics hardware acceleration. In *2008 IEEE/AIAA 27th Digital Avionics Systems Conference*, pages 4–D. IEEE, 2008.

[21] M. Pellkofer. *Verhaltensentscheidung fur autonome Fahrzeuge mit Blickrichtungssteuerung (English title: Behavior decision for autonomous vehicles with saccadic vision)*. PhD thesis, 20003.

[22] F. Rosique, P. J. Navarro, C. Fernández, and A. Padilla. A systematic review of perception system and simulators for autonomous vehicles research. *Sensors*, 19(3):648, 2019.

[23] SAE. Taxonomy and definitions for terms related to driving automation systems for on-road motor vehicles, 2014.

[24] SAKURA. Scenario database, 2020. Accessed July 30, 2022. https://www.sakura-prj.go.jp/project_info/tabid84.html.

[25] M. Scholtes, L. Westhofen, L. R. Turner, K. Lotto, M. Schuldes, H. Weber, N. Wagener, C. Neurohr, M. H. Bollmann, F. Körtke, et al. 6-layer model for a structured description and categorization of urban traffic and environment. *IEEE Access*, 9:59131–59147, 2021.

[26] F. Schuldt, F. Saust, B. Lichte, M. Maurer, and S. Scholz. Effiziente systematische testgenerierung für fahrerassistenzsysteme in virtuellen umgebungen. *Automatisierungssysteme, Assistenzsysteme und Eingebettete Systeme Für Transportmittel*, 2013.

[27] M. Thomason and R. Gonzalez. Data structures and databases in digital scene analysis. In *Advances in Information Systems Science*, pages 1–47. Springer, 1985.

[28] S. Ulbrich, T. Menzel, A. Reschka, F. Schuldt, and M. Maurer. Defining and substantiating the terms scene, situation, and scenario for automated

driving. In *2015 IEEE 18th international conference on intelligent transportation systems*, pages 982–988. IEEE, 2015.

[29] A. Von Bernuth, G. Volk, and O. Bringmann. Simulating photo-realistic snow and fog on existing images for enhanced cnn training and evaluation. In *2019 IEEE Intelligent Transportation Systems Conference (ITSC)*, pages 41–46. IEEE, 2019.

[30] N. Wagner, P. Weissensteiner, J. Coget, L. Eckstein, and A. Bracquemond. Common methodology for data-driven scenario-based safety assurance in the headstart project. In *ITS European Congress*, 2020.

[31] J. Wu and H. Ma. Error analysis on laser measurement device of airborne lidar. In *MIPPR 2007: Automatic Target Recognition and Image Analysis; and Multispectral Image Acquisition*, volume 6786, pages 1331–1339. SPIE, 2007.

[32] C. Zhou and S. K. Nayar. Computational cameras: convergence of optics and processing. *IEEE Transactions on Image Processing*, 20(12):3322–3340, 2011.

Authors' Biography

Arpan Kusari is an Assistant Research Scientist in the Engineering Systems Group at UMTRI. He is primarily interested in refining the Automated Driving Systems (ADS) pipeline with a broad impact in terms of safety, efficiency and robustness. He has interests in the entire ADS pipeline spanning from mapping and localization, to perception and finally to robust decision making. He previously spent five years at Ford Motor Company researching exclusively on making autonomous vehicles safe and viable, working collaboratively with researchers from MIT and University of Michigan to advance the state-of-the-art knowledge in autonomous vehicles. Dr. Kusari has a masters in Civil Engineering from Purdue University and a Ph.D. in Civil Engineering from University of Houston.

12

Path Following Control System Design and Analysis of Autonomous Vehicles

Mümin Tolga Emirler

Yildiz Technical University, Turkey
E-mail: emirler@yildiz.edu.tr

Abstract

The research on autonomous vehicles is highly multidisciplinary and includes many components from different domains of engineering. Besides perception, localization, and path planning, the control system design and implementation is vital for a successful autonomous vehicle operation. From the perspective of control systems, autonomous vehicle control can be categorized into longitudinal and lateral automation. Longitudinal automation is related to vehicle speed control. On the other hand, lateral automation considers path following control with steering actuator. This chapter focuses on the path following the control system design which is used to minimize the lateral deviation error from the reference path while traveling at the desired speed. Firstly, a linear path following autonomous vehicle model is introduced in state space representation. This model is used to design feedback linear control systems such as PID and linear quadratic optimal controllers. Simulations are conducted for different scenarios. In the simulation studies, a high degree-of- freedom vehicle model is employed to get realistic results. Finally, the results are obtained and analyzed in terms of lateral deviation, vehicle speed, and front wheel steering angle. The performances of different control systems are also compared numerically by using statistical error values.

Keywords: Autonomous vehicles, control system design, linear quadratic optimal control, path following control, simulation.

12.1 Introduction

Autonomous vehicle technology is developing very rapidly to change the way of transportation. Many successful theoretical and practical improvements have been obtained in the last two decades [1]. It is a very attractive multidisciplinary research area including different topics of engineering. Control system design, analysis, and implementation important part of this technology along with perception, localization, and path planning.

Autonomous vehicle control systems can be categorized as longitudinal and lateral automation. Path following control is the main part of lateral automation and it is investigated from the perspective of control engineering in this chapter.

Path following control system automatically steers the wheels of the autonomous vehicle to track the desired path by minimizing the lateral deviation of the vehicle from the reference path at a preview distance. This lateral deviation can be obtained by using magnetic marker reference systems, vision-based sensors, and GNSS-based calculations.

Early works on path following control systems can be traced back to the 990s. In 1990, Ackermann and Guldner proposed a robust $PIDD^2$ automatic steering control system for a bus which uses a displacement sensor to measure the lateral deviation of the bus from a reference path (a magnet trail). Only a front displacement sensor is used in this study [2]. A robust parameter space approach-based control method in an invariance plane is applied to a similar problem in [3] using the front and tail displacement measurements of the vehicle from the reference path. The proposed method is tested with experiments on magnets installed test track successfully. In 2002, Aksun Guvenc and Guvenc proposed a discrete-time disturbance observer-based control system for the benchmark problem on automatic steering control of a city bus. The details of the disturbance observer design in discrete-time including alternative implementations are given in this study [4].

The research on autonomous vehicle path following control has been accelerated again in the last decade due to the rapid development in microprocessor and sensor technologies. As a result, various research results have been published on this subject. For example, Choi proposed a LMI-based H_∞ controller that uses the lateral deviation and yaw angle error at a preview distance as feedback signals. It is tested by simulations against the modelling uncertainties and random sensing noises [5]. In [6, 7], a two degrees of freedom control structure is proposed which includes a feedforward controller, a tracking controller, and a disturbance observer. Parameter space approach is used in the design phase and the proposed control structure is tested with a simulated sudden crosswind, a simulated lane change, and an experimental lane change scenarios.

A robust PID automatic steering control system is designed by using parameter space approach in [8, 9]. Vehicle mass, vehicle velocity, and tire-road friction coefficient are taken as uncertain parameters in this design. In [10], the required feedback signals for path following controller are obtained by comparing GNSS-based measured vehicle position with the generated digital map trajectory. The proposed robust PID control system is tested by simulations using these calculated feedback signals. In [11], an elastic band method-based collision avoidance system is added to aforementioned path following control system to follow a collision-free path. The simulation results are given. In [12], real-time implementation of the elastic band method-based collision avoidance algorithm and path following control system is realized. Hardware-in-the-loop simulations and road experiments are used to show the effectiveness of the proposed approach.

A comparative study of path-following controllers is conducted in [13]. PID, LQG, and H_∞ controllers are designed and compared in terms of tracking performance, stability robustness, and passenger comfort. A speed-dependent gain scheduled LQI controller is designed and compared with PID and LQR controllers in [14]. Parameter space-based speed scheduled path-following control system is proposed in [15] and it is tested by hardware-in-the-loop simulations and road tests succesfully. In [16, 17], a discrete-time preview optimal path tracking control algorithm is designed. It includes a feedback controller to compensate tracking errors and a feedforward controller considering the upcoming road curvatures. The proposed system is tested using simulations and road experiments.

Frequently used and well-developed control system design methods such as PID, linear quadratic optimal controllers are applied to autonomous vehicle path-following control in this chapter. Performance improvements are achieved by using LQI (linear quadratic integral) control over LQR (linear quadratic regulator) optimal control and PID control. The proposed control methods are compared with each other in different scenarios.

This chapter is the extended version of the journal paper [14] written in Turkish by the author of this chapter. The rest of this chapter is organized as follows. In Section 12.1, path-following autonomous vehicle model is introduced which is employed for linear control system design. Section 12.3 presents different control system design methods for path-following control. In Section 12.4, the details of a nonlinear autonomous vehicle model are given which is a more realistic model for testing and comparing of control systems. In Section 12.5, the simulation studies are described, and the results are given including numerical comparisons of control systems. The chapter ends with the conclusion part in Section 12.6.

12.2 Path Following Autonomous Vehicle Model

Modeling is the first and the fundamental step of control system design. Single-track vehicle model is used frequently for representing vehicle dynamics in control applications. It is a well-established model in the literature of vehicle dynamics [18]. In this model, two front and rear tires of the vehicle are lumped together to ease the model representation and to focus on the lateral dynamics. Because of this representation, it is sometimes called the bicycle model. The modeling details of the single track vehicle model can be found in the literature [19–21].

Path following autonomous vehicle model is obtained based on the single track model by adding the equations which describe the vehicle's motion relative to a reference path [2, 4, 8–10]. The path-following autonomous vehicle model shown in Figure 12.1 is described in state-space form as follows:

$$\dot{x} = Ax + B_1 u + B_2 u_d \tag{12.1}$$

where the states are $x = [\beta \; r \; \Delta\psi \; e_p]^T$, the control input is $u = \delta_u$, and the disturbance is $u_d = \rho$. Here, β is the vehicle side slip angle, r is the vehicle yaw rate, $\Delta\psi$ is the yaw angle error, e_p is the lateral deviation at the preview distance, δ_u is the steering control input generated by an electric motor actuator, and ρ is the road curvature.

The matrices A, B_1, and B_2 in eqn 12.1 are defined as follows:

$$A = \begin{bmatrix} a_{11} & a_{12} & 0 & 0 \\ a_{21} & a_{22} & 0 & 0 \\ 0 & 1 & 0 & 0 \\ V & l_s & V & 0 \end{bmatrix}, \; B_1 = \begin{bmatrix} b_{11} \\ b_{21} \\ 0 \\ 0 \end{bmatrix}, \; B_2 = \begin{bmatrix} 0 \\ 0 \\ -V \\ -l_s V \end{bmatrix} \tag{12.2}$$

where V is the vehicle speed and the remaining coefficients are

$$\begin{aligned} a_{11} &= -\frac{C_f + C_r}{mV}, & a_{12} &= -1 + \frac{C_r l_r + C_f l_f}{mV^2}, & a_{21} &= \frac{C_r l_r - C_f l_f}{I_z}, \\ a_{22} &= -\frac{C_r l_r^2 + C_f l_f^2}{I_z V}, & b_{11} &= \frac{C_f}{mV}, & b_{21} &= \frac{C_f l_f}{I_z} \end{aligned} \tag{12.3}$$

The autonomous vehicle parameter values are given in Table 12.1.

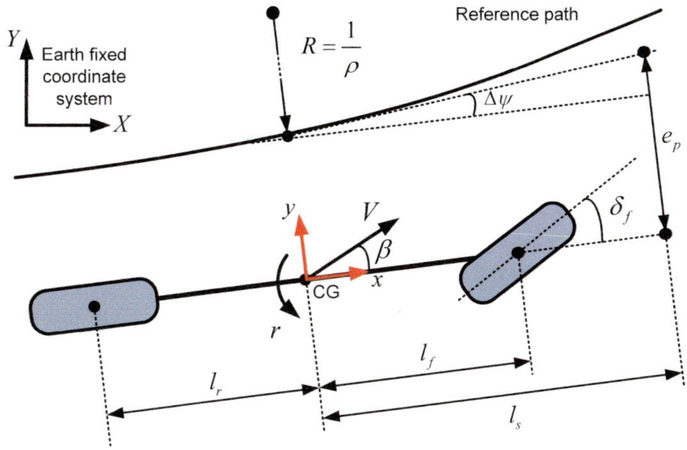

Figure 12.1 Path following autonomous vehicle model.

12.3 Control System Design

12.3.1 PID control

Conventional proportional-integral-derivative (PID) controllers are used frequently in industrial applications such as in vehicle control systems due to their reliability, simplicity, ease of implementation, and tuning. In this chapter, PID controller is employed for path-following control as a baseline control system shown in block diagram in Figure 12.2. In this control structure, only the lateral deviation at the preview distance (e_p) is used as a feedback signal. PID controller generates the control signal δ_u to decrease the road curvature disturbance on the system by minimizing the lateral deviation at the preview distance. An example of PID control system design considering robustness against the parametric uncertainties in vehicle parameters can be

Table 12.1 Parameters used in path-following autonomous vehicle model.

Symbol	Description	Value
m	Vehicle mass	1550 [kg]
I_z	Vehicle moment of inertia about the z axis	2400 [kg/m^2]
C_f	Cornering stiffness of the front tires	145,000 [N/rad]
C_r	Cornering stiffness of the rear tires	185,000 [N/rad]
l_f	Distance from the center of gravity to the front axle	1.07 [m]
l_r	Distance from the center of gravity to the rear axle	1.53 [m]
l_s	Preview sensing distance	20 [m]

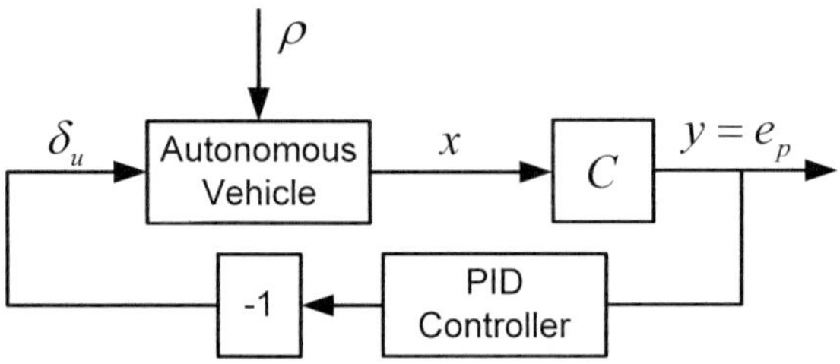

Figure 12.2 Autonomous vehicle PID path-following control system block diagram.

found in [8, 9]. In this chapter, the gain of the PID controller is determined by using MATLAB's PID Tuner. The gains are optimized for the parameters given in Table 12.1 with a constant vehicle speed of 72 km/h. Since LQR and LQI controllers are in optimal fashion, PID controller is designed here based on optimization to make a fair comparison. By using MATLAB's PID Tuner, PID gains k_p, k_i, and k_d are calculated as $(0.067, 0.045, \text{and } 0.008)$, respectively.

12.3.2 Linear quadratic optimal control

Linear quadratic regulator (LQR) and linear quadratic integral (LQI) control are optimal state feedback control methods used frequently in engineering applications. They realize control action by driving the state variables of a dynamic system from arbitrary initial conditions to zero or to operating conditions. By proper selection of weighting matrices in quadratic performance index of these controllers, asymptotic stability, and performance requirements of the closed-loop control system are provided [22, 23].

LQR and LQI control have been applied in a wide range of problems. Some examples are given in this part. In [24], speed scheduled LQR controller is designed for optimal vehicle yaw moment control. Luong and Tsao proposed a LQI-based control system to control an organic Rankine cycle for waste heat recovery in heavy-duty diesel powertrain. In this study, PI and LQI controllers are compared in the presence of automotive transient disturbances [25]. In [26], LQR control with integral action is employed for optimal control of a twin rotor multi-input multi-output helicopter system. LQI control system is designed for a fixed-wing micro aerial vehicle in [27].

LQR, LQI, and disturbance observer augmented LQR are designed and compared for a small fixed-wing-unmanned aerial vehicle in [28]. Dal Poggetto and Serpa proposed a gain-scheduled LQR control system for vehicle rollover avoidance control. LQR controller gains are calculated as a function of vehicle speed and the closed loop control system is analyzed considering changes in speed [29]. In [30], robust LQR and LQI controllers are designed for unmanned bicycle robot stabilization. Control strategies are compared experimentally. LQI control is utilized in order to control a small-scale turbojet engine with variable exhaust nozzle in [31]. A robust LQR controller are proposed for vehicle direct yaw moment control in [32]. Robustness properties of LQR controller are improved by an additional tuned feedback controller.

A general linear time invariant (LTI), multi-input multi-output system is written in state space representation as follows:

$$\dot{x} = Ax + Bu \tag{12.4}$$

$$y = Cx \tag{12.5}$$

where x, u, and y are state, input, and output variables, and A, B, and C denote state, control, and output matrices, respectively. For the state space system given in eqn (12.1)- (12.3), A is the same in eqn (12.4), $B_1 = B, C = [0\ 0\ 0\ 1]$.

The aim of LQR optimal control is to minimize the following performance index J_{lqr} [23]:

$$J_{\text{lqr}} = \int_0^{\infty} \left(x^T Q x + u^T R u\right) dt \tag{12.6}$$

where Q is a symmetric positive semidefinite weighting matrix, and R is a symmetric positive definite weighting matrix. Optimal control law which minimizes eqn (12.6) and guarantees the closed-loop system stability can be written as follows:

$$u = -R^{-1} B^T P x = K_{\text{lqr}} x \tag{12.7}$$

where K_{lqr} is the optimal feedback gain matrix, and P is the unique positive semidefinite solution of the Algebraic Riccati equation given below:

$$PA + A^T P - PBR^{-1} B^T P + Q = 0 \tag{12.8}$$

LQR optimal control method has not an integral effect in its standard structure. Therefore, steady state error problem is frequently seen in LQR control. In order to overcome this problem, LQI control method can be used. In LQI control, an augmented system representation is formed by defining a new state variable which adds an integral effect to the system. Then, the problem solution is arranged as in the LQR control method solution, and the optimal control law, which minimizes the performance index, is obtained by solving the related algebraic Riccati equation.

The advantages of LQI control method over the other linear control methods can be summarized as follows [31]:

- It provides quadratic optimal control.
- It globally stabilizes LTI systems.
- It has good reference tracking property through integrator in its control structure.
- It has good robustness property by providing positive infinity gain margin to the closed loop system.

In this part, LQI control method equations are derived. The difference between reference input r and system output y is defined as error e and the integral of this error can be defined as a new (extra) state variable for LQI structure [23, 28]:

$$x_{ei} = \int_0^\infty (r-y)\,dt = \int_0^\infty e\,dt \tag{12.9}$$

At steady state conditions, the following equations can be written as related to steady state values of x and u:

$$Ax_\infty + Bu_\infty = 0 \tag{12.10}$$

$$Cx_\infty = r\ (r=y) \tag{12.11}$$

Steady state value of x_{ei} can be considered as a design parameter. Here, it is selected as zero $(x_{ei\infty} = 0)$.

The difference of x, u, x_{ei} values from their steady state values can be defined as

$$\tilde{x} = x - x_\infty \tag{12.12}$$

$$\tilde{u} = u - u_\infty \tag{12.13}$$

$$\tilde{x}_{ei} = x_{ei} - x_{ei\infty} \tag{12.14}$$

By using these defined variables, the augmented state space form of the system can be written as below:

$$\underbrace{\begin{bmatrix} \dot{\tilde{x}} \\ \dot{\tilde{x}}_{ei} \end{bmatrix}}_{\dot{\tilde{z}}} = \underbrace{\begin{bmatrix} A & 0 \\ -C & 0 \end{bmatrix}}_{A_z} \tilde{z} + \underbrace{\begin{bmatrix} B \\ 0 \end{bmatrix}}_{B_z} \tilde{u} + \underbrace{\begin{bmatrix} 0 \\ I \end{bmatrix}}_{=0} \tilde{r} \tag{12.15}$$

Eqn (12.15) is written in standard form by considering zero reference input ($\tilde{r} = 0$):

$$\dot{\tilde{z}} = A_z\tilde{z} + B_z\tilde{u} \tag{12.16}$$

Similar to LQR system, LQI control performance index J_{lqi} for the augmented system can be described as

$$J_{\mathrm{lqi}} = \int_0^{\infty} \left(\tilde{z}^T Q_z\tilde{z} + \tilde{u}^T R_z u\right) dt \tag{12.17}$$

The optimal control law is to minimize the performance index given in eqn (12.17) can be found as below:

$$\tilde{u} = -R_z^{-1} B_z^T P_z\tilde{z} = K_x\tilde{x} + K_{x_{ei}}\tilde{x}_{ei} \tag{12.18}$$

P_z matrix is obtained by solving the following algebraic Riccati equation:

$$P_z A_z + A_z^T P_z - P_z B_z R_z^{-1} B_z^T P_z + Q_z = 0 \tag{12.19}$$

where the partitions of P_z positive semidefinite matrix is

$$P_z = \begin{bmatrix} P_{11} & P_{12} \\ P_{12}^T & P_{22} \end{bmatrix} \tag{12.20}$$

The unique solution of algebraic Riccati equation given in eqn (12.19) results in optimal LQI control gains as follows:

$$K_x = -R^{-1} B^T P_{11} \tag{12.21}$$

$$K_{x_{ei}} = -R^{-1} B^T P_{12} \tag{12.22}$$

By using eqn (12.10)-(12.14) and (12.18), the final optimal control law is written as below:

$$u = K_x x + K_{x_{ei}} x_{ei} + K_r r \tag{12.23}$$

where

$$K_r = -\left[C\left(A + BK_x\right)^{-1} B\right]^{-1} \tag{12.24}$$

The optimal control gain matrix of LQI control system is also expressed as follows:

$$K_{\mathrm{lqi}} = [K_x \; K_{x_{ei}}] \tag{12.25}$$

A general LQI control system block diagram containing optimal gain blocks is shown in Figure 12.3.

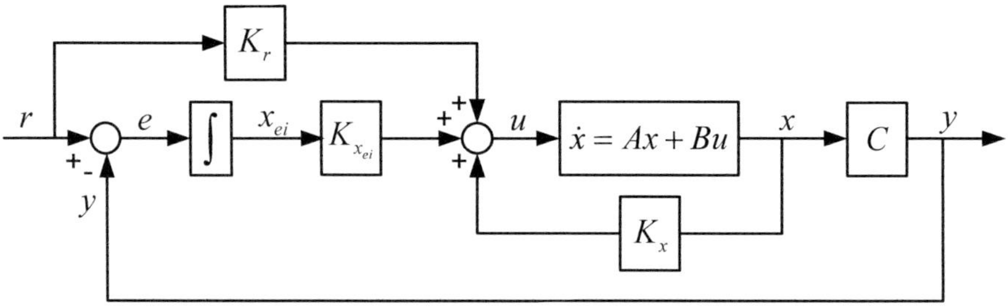

Figure 12.3 LQI control system block diagram.

12.3.3 LQR and LQI controllers for autonomous path following

After some theoretical background given in Section 12.3.2, LQR and LQI control design details are given in this part. Autonomous vehicle LQR path-following control system block diagram can seen from Figure 12.4. For the linear path-following vehicle model given in eqn (12.1)-(12.3), the optimal control gain matrix $K_{\text{lqr}} = -\left[k_\beta \; k_r \; k_{\Delta\psi} \; k_{e_p}\right]$ can be obtained using eqn (12.4)-(12.8) including the solution of algebraic Riccati equation. Here, k_β, k_r, $k_{\Delta\psi}$, and k_{e_p} show the optimal control gains of the corresponding vehicle states, β, r, $\Delta\psi$, and e_p, respectively.

LQI control system for autonomous path-following control is depicted in Figure 12.5. In LQI control system, a new state variable is defined as $x_{ei} = \int_0^\infty (e_{p,\text{des}} - e_p)\,dt$ to add the integral effect into the closed-loop control system. Here, $e_{p,\text{des}}$ shows the desired lateral deviation at the preview distance, and it is equal to zero. LQI control system optimal gain matrix $K_{\text{lqi}} = [K_x \; K_{x_{ei}}]$ can be calculated by following eqn (12.9)-(12.24) and determined as $K_x = -\left[k_\beta \; k_r \; k_{\Delta\psi} \; k_{e_p}\right]$ and $K_{x_{ei}} = -k_{xei}$. Here, k_{xei} is the optimal control

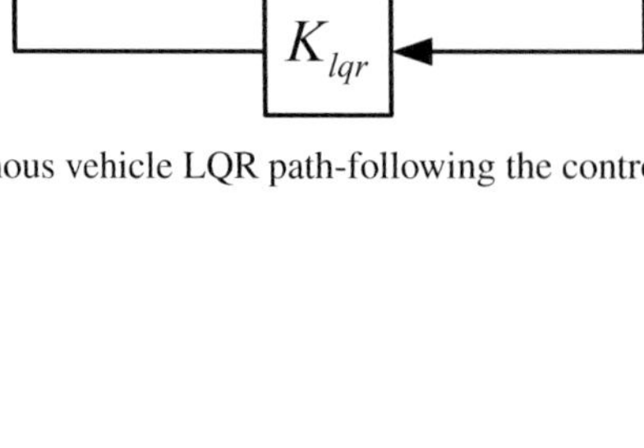

Figure 12.4 Autonomous vehicle LQR path-following the control system block diagram.

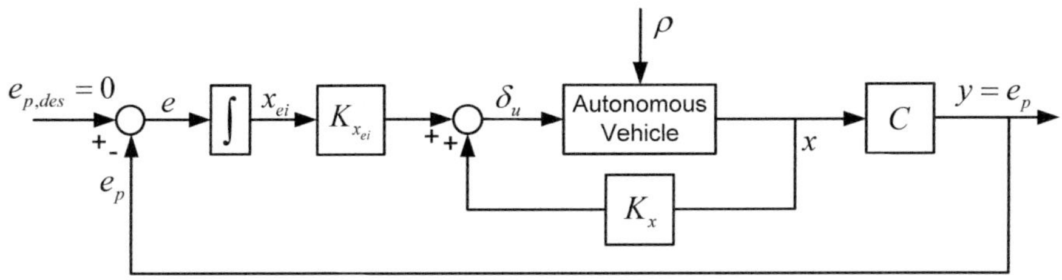

Figure 12.5 Autonomous vehicle LQI path-following control system block diagram.

gain which is multiplied by the state variable x_{ei}. Note that there is no need to calculate K_r defined in eqn (12.24) for this problem because $e_{p,\text{des}} = 0$.

For the calculation of LQR control optimal gain matrix K_{lqr}, the weighting matrices are taken as $Q = q_w (I_{4\times 4})$, and $R = r_w$. Similarly, the weighting matrices are considered as $Q = q_w \ (I_{5\times 5})$, $R = r_w$ for LQI control optimal gain matrix K_{lqi} calculation. q_w and r_w are chosen as same for both LQR and LQI control system design to provide a fair comparison. For this problem, q_w is selected as 8, and r_w is determined as 150 by intensive simulation studies. $I_{n\times n}$ shows n-dimensional identity matrix.

Dynamic behavior of a vehicle is strongly dependent on vehicle speed V. From the perspective of system modeling, the location of the eigenvalues of the system is changing by differing vehicle speed. For this reason, vehicle speed effect is considered in LQR and LQI control system design. The matrices of state space model of the path-following autonomous vehicle given in eqn (12.1)-(12.3) can be considered as $A = A(V)$, $B_1 = B_1(V)$, and $B_2 = B_2(V)$. For each vehicle speed, LQR and LQI controller optimal gain matrices can be calculated. Here, $K_{\text{lqr}} = K_{\text{lqr}}(V)$ and $K_{\text{lqi}} = K_{\text{lqi}}(V)$ are computed for speed values from 1 km/h to 40 km/h with a 1 km/h step. This kind of speed-dependent gain scheduled approach has been used before in vehicle yaw stability control [24] and vehicle rollover avoidance control [29]. Figure 12.6 shows the variations of LQR and LQI controller gains by vehicle speed. The inner points of the curves in Figure 12.6 are filled by interpolation. Then, 2D look-up tables are prepared for each control gain and used in Simulink simulation model.

Closed loop system stability is investigated for differing vehicle speeds. Figure 12.7 shows the maximum real part of eigenvalues of the closed-loop system for varying vehicle speeds. For both LQR and LQI-controlled systems, the real parts of eigenvalues are negative, so they are at the left half of the complex plane. This verifies the closed-loop system stability.

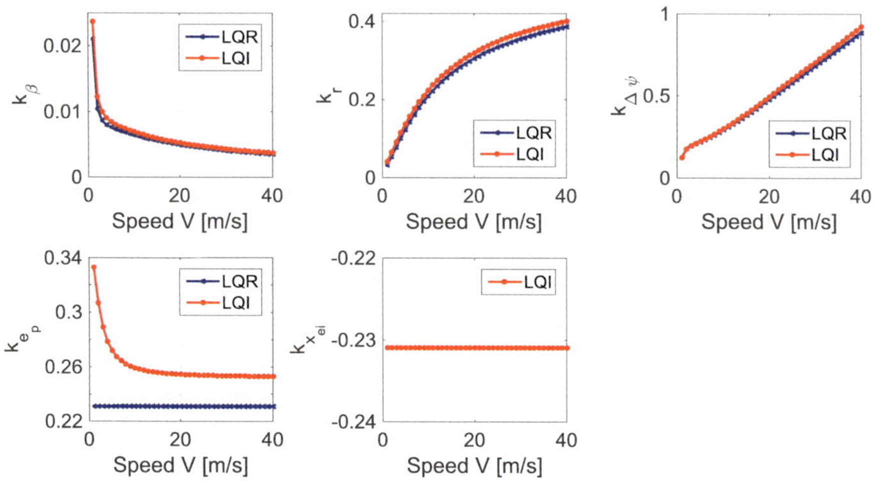

Figure 12.6 Variations of LQR and LQI controller gains as a function of vehicle speed.

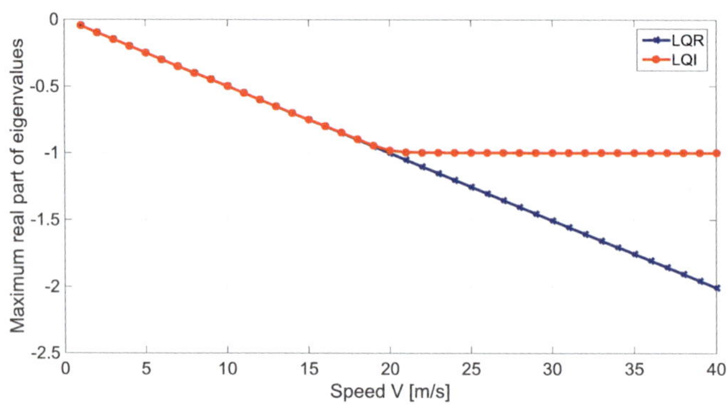

Figure 12.7 Closed-loop system maximum real part of eigenvalues varying with vehicle speed.

12.4 Simulation Model

This section provides a nonlinear autonomous vehicle simulation model. This model is used for testing the proposed control systems. It has seven degrees of freedom including longitudinal, lateral, and yaw motions and the rotations of four wheels. The vertical, roll, and pitch dynamics of the vehicle and the vertical motion of the tires are neglected due to the investigation of path-following problem of the autonomous vehicle.

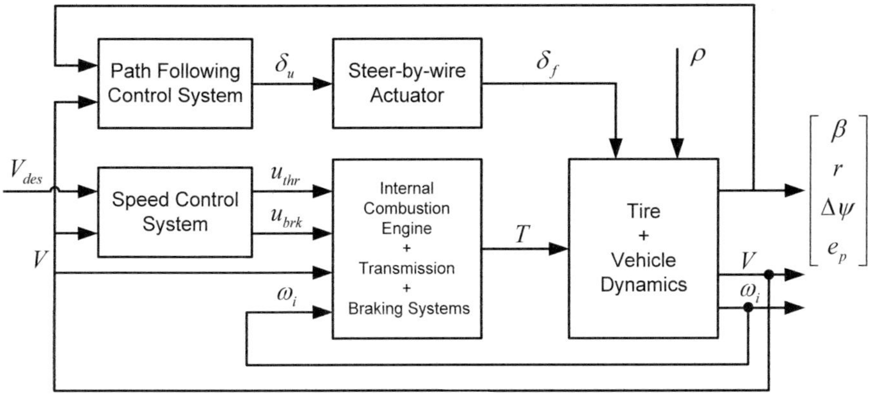

Figure 12.8 Autonomous vehicle simulation model block diagram.

The main block diagram of the nonlinear autonomous vehicle simulation model can be seen from Figure 12.8 with input-output signals of blocks. Autonomous vehicle simulation model consists of speed control system, internal combustion engine (ICE), transmission and braking system, tire and vehicle dynamics systems, path-following control system, and steer-by-wire actuator model. In Figure 12.8, V_{des} denotes desired vehicle speed, δ_f is the front wheel steering angle, u_{thr} is the throttle control signal, u_{brk} is the braking control signal, ω_i is angular speed of the ith wheel, T is the traction or braking torque transmitted to the wheels.

In Figure 12.8, speed control system block contains throttle and braking switching decision algorithm, and PID type throttle and braking lower controllers. These controllers are generate u_{thr} throttle and u_{brk} braking control signals in percentages. Internal combustion engine, transmission, and braking system block includes engine map and equations, torque convertor and transmission equations, automatic gear shift logic based on vehicle speed, and braking torque calculation blocks. By using this block, torque transmitted to the wheels (T) is calculated considering traction (acceleration) or braking situations.

The equations governing the dynamics of the nonlinear vehicle model are introduced in the following part. This section is based on the modeing parts in [8, 10, 11].

The equations of motion for nonlinear autonomous vehicle model shown in Figures 12.1 and 12.9 can be written in the lateral, longitudinal directions

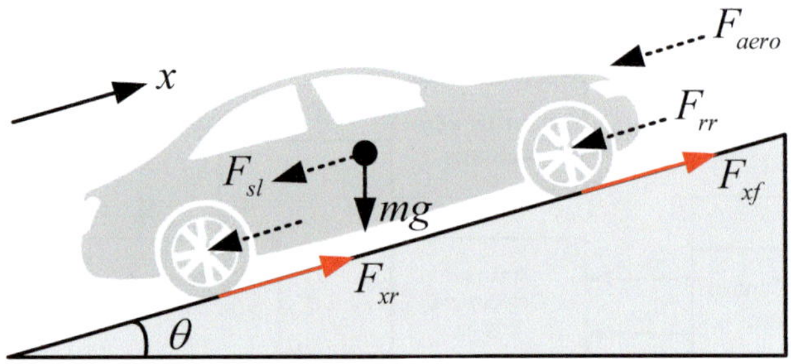

Figure 12.9 Longitudinal forces acting on the vehicle.

and about the yaw axis as follows:

$$m\left(\dot{V}_y + rV_x\right) = \sum_{i=f,r} F_{xi}\sin\delta_i + F_{yi}\cos\delta_i \qquad (12.26)$$

$$m\left(\dot{V}_x - rV_y\right) = \sum_{i=f,r} F_{xi}\cos\delta_i - F_{yi}\sin\delta_i + \left(F_{\text{aero}} + F_{\text{rr}} + F_{\text{sl}}\right) \qquad (12.27)$$

$$I_z\dot{r} = l_f\left(F_{yf}\cos\delta_f + F_{xf}\sin\delta_f\right) - l_r\left(F_{yr}\cos\delta_r + F_{xr}\sin\delta_r\right) \qquad (12.28)$$

where F_{xi} and F_{yi} are the longitudinal and lateral tire forces. f and r denote the front and rear tires. V_x, V_y, and I_z are the longitudinal velocity, lateral velocity, and the moment of inertia about the yaw axis, respectively. Here, steering is considered from the front wheels so $\delta_r = 0$ in this chapter.

The longitudinal forces acting on the vehicle are shown in Figure 12.9. The resistive forces are the aerodynamic drag force, the rolling resistance force, and the gravitational slope resistance force.

The aerodynamic drag force F_{aero} can be represented by

$$F_{\text{aero}} = \frac{1}{2}\rho C_d A V^2 \qquad (12.29)$$

where ρ is the air density, C_d is the drag coefficient, and A is the frontal area of the vehicle.

The rolling resistance force F_{rr} is given by

$$F_{\text{rr}} = C_{\text{rr}}\text{mg}\cos\theta \qquad (12.30)$$

where C_{rr} is the rolling resistance coefficient, and θ is the road inclination angle.

The gravitational slope resistance force F_{sl} is determined as

$$F_{\text{sl}} = \text{mg}\cos\theta \qquad (12.31)$$

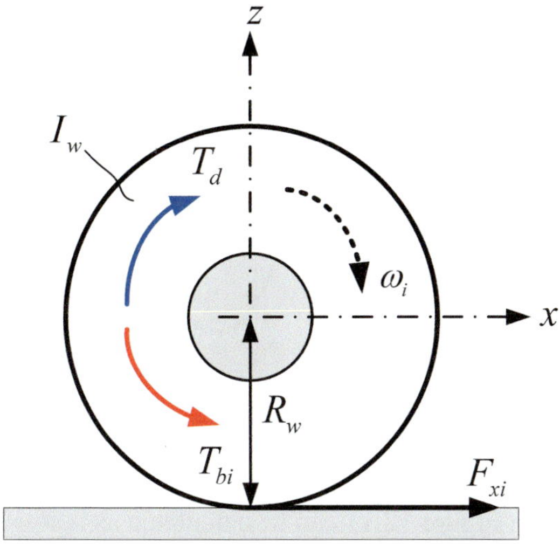

Figure 12.10 Forces and torques acting on the wheel.

Internal combustion engine (ICE) model is represented by using a static engine map. Look-up table type engine map sets a relationship between the inputs of throttle position α_{pos}, the engine speed ω_e, and the engine torque output $T_{\text{ice}}\left(\alpha_{\text{pos}}, \omega_e\right)$.

The engine torque output is transmitted to the wheels through the driveline and the transmitted torque T_d can be calculated as follows:

$$T_d = \eta_t i_t T_{\text{ice}}\left(\alpha_{\text{pos}}, \omega_e\right) \tag{12.32}$$

where η_t is a static efficiency factor to represent mechanical loses, and i_t is the transmission ratio.

The forces and the torques action on a single wheel is shown in Figure 12.10. The moment balance equation with respect to the center of the wheel can be written as follows:

$$I_w \dot{\omega}_i = T_d - T_{bi} - F_{xi} R_w \tag{12.33}$$

where I_w denotes the moment of inertia of a wheel, ω_i is the angular speed of the ith wheel. T_{bi} denotes the braking torque applied to the ith wheel. F_{xi} is the longitudinal tire force of the ith wheel and R_w is the effective tire radius.

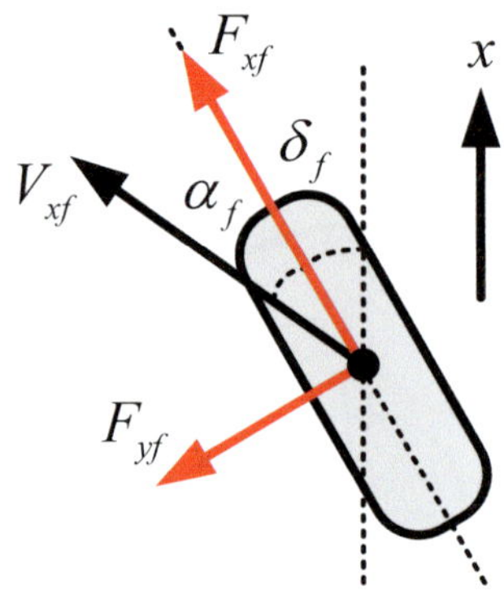

Figure 12.11 Velocity, angles, and forces at the front wheel.

The longitudinal velocities at the center of the front and rear wheels shown for one wheel in Figure 12.11 can be calculated as below:

$$V_{xf} = \sqrt{V_x^2 + (V_y + l_f r)^2} \cos\alpha_f \tag{12.34}$$

$$V_{xr} = \sqrt{V_x^2 + (V_y - l_r r)^2} \cos\alpha_r \tag{12.35}$$

where tire slip angles are

$$\alpha_f = \delta_f - \arctan\left(\tan\beta + \frac{l_f r}{V_x}\right) \tag{12.36}$$

$$\alpha_r = \delta_r - \arctan\left(\tan\beta - \frac{l_r r}{V_x}\right) \tag{12.37}$$

The longitudinal slip ratio is defined as

$$s_i = \begin{cases} \frac{R_w \omega_i - V_{xi}}{V_{xi}}, & R_w \omega_i < V_{xi} \text{ (braking)} \\ \frac{R_w \omega_i - V_{xi}}{R_w \omega_i}, & R_w \omega_i > V_{xi} \text{ (acceleration)} \end{cases} \tag{12.38}$$

Longitudinal and lateral tire forces are calculated by using the Dugoff tire model as follows:

$$F_{xi} = f_i \, C_{xi} \, s_i \tag{12.39}$$

$$F_{yi} = f_i \, C_{yi} \, \alpha_i \tag{12.40}$$

where C_{xi} and C_{yi} denote the longitudinal and the lateral cornering stiffness of the ith wheel, respectively.

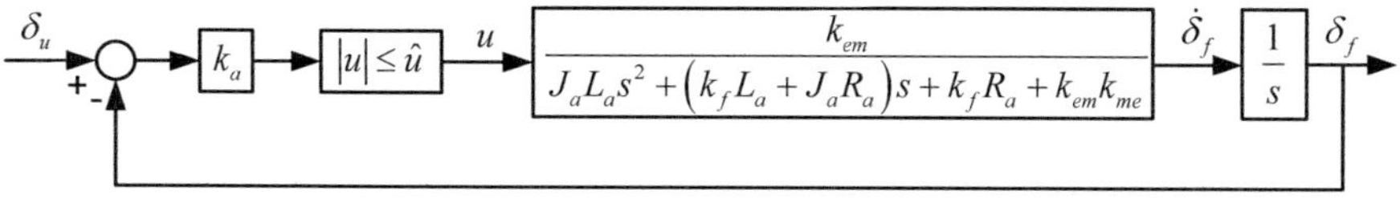

Figure 12.12 Electric motor-based steer-by-wire actuator model for automatic steering.

The coefficient f_i is determined as

$$f_i = \begin{cases} 1, & F_{Ri} \leq \frac{\mu F_{zi}}{2} \\ \left(2 - \frac{\mu F_{zi}}{2F_{Ri}}\right) \frac{\mu F_{zi}}{2F_{Ri}}, & F_{Ri} > \frac{\mu F_{zi}}{2} \end{cases} \tag{12.41}$$

where

$$F_{Ri} = \sqrt{(C_{xi} s_i)^2 + (C_{yi} \alpha_i)^2} \tag{12.42}$$

An experimental validation study has been conducted in [8, 9] for the aforementioned nonlinear vehicle simulation model. The simulated velocity and yaw rate results have been compared with the experimental results obtained from the same steering angle and desired velocity inputs. The simulation results have been found to be consistent with the collected experimental data. The model has been validated for both longitudinal and lateral dynamics of the vehicle.

An electric motor-based steer-by-wire actuator can be used for steering the wheels of the vehicle [19]. Figure 12.12 shows the closed-loop position controlled electric motor actuator block diagram. The control signal δ_u, which is obtained from path-following controller, is applied to steer-by-wire actuator to set required front wheel steering angle δ_f. The actuator model can be represented by a third-order system with proportional (P) position control while neglecting the nonlinear voltage limitation term. Additionally, by

Table 12.2 Electric motor actuator model parameters [19].

Symbol	Description	Value
J_a	Electric motor inertia	0.004053 [kgm^2]
k_f	Electric motor damping coefficient	0.01625 [Nms/rad]
L_a	Electric motor inductance (neglected)	0 [H]
R_a	Electric motor resistance	5 [Ω]
k_{me}	Electric motor torque constant (back EMF constant = k_{em})	0.9 [N/m]
k_a	Position control gain of the actuator	22.22 [V/rad]
$\hat{u}$	Voltage limitation level	12 [V]

neglecting the electric motor inductance (L_a), actuator model can be reduced to a second-order system [19]. In the simulation model, the voltage limitation is considered under the assumption of a $12V$ power supply $(|u| \leq 12V)$ and the structure given in Figure 12.12. The electric motor actuator model parameters are given in Table 12.2.

12.5 Simulation Studies

After designing control system using linear autonomous path-following model, the proposed control systems are evaluated using high degree of freedom nonlinear simulation model given in Section 12.4. In simulation 1, autonomous vehicle tries to follow a constant road curvature (ρ) of 0.005 1/m with a constant speed of 72 km/h (= 20m/s) in a 30 s simulation. It is equivalent to turn a road radius (R) of 200 m. The road curvature for simulation 1 is depicted in Figure 12.13.

Figure 12.14 shows the vehicle speed during simulation 1. It is kept around desired speed value of 72 km/h successfully by the speed control system. In Figure 12.15, the lateral deviation at the preview distance is shown for three different controllers. PID control system results in 21.94 cm maximum lateral deviation and the deviation approaches zero after approximately 15 s of the simulation. By using LQR-control system, the maximum

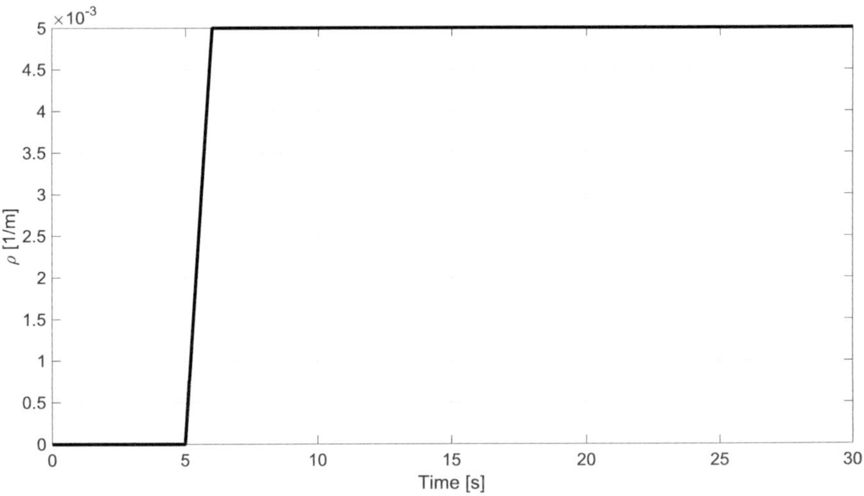

Figure 12.13 Simulation 1: Road curvature input.

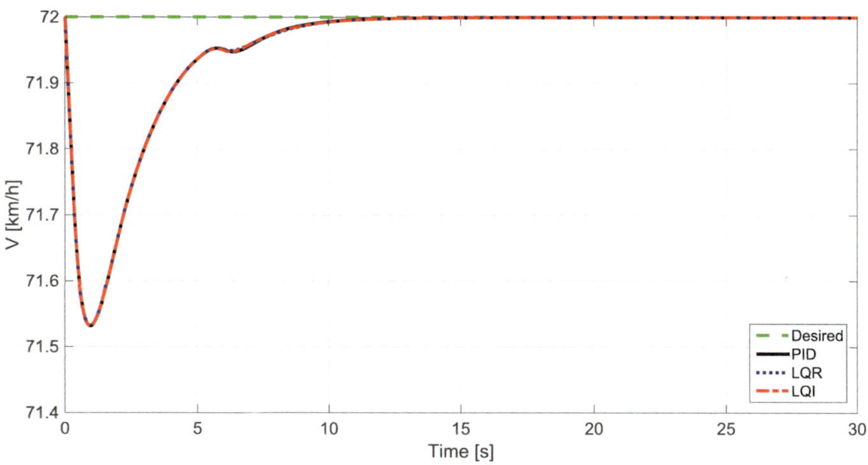

Figure 12.14 Simulation 1: Vehicle speeds.

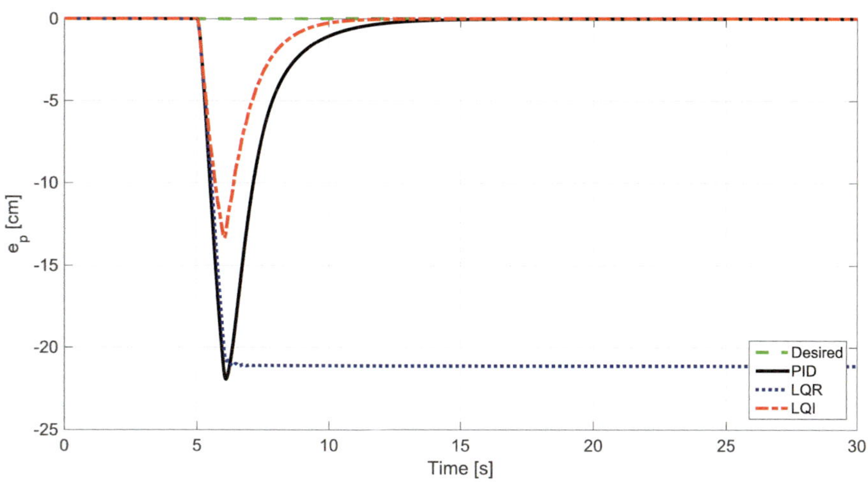

Figure 12.15 Simulation 1: Lateral deviations at the preview distance.

lateral deviation is reduced to 21.15 cm. However, there is a steady state error in LQR-controlled system due to the lack of integral component in the controller. In the case of LQI control system, there is no steady state error because of integral effect, and it approaches zero lateral deviation before the simulation time of 10 s. The maximum lateral deviation of 13.36 cm is obtained with LQI control system. The front wheel steering angle

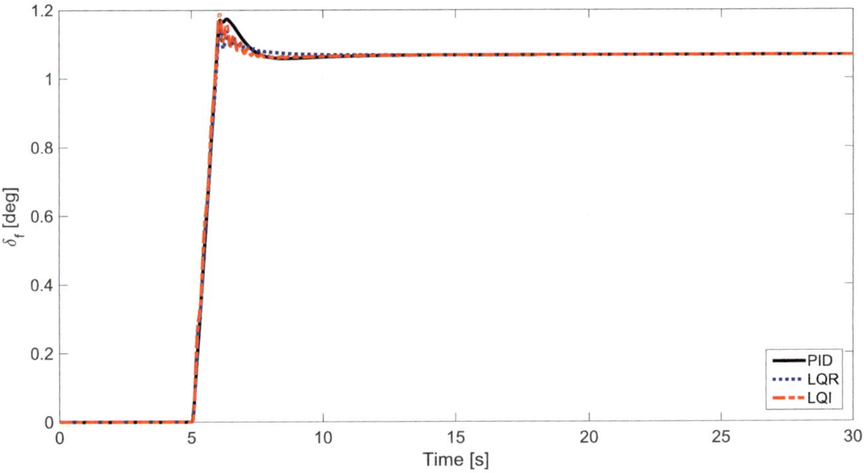

Figure 12.16 Simulation 1: Front wheel steering angles.

during simulation 1 is shown in Figure 12.16. The steering angle values are convenient with autonomous vehicle operation and physical realization.

In simulation 2, autonomous vehicle tries to track a varying road curvature including left and right turns with a varying speed profile containing acceleration and braking regions. Figure 12.17 shows the varying road curvature input for simulation 2. Also, the speed profile is given in Figure 12.18. Here,

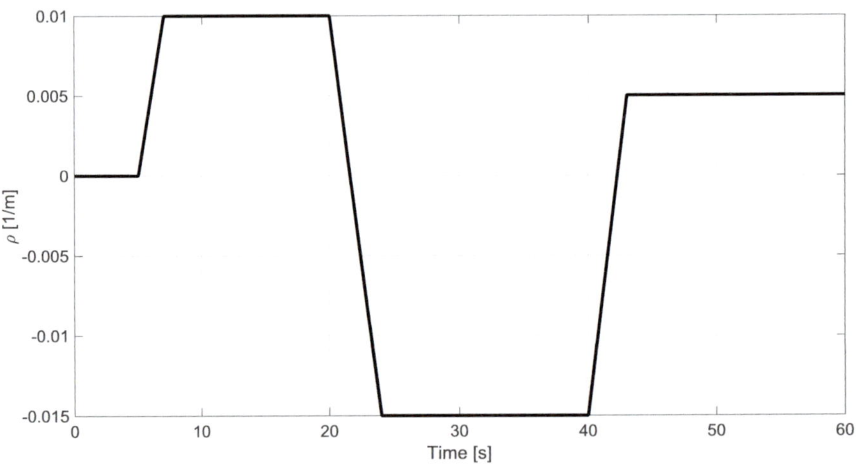

Figure 12.17 Simulation 2: Road curvature input.

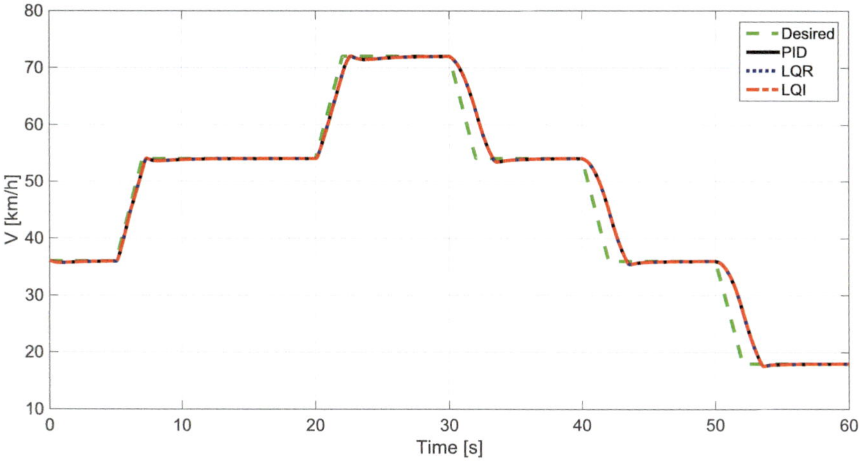

Figure 12.18 Simulation 2: Vehicle speeds.

the vehicle speed is taken as a varying profile in order to test the speed-dent gain scheduled control strategy. It is seen from Figure 12.18 that the speed control system in the simulation model is working satisfactorily and desired speed profile is followed by the autonomous vehicle successfully.

Figure 12.19 shows the lateral deviation values of three different control approach in simulation 2. In PID controlled system, there are sharp lateral deviation values around zero with a maximum deviation of 49.07 cm. PID

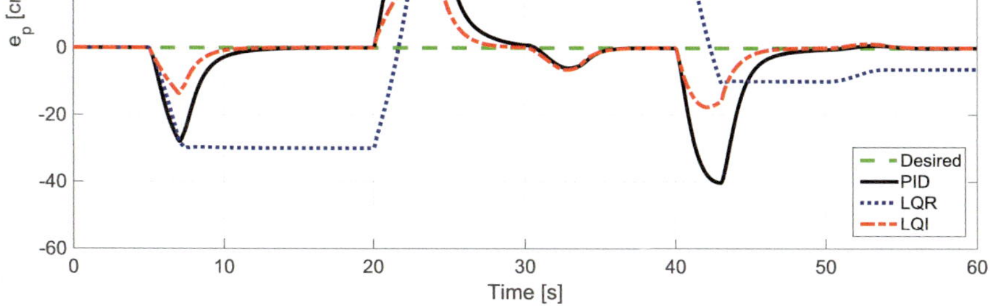

Figure 12.19 Simulation 2: Lateral deviations at the preview distance.

control system tries to obtain zero steady state error with integral effect but it shows large deviations around the zero. The maximum lateral deviation at the preview distance is obtained as 63.96 cm with LQR control. LQR-controlled system shows large steady state errors, and it cannot prevent this errors due to the absence of integral effect. By using LQI control system, the maximum lateral deviation is reduced to 26.31 cm. It converges to zero lateral tracking error without large deviations. It can be seen from both simulation studies that LQI path-following system shows better performance than LQR and PID control systems in different conditions.

Different kind of control methods for simulations 1 and 2 are compared numerically, and the results are given in Table 12.3. For numerical comparison, maximum absolute value of lateral deviation from the reference path at the preview distance ($\max|e_p|$), and root mean square error value of the same variable ($\text{RMSE}(e_p)$) are employed as statistical error values. $\text{RMSE}(e_p)$ is

Table 12.3 Numerical comparison of different controllers.

	$max\|e_p\|$ [cm]			$RMSE(e_p)$ [cm]		
	PID	LQR	LQI	PID	LQR	LQI
Simulation 1	21.9422	21.1487	13.3550	4.2258	18.9925	2.3963
Simulation 2	49.0694	63.9621	26.3092	14.2211	33.3663	6.9662
Average	35.5058	42.5554	19.8321	9.2234	26.1794	4.6813

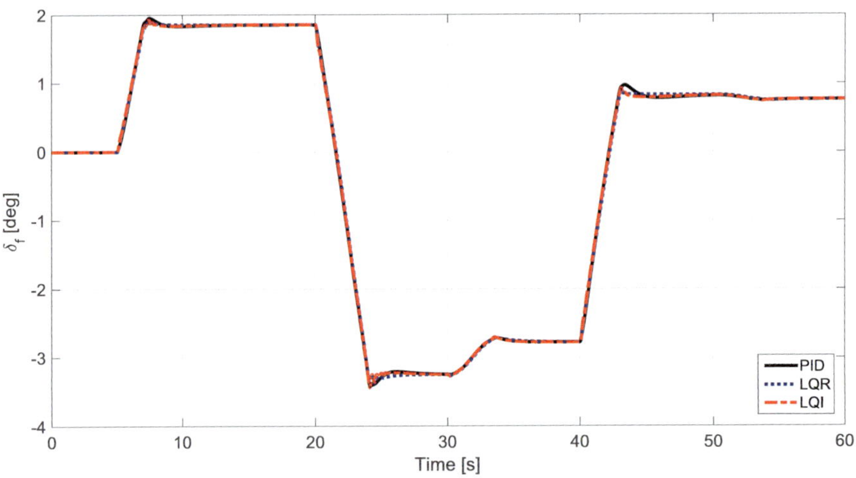

Figure 12.20 Simulation 2: Front wheel steering angles.

calculated as follows:

$$\mathrm{RMSE}(e_p) = \sqrt{\frac{\sum_{i=1}^{n}\left(e_p - e_{p,\mathrm{des}}\right)^2}{n}} \tag{12.43}$$

where n is the sample size.

It can be seen from Table 12.3, LQI controlled system has minimum error values for both indicators. The average performance of LQI control system is 53.40% better in terms of $\max|e_p|$ and 82.12% better in terms of $\mathrm{RMSE}(e_p)$ than LQR control system performance. It is also 44.14% better in terms of $\max|e_p|$ and 49.25% better in terms of $\mathrm{RMSE}(e_p)$ than PID control system performance.

12.6 Conclusion

In this chapter, different control systems are designed and compared with each other for autonomous vehicle path-following control. Linear path-following vehicle model is employed for designing PID, LQR, and LQI control systems. The effect of vehicle speed is considered in LQR and LQI control system design, and gain scheduled control methodology is followed. Proposed control systems are tested with simulations using seven degrees of freedom vehicle model including steer-by-wire electric motor actuator model. For comparison purpose, two different simulation studies are conducted. In simulation 1, constant curvature reference path is followed with a constant vehicle speed. In simulation 2, varying curvature reference path is tracked with a varying vehicle speed profile. A comparative analysis of simulation results is given. The maximum absolute value of the lateral deviation from the reference path at the preview distance and its root mean square error value are used as statistical error values. According to simulation results and numerical comparison, speed-dependent gain scheduled LQI control system is found to be more successful than PID and LQR control systems for autonomous path-following control.

References

[1] Levent Güvenç, Bilin Aksun Güvenç, and Mümin Tolga Emirler. Connected and autonomous vehicles. In Hwaiyu Geng, editor, *Internet of Things and Data Analytics Handbook*, chapter 35, pages 581–595. John Wiley & Sons, Ltd, 2017.

[2] Jürgen Ackermann and Wolfgang Sienel. Robust control for automatic steering. In *1990 American Control Conference*, pages 795–800. IEEE, 1990.

[3] Jürgen Guldner, Wolfgang Sienel, Han-Shue Tan, Jürgen Ackermann, Satyajit Patwardhan, and Tilman Bünte. Robust automatic steering control for look-down reference systems with front and rear sensors. *IEEE Transactions on Control Systems technology*, 7(1):2–11, 1999.

[4] Bilin Aksun Güvenç and Levent Güvenç. Robust two degree-of-freedom add-on controller design for automatic steering. *IEEE Transactions on Control Systems Technology*, 10(1):137–148, 2002.

[5] Ju Yong Choi. Robust controller for an autonomous vehicle with look-ahead and look-down information. *Journal of Mechanical Science and Technology*, 25(10):2467–2474, 2011.

[6] Christian Rathgeber, Franz Winkler, Dirk Odenthal, and Steffen Müller. Lateral trajectory tracking control for autonomous vehicles. In *2014 European Control Conference (ECC)*, pages 1024–1029. IEEE, 2014.

[7] Christian Rathgeber, Franz Winkler, Dirk Odenthal, and Steffen Müller. Disturbance observer for lateral trajectory tracking control for autonomous and cooperative driving. *International Journal of Mechanical and Mechatronics Engineering*, 9(6):921–928, 2015.

[8] Mümin Tolga Emirler, İsmail Meriç Can Uygan, Bilin Aksun Güvenç, and Levent Güvenç. Robust pid steering control in parameter space for highly automated driving. *International Journal of Vehicular Technology*, 2014, 2014.

[9] Mümin Tolga Emirler. *Advanced Control Systems for Ground Vehicles*. PhD thesis, İstanbul Technical University, İstanbul, Turkey, 2015.

[10] Mümin Tolga Emirler, Haoan Wang, Bilin Aksun Güvenç, and Levent Güvenç. Automated robust path following control based on calculation of lateral deviation and yaw angle error. In *Dynamic Systems and Control Conference*, volume 57267, page V003T50A009. American Society of Mechanical Engineers, 2015.

[11] Mümin Tolga Emirler, Haoan Wang, and Bilin Aksun Güvenç. Socially acceptable collision avoidance system for vulnerable road users. *IFAC-PapersOnLine*, 49(3):436–441, 2016.

[12] Haoan Wang, Antonio Tota, Bilin Aksun Güvenç, and Levent Güvenç. Real time implementation of socially acceptable collision avoidance of a low speed autonomous shuttle using the elastic band method. *Mechatronics*, 50:341–355, 2018.

[13] Kibeom Lee, Shengbo Eben Li, and Dongsuk Kum. Synthesis of robust lane keeping systems: impact of controller and design parameters on

system performance. *IEEE Transactions on Intelligent Transportation Systems*, 20(8):3129–3141, 2018.

[14] Mümin Tolga Emirler. Otonom taşıtlar için hıza bağlı kazanç uyarlamalı lqi tabanlı yol takip kontrol sistemi tasarımı. *Gazi University Journal of Science Part C: Design and Technology*, 7(4):855–868.

[15] Sheng Zhu, Şükrü Yaren Gelbal, Bilin Aksun Güvenç, and Levent Güvenç. Parameter-space based robust gain-scheduling design of automated vehicle lateral control. *IEEE Transactions on Vehicular Technology*, 68(10):9660–9671, 2019.

[16] Shaobing Xu and Huei Peng. Design, analysis, and experiments of preview path tracking control for autonomous vehicles. *IEEE Transactions on Intelligent Transportation Systems*, 21(1):48–58, 2020.

[17] Shaobing Xu, Huei Peng, and Yifan Tang. Preview path tracking control with delay compensation for autonomous vehicles. *IEEE Transactions on Intelligent Transportation Systems*, 22(5):2979–2989, 2020.

[18] Paul Riekert and Theo-Ernst Schunck. Zur fahrmechanik des gummibereiften kraftfahrzeugs. *Ingenieur-Archiv*, 11(3):210–224, 1940.

[19] Jürgen Ackermann, Paul Blue, Tilman Bünte, Levent Güvenc, Dieter Kaesbauer, Michael Kordt, Michael Muhler, and Dirk Odenthal. *Robust control: the parameter space approach*, volume 2. Springer, 2002.

[20] Uwe Kiencke and Lars Nielsen. *Automotive Control Systems: For Engine, Driveline, and Vehicle*. Springer Science & Business Media, 2005.

[21] Rajesh Rajamani. *Vehicle dynamics and control*. Springer Science & Business Media, 2011.

[22] Alena Kozáková and Mária Hypiusová. Lqg/ltr based reference tracking for a modular servo. *Journal of Electrical Systems and Information Technology*, 2(3):347–357, 2015.

[23] Yoshio Ebihara, Tomomichi Hagiwara, and Mituhiko Araki. Sequential tuning methods of lq/lqi controllers for multivariable systems and their application to hot strip mills. *International Journal of Control*, 73(15):1392–1404, 2000.

[24] Ebrahim Esmailzadeh, Avesta Goodarzi, and GR Vossoughi. Optimal yaw moment control law for improved vehicle handling. *Mechatronics*, 13(7):659–675, 2003.

[25] David Luong and Tsu-Chin Tsao. Linear quadratic integral control of an organic rankine cycle for waste heat recovery in heavy-duty diesel powertrain. In *2014 American control conference*, pages 3147–3152. IEEE, 2014.

[26] Andrew Phillips and Ferat Sahin. Optimal control of a twin rotor mimo system using lqr with integral action. In *2014 World Automation Congress (WAC)*, pages 114–119, 2014.

[27] BS Anjali, A Vivek, and JL Nandagopal. Simulation and analysis of integral lqr controller for inner control loop design of a fixed wing micro aerial vehicle (mav). *Procedia Technology*, 25:76–83, 2016.

[28] Jean Smith, Jinya Su, Cunjia Liu, and Wen-Hua Chen. Disturbance observer based control with anti-windup applied to a small fixed wing uav for disturbance rejection. *Journal of Intelligent & Robotic Systems*, 88(2):329–346, 2017.

[29] Vinicius F Dal Poggetto and Alberto L Serpa. Vehicle rollover avoidance by application of gain-scheduled lqr controllers using state observers. *Vehicle System Dynamics*, 54(2):191–209, 2016.

[30] Adam Owczarkowski and Dariusz Horla. Robust lqr and lqi control with actuator failure of a 2dof unmanned bicycle robot stabilized by an inertial wheel. *International Journal of Applied Mathematics and Computer Science*, 26(2):325–334, 2016.

[31] István Kisszölgyémi, Károly Beneda, and Zsolt Faltin. Linear quadratic integral (lqi) control for a small scale turbojet engine with variable exhaust nozzle. In *2017 International Conference on Military Technologies (ICMT)*, pages 507–513. IEEE, 2017.

[32] Zhengyuan Wang, Umberto Montanaro, Saber Fallah, Aldo Sorniotti, and Basilio Lenzo. A gain scheduled robust linear quadratic regulator for vehicle direct yaw moment control. *Mechatronics*, 51:31–45, 2018.

Authors' Biography

Mümin Tolga Emirler received B.S. degrees in mechanical engineering and manufacturing engineering in 2007 and 2008, the M.S. degree in mechatronics engineering in 2010, and the Ph.D. degree in mechanical engineering from Istanbul Technical University, Turkey, in 2014. He worked previously in the Ohio State University as a Postdoctoral Researcher and in Istanbul Okan University as an Assistant Professor. He is currently an Assistant Professor with Control and Automation Engineering, and Aviation Electrics and Electronics Departments, Yildiz Technical University, Istanbul, Turkey. His current research interests include control of automated and connected vehicles, vehicle dynamics control, and applied robust control.

13

Public Transport Travel Time Prediction using Sequential Forward Floating Selection Algorithm and Stacked Autoencoder

Makram Soui[1], Ines Gasmi[2], Nesrine Mansouri[2], Imen Ben Rjeb[2], Marouane Kessentini[1], and Karthic Srinivasin[3]

[1]School of Engineering and Computer Science Oakland University, USA
[2]Artificial Intelligence Research Unit, National School of Computer Sciences, University of Manouba, Tunisia
[3]College of Computing and Informatics, Saudi Electronic University, Saudi Arabia
E-mail: msoui@oakland.edu; kessentini@oakland.edu; ines.gasmi@ensi-uma.tn; nesrine.mansouri@isimg.tn; manelbenrejab@gmail.com; k.parimala@seu.edu.sa

Abstract

Travel time prediction plays a major role in intelligent transportation system (ITS). This time prediction is valuable for passengers to organize future trips to plan their schedules effectively. However, travel time prediction is difficult due to various factors such as weather conditions, traffic demand, and road accidents. In this work, we propose an approach based on deep learning to predict travel time. In this context, we used Sequential Forward Floating Selection algorithm (SFFS) to select the most relevant features, and then autoencoder algorithm was applied to build a model for travel time prediction. We assess our proposed model using the historical global positioning systems

(GPS) dataset of Dublin City in Ireland. The obtained results show that our proposed model provides an average of 90.5% accuracy compared to the other studied algorithms.

Keywords: Intelligent transportation systems, travel time prediction, SFFS, autoencoder, feature selection.

13.1 Introduction

Travel time prediction is essential for passengers to schedule their plans and choose an optimal itinerary for their journeys. Indeed, an accurate and reliable travel time estimation can minimize the uncertainty and the stress of the passengers [1], and also it allows them to select more comfortable travel trip [2]. The travel time prediction has a positive impact on the quality of public transport services in terms of reducing traffic and minimizing travel time [3].

In this context, travel time estimation is a big concern in the Intelligent Transportation System (ITS) field in order to provide new services to satisfy the customers [11]. The travel time prediction aims to motivate a passenger to select a suitable itinerary by taking into consideration many constraints such as price, travel time, and comfort of the trip. [5] and [6] underlined the importance of time attribute in passengers' travel life. The idea is to minimize the travel time as much as possible between the departure and destination points. However, the travel time can be affected by several factors such as vehicle speed, distance between stops, weekdays, weekends, etc. [21]. In addition, the authors [16] mentioned that peak and nonpeak hours change travel time during various times of the day.

Most of the existing works did not reveal the importance of attributes having an impact on the time prediction process. The authors [10] give equal importance to all attributes of the Dublin City travel dataset. However, some features are not relevant to predicting the time estimation of the travel and may affect the model's performance.

[25] mentioned that the irrelevant features in the dataset negatively affect the execution time. To this end, a random forest based on near neighbor (RFNN) approach is suggested to calculate the travel time of a bus between two adjacent stops by using GPS data. The obtained results are promising. However, the predictive model takes a high computation time. Similarly, the authors [10] applied a neural network algorithm to design a model that works based on original destination (OD) matrix to predict the travel time.

In this work, we propose an approach based on a deep learning algorithm for both feature selection and model construction to increase the accuracy rate of travel time prediction. To achieve this task, we aim to apply three feature selection algorithms that are considered the most widely used to choose the optimal subset of features: sequential forward selection (SFS), sequential backward selection (SBS), and sequential forward floating selection (SFFS).

SFS is a wrapper-based algorithm based on a bottom-up search procedure that allows generating an optimal subset of features [22]. Initially, it is empty and gradually adds features that are selected upon an evaluation criterion function, one by one, until reaching the best subset. In SBS [18], features are sequentially eliminated from a full dataset until the elimination of additional features raise the criterion. SFFS algorithm is based on forward and backward steps [18]. It aims to adjust the trade-off between these two steps. SFFS allows a "self-controlled backtracking" that can eventually find better solutions. SFFS algorithm as features selection technique to select the features which are most relevant.

The power of deep learning lies in its ability to learn multiple expressions of raw data layer-by-layer. Each layer is based on the expression features of the previous layer, extracting more abstract and more suitable for complex features and then doing some classification tasks. Stacked autoencoder (SAE) actually does such a thing [4]. An autoencoder is a type of neural network, which can be seen as consisting of two parts: an encoder function $h = f(x)$ and a decoder that generates reconstruction $r = g(h)$.

In this work, we applied the features selection algorithm with autoencoder to build a model for travel time prediction. The rest part of the paper is structured as follows: Section 13.2 deals with the related works. Our proposed approach is described in section 13.3. The experimental framework and validations are given in section 13.4. Finally, the conclusion and future directions are presented in section 13.5.

13.2 Related Works

Travel time prediction represents the time that a vehicle takes to travel from a source to a destination. In this context, various models have been proposed to estimate the travel time. The authors [17] introduced a historical trajectory-based time prediction (HTTP) method to forecast travel time. This framework can make a real-time estimation of a journey duration using large data collected from historical trajectories of an active bus journey. It first samples the data, then explores variant estimation methods to trim the sample set of identical trajectories, and finally returns the average driving time.

The authors [23] focus on an autoregressive integrated moving average model (ARIMA), which combines traffic data from neighboring links to predict short-term travel time when an incident happens. The ARIMA model analyzed the spatiotemporal data by using prewhitening cross-correlation functions. The model was developed based on the city of Charlotte, North Carolina, and includes 181 "vehicle accident" type incidents. The obtained results of the validation model show that the predicted incidents of travel time are near to the real-time observations.

The authors [12] proposed an approach that aims to combine the support vector regression (SVR) algorithm and immune genetic algorithm (IGA) to create a model. The SVR is used primarily for travel time estimation, and IGA is used for optimizing the parameters in comparison with multivariate linear regression. This model was built based on traffic data from Langfang City, China, and performed well as compared to the historical averages method, multi-linear regression, and back propagation neural networks.

The authors [20] proposed a modified K-means clustering approach (MKC) for the estimation of travel time by discovering hidden knowledge from a collection of historical data that are divided into a group of clusters. These clusters are based on the frequency of travel time, duration of travel time and speed for a given road segment. This model is compared to the other three methods such as naïve Bayesian classification (NBC), chain average (CA), and successive moving average (SMA), and the results are promising.

The authors [8] proposed a model for travel time estimation for road-based mass transit systems. This approach calculates travel time by considering historical behavior from historical profiles and the recent behavior recorded on the public transportation system for which the estimation is to be made. This model was built by k-medoids clustering algorithm to derive historical travel time information. This model particularly focused on bus travel time and did not consider other vehicles' travel time. The obtained results of the model provide significant improvement in travel time prediction.

The authors [13] recommended an approach to estimate the travel time using artificial neural networks (ANN) and global positioning systems (GPS). In this work, the ANN classifier was first trained offline. After that, the GPS provides the necessary data to predict real-time travel information to the model. This model is based on examining both the real-time and historical patterns of arrival and departure times at various stops in the chosen route. The neural network (NN) model has one hidden layer with 15 neurons and was trained based on the Levenberg–Marquardt algorithm. The outcome demonstrated that the model makes better predictions 70% of the time.

The authors [9] applied an alternate neural network approach called long short-term memory (LSTM) to resolve the travel time estimation issue. The issue has been translated into a time series, and the LSTM model was utilized to predict travel time. The LSTM can automatically store historical sequences that help make exact travel time expectations. In addition, LSTM is viewed as a specific sort of recurrent neural network, where the hidden layer of the NN in various time sequences are associated with one another. Each hidden layer has an LSTM cell with various gate types, in particular: input gate, forget entryway, and yield entryway.

Furthermore, the authors [26] developed a mixed neural network model for travel time estimation. The model was made as a combination of a two-dimensional convolution neural network (CNN) and LSTM. The idea was motivated by the dominant performance of both CNN and LSTM. The proposed model was trained using an RMSprop optimizer and Relu activation function. The proposed model is used to forecast the running time on a 35 km urban expressway in Beijing, China. The forecasting accuracy was then compared with other classical prediction models like naive K-nearest neighbor (KNN), instantaneous travel time, and historical average travel time. The outcome indicated that the model outperformed the classical algorithms during peak and nonpeak hours. However, the model made most-accurate predictions when making short-term travel time forecasts.

13.3 Proposed Approach

Our proposed approach aims to predict the travel time using a deep learning model based on sequential forward floating selection algorithm (SFFS) and stacked autoencoder classifier. It includes three phases: (1) filtering phase, which consists of selecting the frequently visited journeys and ignoring the rarely visited ones, (2) feature selection phase which consists of selecting the relevant features of the studied dataset, and (3) the classification phase, which is used to create a deep neural network model for predicting travel time of journey. Figure 13.1 shows the overview of the proposed approach.

The filtering phase is mainly used to select the most frequently visited journeys. The idea is to keep only the journeys (instances) having high occurrences. Then, the SFFS algorithm is applied to select only the relevant features that will be used to create a model. This phase takes all the features of the dataset as input and generates only the most important ones. The aim is

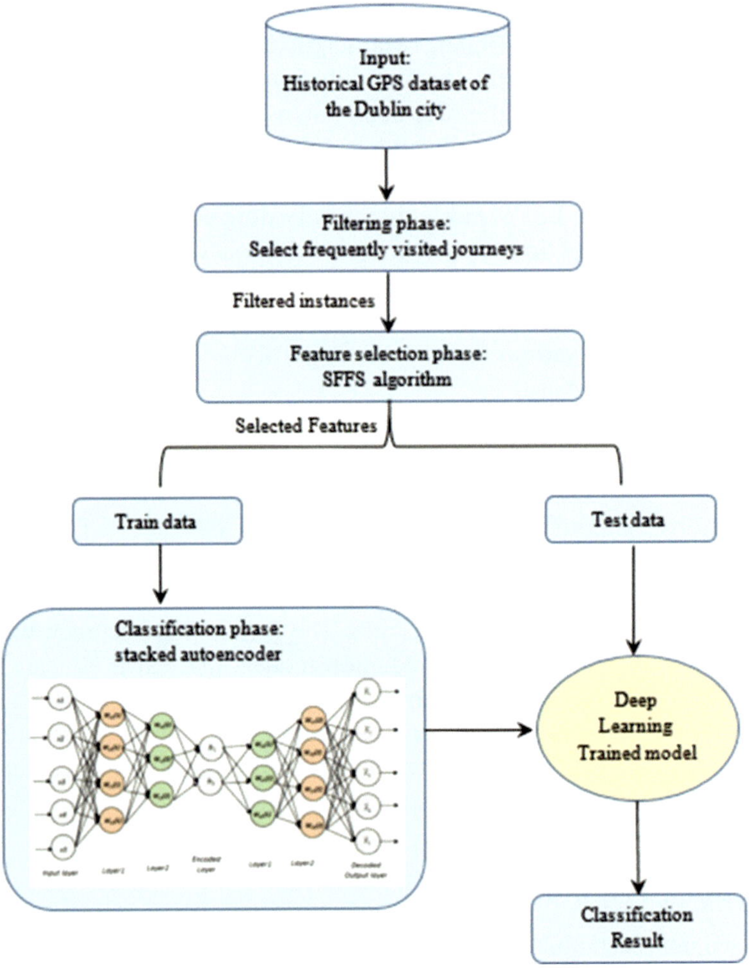

Figure 13.1 Proposed approach overview.

to reduce the high dimensionality of the dataset. The SFFS algorithm has an additional exclusion or inclusion step once it is excluded or included so that a large number of feature subsets combinations can be sampled.

As shown in Algorithm 13.1, the SFFS algorithm [18] takes several features as input and produces number of selected subfeatures (k) as output. The SFFS algorithm initially begins with an empty set. For every forward step, whenever an objective function increases, the SFFS algorithm executes backward steps. Line (3) aims at initializing the algorithm with an empty set X

and the number of selected features $k = 0$. The algorithm involves two steps: The first one is the inclusion step, shown in line (4) to line (7), which use the simple SFS method to retrieve the most significant features with respect to X and include it in the travel subset X_k. Step 1 stops if k features have been selected; otherwise, go to step 2. The second step is the conditional exclusion step, shown from line (8) to line (12), which is used to find the least significant feature k in X. If the feature is added, then retain it and go back to step 1. Otherwise, if the condition is false, exclude feature k. Keep in mind that X is now slightly better than it was before in step 1. Line (15) indicates the termination step, which stops the algorithm when k is equal to the number of desired travel features.

Algorithm.1: Sequential Forward Floating selection (SFFS)

1.Input: travel features, dataset
2.Output: selected travel features
3. Initialization: $\boldsymbol{X_0} = \emptyset$, k=0
4. Step 1(inclusion):
5. $\boldsymbol{x}^+$ = arg max j($\boldsymbol{x_k}$+x), where x ∈ Y-$\boldsymbol{X_k}$
6. $\boldsymbol{X_{k+1}} = \boldsymbol{X_k} + \boldsymbol{x}^+$
7. K=k+1
8. Step 2 (conditional exclusion):
9. $\boldsymbol{x}^-$ = arg max j($\boldsymbol{x_k}$- x), where x ∈ $\boldsymbol{X_k}$
10.If j($\boldsymbol{x_k}$- x)> j($\boldsymbol{x_k}$- x):
11. $\boldsymbol{X_{k-1}} = \boldsymbol{X_k} - \boldsymbol{x}^-$
12. K=k-1
13. Go to step 1
14. Step 1 and 2 are repeated until the termination is reached
15. Termination: stop when k equals the number of desired features

---

After the feature selection phase, the classification phase is executed. In this phase, we used an autoencoder, an unsupervised artificial neural network (ANN) type. In our work, we used it to estimate the travel time by providing travel data. Autoencoder includes three layers: input layer, hidden layers, and output layer. First, the input layer receives travel data as input and passes it to the hidden layers. Next, one or more hidden layers are present to process and encode the data.

An autoencoder is an underlying unit of stacked autoencoders. In this work, we built a stacked autoencoder (SAE) with three hidden layers and a softmax layer. It is a deep learning model which includes three layers of sparse autoencoders where the output of each layer acts as an input of the successive layer. Furthermore, it groups two or more autoencoders and uses the greedy layer-wise for classification.

In a stacked autoencoder, encoding and decoding are implemented in a stack of hidden layers. More hidden layers help the encoders to learn better

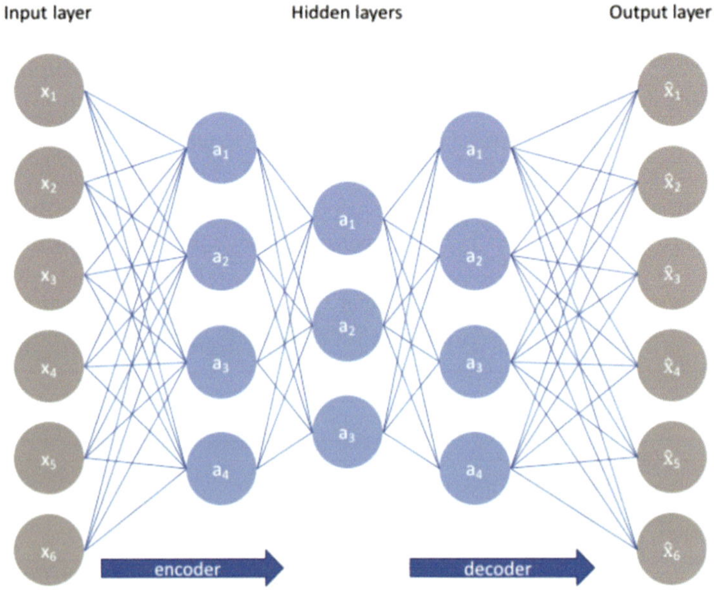

Figure 13.2 Structure of stacked autoencoder.

and understand the data. The encoder compresses and generates the code, while the decoder decompresses and reconstructs the data from the code. The structure of the stacked autoencoder with three hidden layers is shown in Figure 13.2. The SAE model is generated by stacking autoencoders forming a deep neural network by taking the output of the trained autoencoder found in the previous layer as the input of the present layer [3].

After training the first hidden layer number, the output of the hidden layer number k is utilized as input to the hidden layer number $k + 1$. Each autoencoder in SAE is first trained based on the backpropagation (BP) algorithm, which enables after each forward pass through a network to adjust the model's parameters. After that, the parameters of the autoencoders are updated. Next, in the layer-wise training, the stacked autoencoder parameters can help to find the local optimum values. The training procedure was inspired from studies in [3, 4, 14].

The encoder receives travel data as input, compresses it, and generates the code; on the other hand, the decoder decompresses and reconstructs the input. This process is based on eqn (13.1) and (13.2)

$$y = s(W_x + b) \tag{13.1}$$

$$z = s(W_{ty} + b'), \tag{13.2}$$

where W and W_t represent the weight matrices; b and b' corresponds to bias vectors; x is the input layer, and y is named as a latent variable that is the input; s is the nonlinearity function. In this work, it is considered as the sigmoid function. Finally, z is the output prediction.

Typically, autoencoders are stacked, which means the output of one hidden layer will become the input of the next autoencoder. The main aim of the autoencoder is to reduce the variation between the input and output according to the given input vectors. To calculate the reconstruction error, the cross entropy function is used, as shown in eqn (13.3), where x_k and z_k signifies the kth element of x and z, respectively

$$L(x, z) = -\sum [x_k ln\ z_k + (1 - x_k) ln\ (1 - Z_k)_{k=1}^{d} \tag{13.3}$$

The gradient descent method is used to reduce the reconstruction error. The weights in eqn (13.1) and (13.2) can be updated based on eqn (13.4)–(13.6), where α represents the learning rate

$$W = W - \alpha \frac{\partial L(x, z)}{\partial W} \tag{13.4}$$

$$b = b - \alpha \frac{\partial L(x, z)}{\partial b} \tag{13.5}$$

$$b' = b' - \alpha \frac{\partial L(x, z)}{\partial b'} \tag{13.6}$$

Initially, the stacked autoencoder is used for the training phase after extracting the relevant features from the dataset. Next, for the classification phase, a softmax classifier is used as a last layer to predict the travel time based on the softmax activation function for our deep learning model. The softmax provides the probability that the current input data belongs to class K as output. The weight matrix among the hidden units of the top layer and the softmax layer (W_{ij}) is determined by reducing the cost function shown in eqn (13.7):

$$J_{\text{softmax}} = -\frac{1}{N} \sum_{n=1}^{N} \sum_{i=1}^{k} y_i^{(n)} \log\left(\tilde{y}_i^{(n)}\right) + \left(1 - y_i^{(n)}\right) \log\left(1 - \tilde{y}_i^{(n)}\right) + \frac{\lambda}{2} \sum_i \sum_j W_{ij}^2, \tag{13.7}$$

where N represents the number of training samples in the small-batch, y_i represents the prediction class i of the input data, and assign y_i to 1, if the input data is related to class i, else assign it to 0. To enhance the performance of our SAE, we fine-tune our model based on backpropagation algorithm for all hidden layers. The output of the softmax units is defined using eqn (13.8):

$$y_i = \frac{\exp \sum_j W_{ij} x_j}{\sum_{i=1}^{k} \exp \sum_j W_{ij} x_j}, \tag{13.8}$$

where k is the number of classes. Similarly, the deep network can be used for a regression task if a regressor is placed on top layer hidden units.

13.4 Experimental Framework and Validation

13.4.1 Dataset description

Our work is based on Dublin's dataset, which is GPS data collected from Dublin City Traffic Control System. The dataset contains 1400 instances and 16 attributes. The GPS information was sampled in intervals of 5–300 s (20 s on average). Line 046A was chosen for its frequent and lengthy journey pattern. Furthermore, the route passes through Dublin City center, where the blue markers correspond to bus stops observed along the route, the green

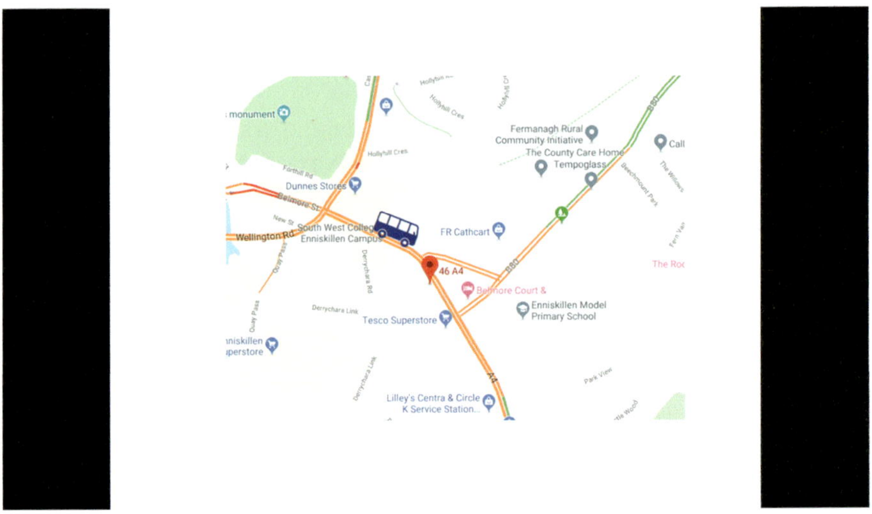

Figure 13.3 Route for bus line 46A.

Table 13.1 Description of the dataset.

No.	Feature	Description	Type	Example of value
1	Timestamp	The time the data was observed	Numerical	1353249926
2	Time	The converted time for the timestamp	Numerical	14:45:26
3	Line ID	The ID of the line that the bus travels	Numerical	46
4	Journey pattern ID	The sequence of bus stop journey	Categorical	046A1001
5	Time frame	Date of the production of the timetable	Date	November 18, 2012
6	Vehicle journey ID	A run on the journey pattern	Numerical	16220
7	Operator	The operator, not the bus's driver	Categorical	D1
8	Longitude	Position of the bus in terms of longitude	Numerical	−6.2315
9	Latitude	Position of the bus in terms of latitude	Numerical	53.317268
10	distance	Distance between the last stop and the current stop	Numerical	0.135459
11	Speed	Speed of the bus between the last stop and the current stop	Numerical	30.478241
12	Delay	Gives negative when the bus is ahead of schedule	numerical	424
13	Block ID	A section ID of the journey pattern	Numerical	46010
14	Vehicle ID	An identifier of the bus	Numerical	43001
15	Stop ID	Defines the bus stop closest to a bus	Numerical	775
16	At_Stop	Specifies if the bus is at a particular stop or not	Binary	0 or 1

marker corresponds to the first station Crofton Road, Railway Station, and the red marker corresponds to the last stop Infirmary Rd, Phoenix Pk Gate as shown in Figure 13.3. The description of the dataset is given in Table 13.1.

13.4.2 Validation

In this validation part, we are going to answer the three following questions:
RQ1. To what extent can the feature selection technique improve the efficiency of our prediction model?

RQ2. How does our proposed approach perform, when compared to similar existing works?

To answer RQ1, we assess the effectiveness of the studied algorithms: autoencoder, K-nearest neighbor (KNN), decision tree, and support vector machine (SVM) in terms of accuracy, AUC, precision, recall, and *F*-score indicators on full dataset and filtered dataset, respectively.

The accuracy metric corresponds to the percentage of test set tuples that are correctly predicted cases on a dataset defined

$$\text{Accuracy} = \frac{\text{TP} + \text{TN}}{\text{TP+FP} + \text{TN} + \text{FN}} \tag{13.9}$$

$$\text{AUC} = \frac{\text{Sensitivity} + \text{Specificity}}{2}, \tag{13.10}$$

where sensitivity is the percentage of correctly predicted travel time

$$\text{Sensitivity} = \frac{\text{TP}}{\text{TP} + \text{FN}}, \tag{13.11}$$

whereas specificity represents the percentage of incorrectly predicted travel time.

$$\text{Specificity} = \frac{\text{TN}}{\text{TN} + \text{FP}} \tag{13.12}$$

The precision is the percentage of tuples that the classifier labeled as positive and are actually positive. It represents the ratio of true positives (number of correct predictions of travel time) and the total of false positives (number of incorrect predictions of travel time) and true positives (number of correct predictions of travel time)

$$\text{Precision}{=}\frac{\text{TP}}{\text{TP+FP}} \tag{13.13}$$

Recall is the percentage of positive tuples detected by the classifier as positive. As shown in eqn (13.14), recall is the ratio of TP (number of correct predictions of travel time) over the total number of predictions

$$\text{Recall}{=}\frac{\text{TP}}{\text{TP+FN}} \tag{13.14}$$

Finally, we used the *F*-score indicator, which measures the average between the sensitivity and the specificity

$$\text{F}-\text{score}{=}\ 2*\frac{\text{sensitivity} * \text{specificity}}{\text{sensitivity+specificity}} \tag{13.15}$$

Table 13.2 Experimental results with full dataset.

Algorithm	Accuracy (%)	Precision (%)	Recall (%)	*F*-score (%)
SVM	72.25	78.45	53.52	63.11
KNN	71.75	76.65	51.82	61.03
Decision tree	76.50	86.50	**76.50**	**80.69**
Autoencoder	**77.50**	**89.25**	75.75	71.34

To answer RQ2, we compare the obtained results of the best algorithm with other similar studies that applied the ANN machine learning algorithm. The comparison is based on two evaluation criteria such as accuracy and root mean square error (RMSE) indicators.

13.4.2.1 Results for research question 1

In our research study, experiments are conducted on dataset of Dublin City Traffic Control System, in which SFS, SBS, and SFFS algorithms are used to select the optimal subset of features for travel time prediction. We respectively present the results of the studied algorithms. In this experiment, we train the studied classifiers using the full and filtered datasets. For both experiments, the dataset was divided into training set (70%) and testing set (30%). In this experiment, we applied four studied algorithms: autoencoder, SVM, KNN, and decision tree.

13.4.2.1.1 Experimental results with full dataset

In this part, we present an empirical evaluation by using full features of the dataset. The experimental results in Table 13.2 show the performance of the four applied algorithms. The achieved results are based on the evaluation metrics of accuracy, precision, recall, and F-score.

As reports in Table 13.2, the decision tree and autoencoder outperform the remaining used algorithms for all 16 features. The autoencoder provides an accuracy of 77% and a precision of 89%. The decision tree provides a recall of 76% and an *F*-score of 80.69%.

13.4.2.1.2 Experimental results with filtered dataset

To select the features in the dataset, we used SFS, SBS, and SFFS algorithms and managed to get the most suitable subset of features for travel prediction. All the achieved results are shown in Table 13.3. As reported in Table 13.3, the performance of the majority of algorithms is enhanced due to the feature selection algorithms application.

Table 13.3 Experimental results of studied feature selection algorithms.

Feature selection + algorithm	Accuracy (%)	Precision (%)	Recall (%)	*F*-score (%)
SFS+SVM	82.35	83.48	74.82	63.11
SFS+KNN	75.41	77.52	70.21	61.03
SFS+decision tree	81.56	87.90	75.30	80.69
SFS+autoencoder	85.82	88.25	80.50	81.34
SBS+SVM	83.58	81.65	75.38	63.11
SBS+KNN	76.10	76.55	71.40	61.03
SBS+decision tree	85.34	86.75	76.28	80.69
SBS+autoencoder	86.29	89.80	81.55	80.72
SFFS+SVM	86.61	90.15	76.85	81.81
SFFS+KNN	79.93	88.50	74.65	82.69
SFFS+decision tree	88.56	89.25	78.30	83.63
SFFS+autoencoder	**90.50**	**91.75**	**83.25**	**85.14**

As shown in Table 13.3, feature selection algorithms combined with studied machine learning algorithms such as SVM, KNN, decision tree, and autoencoder. The obtained results show that autoencoder performs better than other algorithms when combined with feature selection algorithms. The first experiment was based on the SFS feature selection algorithm. In this case, SFS+autoencoder provides the best results in terms of accuracy (85.82%), precision (88.25%), recall (8050%), and *F*-score (81.34%). The second experiment is based on the SBS feature selection algorithm. In this case, SBS+autoencoder provides the best results in terms of accuracy (86.29%), precision (89.80%), recall (81.55%), and F-score (80.72%). The third experiment is based on the SFFS feature selection algorithm. In this case, SFFS+autoencoder provides the best results in terms of accuracy (90.50%), precision (91.75%), recall (83.25%), and *F*-score (85.14%). Hence, the SFFS+autoencoder model outperforms the remaining models in terms of accuracy, precision, recall, and *F*-score.

SFFS+autoencoder proved its efficiency compared to the other evaluated models. To statistically confirm the hypothesis, a paired t-test is used to evaluate the importance of the obtained results. The null hypothesis H0 considers that there is no statistically significant difference between the two models means "The accuracy of model A = accuracy of model B." H0 is complimented by the alternate hypothesis H1 which deems that there is a statistically significant difference between the two models means "accuracy of model A $\neq$ accuracy of model B." From Table 13.4, it is observed that the p-value is less than 0.05 for all the cases. Therefore, the null hypothesis is rejected, and it is

Table 13.4 Significant test results of paired t-test ($\alpha = 0.05$).

Model A	Model B	t-test	p-value
SFFS+autoencoder	SFS+SVM	4.764	0.0043
	SFS+KNN	3.154	0.0026
	SFS+decision tree	14.362	1.04E−06
	SFS+autoencoder	5.374	0.0024
	SBS+SVM	14.354	2.53E−07
	SBS+KNN	11.523	1.30E−06
	SBS+ decision tree	5.213	0.0034
	SBS+autoencoder	4.350	0.0037
	SFFS+SVM	4.768	0.0004
	SFFS+KNN	3.967	0.0043
	SFFS+ decision tree	8.989	3.45E−07

wrapped up that there is a statistically significant improvement in predicting the travel time using SFFS feature selection with the autoencoder algorithm. Therefore, it is drawn to conclude that SFFS+Autoencoder has outstanding performance in travel time prediction.

13.4.2.2 Results for research question 2

In this experiment, we compare our proposal with other existing approaches ANN model [10] that used the Dublin City travel based on the GPS dataset. The conventional ANN attempts to predict a target variable y from the input features x. Conventional ANN is an example of a feed-forward neural network. However, in this type of network, there are no feedback connections (backpropagation), where the output of the model is feedback into itself. Typically, backpropagation is not supported by a feed-forward neural network to minimize the error. Autoencoder is a type of neural network which supports backpropagation to minimize error. The autoencoder aims to learn compressed, distributed representation for the given data for dimensionality reduction.

In our work, in each loop of execution, every hidden layer of the autoencoder attempts to increase the quality of the output data. In addition, the filtered dataset derived by the SFFS algorithm contains the most important features to help predict the model. Table 13.5 compares the proposed model with the ANN model. As shown in Table 13.3, our proposed model provides the best result with an average of 90.5% accuracy.

In this chapter, we configured two important parameters to boost our model such as the number of hidden layers and the number of units in each hidden layer. In our experiments, we select 1 to 3 hidden layers in the deep

Table 13.5 Comparison of the best solution with ANN.

Compared models	Accuracy without feature selection (%)	Accuracy with feature selection (%)
Proposed model	77.5	90.5
ANN model	70	79

network. The evaluation results of the Dublin City travel datasets proved that the SAE has high accuracy when the count of hidden layers is three. At this level, the SAE is applied with three hidden layers. After that, we adjusted the number of neurons (units) per each hidden layer to experiment again. The obtained results confirm that autoencoder achieves the best accuracy when the number of neurons in the first, second, and third hidden layers is 14, 12, and 9, respectively. Sparse Regularization is an essential parameter that supervises the regularization value of an objective function. L2 regularization aims to reduce the overfitting. In this work, we set 0.02 as the L2 regularization value and set the Sparse Regularization value is set to 3. The autoencoder is trained with fewer neurons in the hidden layer than in the input layer. The selected features are sufficient to build the model. This paper uses a layer-by-layer greedy training method in the SAE model. This technique gave very encouraging results in terms of AUC.

In addition, the efficiency of the proposed model is evaluated based on root mean square error (RMSE). The aim is to assess how efficiently our proposed model predicts travel time. To achieve this, predicted travel time ($tt_{a,b}$) between two stops a and b of our proposed model is compared against the actual travel time ($TT_{a,b}$) as recorded in the dataset

$$\text{RMSE} = \sqrt{\frac{\sum_{i=1}^{n}\left(tt^{i}_{a,b} - TT^{i}_{a,b}\right)^2}{n}}, \tag{13.16}$$

where $tt^{i}_{a,b}$ is the predicted travel time from stop a to stop b for a journey i in the dataset; $TT^{i}_{a,b}$ is the actual travel time (ground truth) from stop a to stop b for a journey i in the dataset, and n is the total number of links in the journey.

The experiment involves querying a particular day of the week and predicting of travel time of the journey. Figure 13.4 represents the model's behavior when used to make a prediction of a journey outside peak hours. We found that the difference in travel time is minimal compared with actual travel time compared to an existing study of the artificial neural network (ANN)

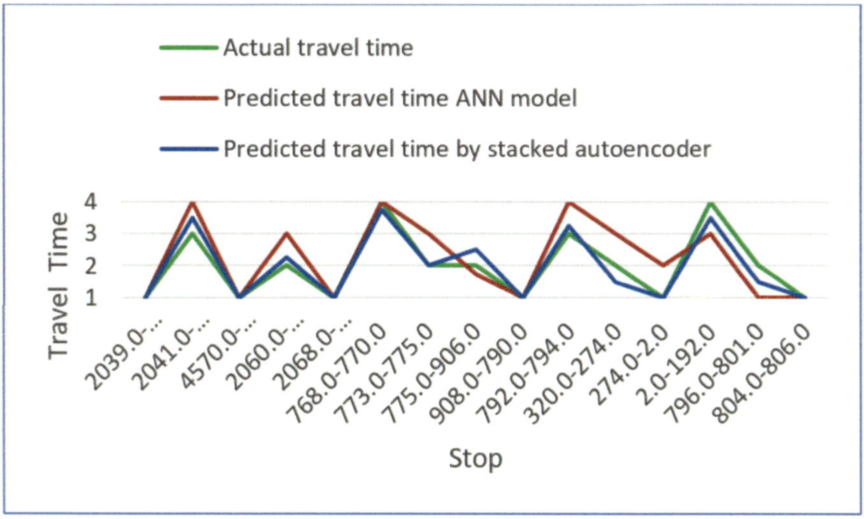

Figure 13.4 Predicted travel time for a journey outside peak hours.

model [10]. The proposed model performed well in time prediction by less than 2 min difference than the ANN model.

Figure 13.5 represents the model's behavior when used to predict a journey in peak hours, for instance, time at 16:00. We found that the proposed model was not much affected by peak hours and performed better than the ANN model.

Table 13.6 shows the RMSE rate value for the predicted trip travel time between the ANN model and the proposed model for a complete week. The proposed autoencoder shows a minimum RMSE value, such as 2.2 in nonpeak-AE and 2.6 in peak-AE. Thus, the proposed autoencoder with SFFS performs better than the ANN model.

Figure 13.6 compares the proposed model and the ANN model in nonpeak hours in all days of the week. The proposed model performed better than the ANN model with a lower error rate.

Figure 13.7 compares the proposed model and the ANN model during peak hours on all days of the week. The proposed model outperformed the ANN model, particularly on Friday, Saturday, and Sunday when more traffic is present. In addition, the error rate of the proposed model is less than the error rate of the ANN model.

We conclude from the results that our proposed autoencoder model performs better prediction than the existing study of the ANN model in both

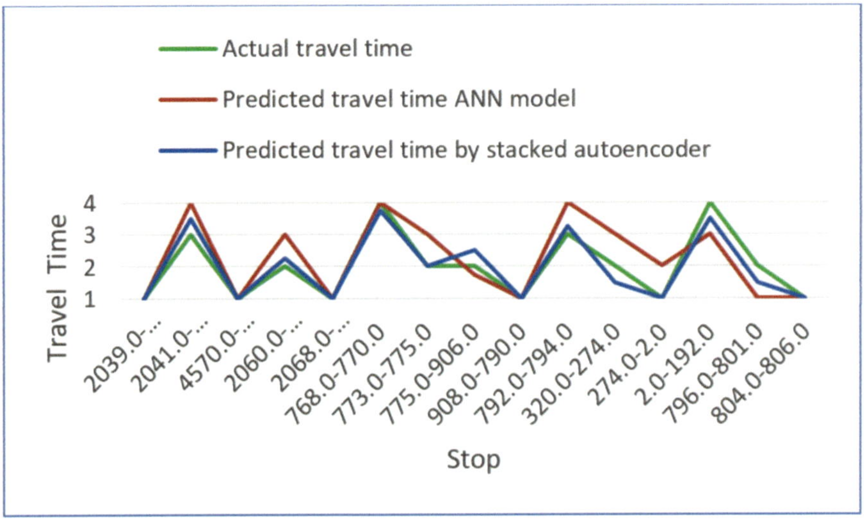

Figure 13.5 Predicted travel time inside peak hours.

Table 13.6 RMSE for predicted trip travel time.

Days	ANN		Autoencoder	
	Nonpeak-ANN	Peak-ANN	Nonpeak-AE	Peak-AE
Sunday	3.2	4.75	2.3	2.9
Monday	2.5	2.76	1.8	2.1
Tuesday	2.55	2.72	2.2	2.2
Wednesday	2.75	3.12	2.1	2.4
Thursday	2.44	3.24	1.9	2.7
Friday	3.01	4.9	2.7	3.2
Saturday	3.1	4.1	2.4	3
Mean of RMSE	2.8	3.7	2.2	2.6

peak and nonpeak time periods. Specifically, during peak periods, the average RMSE value of ANN is more than 3.5 on weekdays; particularly, on Sunday and Friday, the RMSE value is very near to 5. At the same time, the RMSE value of the proposed autoencoder model is almost less than 3 on all the days of the week. In addition, the prediction accuracy of our proposed encoder model is very near to the ground truth.

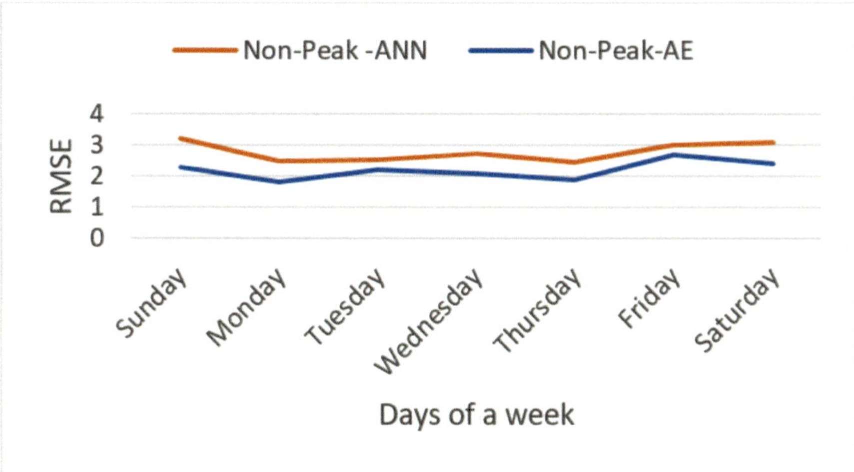

Figure 13.6 Comparison of observed RMSE between ANN and AE model for nonpeak hour periods.

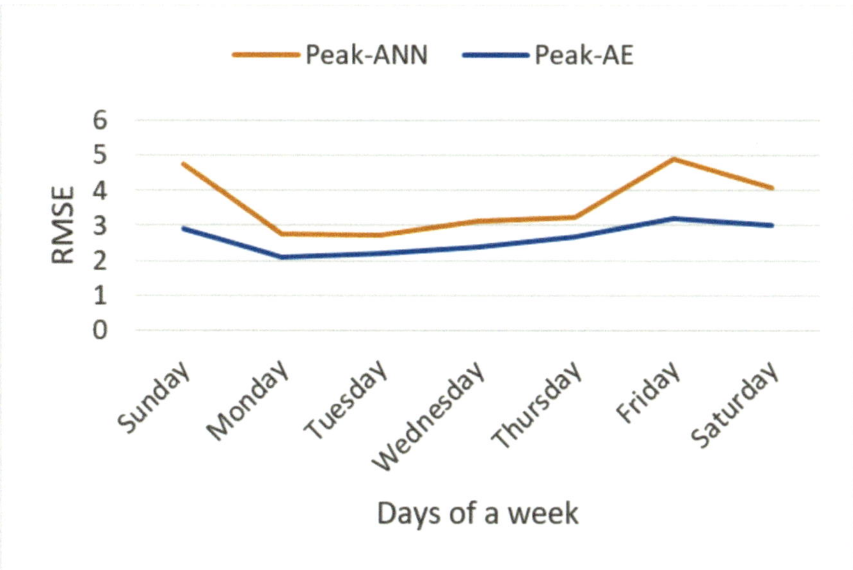

Figure 13.7 Comparison of observed RMSE between ANN and AE models for peak hour periods.

13.5 Conclusion and Future Work

Travel time prediction is an essential requirement for passengers [27]. The main goal of this paper is to propose a model for bus travel time prediction. To this end, we applied SVM, KNN, decision tree, and autoencoder algorithms with feature selection algorithms such as SFS, SBS, and SFFS. The predictive models were constructed and applied to the GPS dataset of route 46A Dublin City. The obtained results show that the efficiency of the proposed model (autoencoder+SFFS) has a 90.5% accuracy which outperformed the other studied algorithms. However, this paper has few limitations which future works can fix. As a future work, the proposed model can be improved by fine-tuning the hyperparameters. Second, the proposed model can be compared with other deep learning algorithms such as convolutional neural networks, deep belief networks, and recursive neural networks. Third, other factors can be investigated such as traffic light waiting, weather conditions, numbers of waiting passengers, and accident time periods. Developing another technique to forecast waiting time at transfer points will also be interesting.

References

[1] Adler, J. L., & Blue, V. J. (1998). Toward the design of intelligent traveler information systems. Transportation Research Part C: Emerging Technologies, 6(3), 157-172.

[2] Ben-Akiva, M., De Palma, A., & Isam, K. (1991). Dynamic network models and driver information systems. Transportation Research Part A: General, 25(5), 251-266.

[3] Bengio, Yoshua, et al. "Greedy layer-wise training of deep networks." Advances in neural information processing systems. 2007.

[4] Bose, T., Majumdar, A., & Chattopadhyay, T. (2018, April). Machine load estimation via stacked autoencoder regression. In 2018 IEEE International Conference on Acoustics, Speech and Signal Processing (ICASSP) (pp. 2126-2130). IEEE.

[5] Bouhana A., Soui, M., Abed, M, A Proposal of Personalized itinerary Search Methods in the Field of Transport, The 11th IFAC/IFIP/IFORS/IEA Symposium on Analysis, Design, and Evaluation of Human-Machine Systems. Valenciennes, France, - 2010.

[6] Soui M, Abed M., Kolski C., Ghédira K., Towards a Dynamic Evaluation of Personalized Information System: Application in Intelligent Transport, IEEE- First International Conference on Mobility,

Security and Logistics in Transport – Hammamet, Tunisia, May, 2011.

[7] Bovy, P. H., & Stern, E. (2012). Route Choice: Wayfinding in Transport Networks: Wayfinding in Transport Networks (Vol. 9). Springer Science & Business Media.

[8] Cristóbal, T., Padrón, G., Quesada-Arencibia, A., Alayón, F., de Blasio, G., & García, C. R. (2019). Bus travel time prediction model based on profile similarity. Sensors, 19(13), 2869.

[9] Duan, Yanjie, Yisheng Lv, and Fei-Yue Wang. "Travel time prediction with LSTM neural network." 2016 IEEE 19th International Conference on Intelligent Transportation Systems (ITSC). IEEE, 2016.

[10] Ephraim. A, "Link Travel Time Prediction Based onO-D Matrix and Neural Networks", Master's thesis, university of TARTU, 2018.

[11] Figueiredo, L., Jesus, I., Machado, J. T., Ferreira, J. R., & De Carvalho, J. M. (2001, August). Towards the development of intelligent transportation systems. In ITSC 2001. 2001 IEEE Intelligent Transportation Systems. Proceedings (Cat. No. 01TH8585) (pp. 1206-1211).

[12] Gao, P., Hu, J., Zhou, H., & Zhang, Y. (2016, June). Travel time prediction with immune genetic algorithm and support vector regression. In 2016 12th World Congress on Intelligent Control and Automation (WCICA) (pp. 987-992). IEEE.

[13] Gurmu, Z. K., & Fan, W. D. (2014). Artificial neural network travel time prediction model for buses using only GPS data. Journal of Public Transportation, 17(2), 3.

[14] Hinton, Geoffrey E., Simon Osindero, and Yee-Whye Teh. "A fast learning algorithm for deep belief nets." Neural computation 18.7 (2006): 1527-1554.

[15] Jain, A., & Zongker, D. (1997). Feature selection: Evaluation, application, and small sample performance. IEEE transactions on pattern analysis and machine intelligence, 19(2), 153-158.

[16] Lee, H., Chowdhury, N. K., & Chang, J. (2008, September). A new travel time prediction method for intelligent transportation systems. In International Conference on Knowledge-Based and Intelligent Information and Engineering Systems (pp. 473-483). Springer, Berlin, Heidelberg.

[17] Lee, Wang-Chien, et al. "HTTP: a new framework for bus travel time prediction based on historical trajectories." Proceedings of the 20th International Conference on Advances in Geographic Information Systems. ACM, 2012.

[18] Miao, J., Niu, L. (2016). A Survey on Feature Selection. Procedia Computer Science, 91(Itqm), 919926.

[19] Moussa, Soumaya, Makram Soui, and Mourad Abed. "User Profile and Multi-criteria Decision Making: Personalization of Traveller's Information in Public Transportation." Procedia Computer Science 22 (2013): 411-420.

[20] Nath, R. P. D., Lee, H. J., Chowdhury, N. K., & Chang, J. W. (2010). Modified K-means clustering for travel time prediction based on historical traffic data. In International Conference on Knowledge-Based and Intelligent Information and Engineering Systems (pp. 511-521). Springer, Berlin, Heidelberg. Cristóbal, Teresa, et al. "Bus travel time prediction model based on profile similarity." Sensors 19.13 (2019): 2869.

[21] Patnaik, J., Chien, S., & Bladikas, A. (2004). Estimation of bus arrival times using APC data. Journal of public transportation, 7(1), 1.

[22] Pohjalainen, J., Rsnen, O., Kadioglu, S. (2015). Feature selection methods and their combinations in high-dimensional classification of speaker likability, intelligibility and personality traits. Computer Speech and Language,29(1), 145171.

[23] Reza, R. M., Pulugurtha, S. S., & Duddu, V. R. (2015). ARIMA model for forecasting short-term travel time due to incidents in spatio-temporal context (No. 15-5553).

[24] Soui M , Abed M., Kolski C., Ghédira K., Towards a Dynamic Evaluation of Personalized Information System: Application in Intelligent Transport, IEEE- First International Conference on Mobility, Security and Logistics in Transport – Hammamet, Tunisia, May, 2011.

[25] Yu, B., Wang, H., Shan, W., & Yao, B. (2018). Prediction of bus travel time using random forests based on near neighbors. Computer-Aided Civil and Infrastructure Engineering, 33(4), 333-350.

[26] Zhang, Z., Chen, P., Wang, Y., & Yu, G. (2017, October). A hybrid deep learning approach for urban expressway travel time prediction considering spatial-temporal features. In 2017 IEEE 20th International Conference on Intelligent Transportation Systems (ITSC) (pp. 795-800).

[27] Soui, M., Kolski, C., Abed, M., & Uster, G. (2008, July). Evaluation of Personalized Information Systems: Application in Intelligent Transport System. In SEKE (pp. 877-880).

[28] Soui, M., Ghédira, K., & Hammadi, S. (2007). Proposal of personalized multimodal information diffusion system. In Proc. of the 1st Intern. Conf. on ICT & Accessibility, ICTA (Vol. 2007, pp. 219-224).

Authors' Biography

Makram Soui received a PhD in Computer Science from the Polytechnic University of Hauts-de-France, in 2010. He holds many academic certificates such as IBM Predictive Analytics Modeler, IBM Business Intelligence Analyst, Oracle Certified Professional JavaProgrammer, IBM Cloud Application Developer, Microsoft 98-367 Security fundamentals. He is currently a visiting professor in Oakland University. His responsibilities include: steering committee member for different branches of SEU, head of research group "Business Intelligence," teaching various courses as the medium of instruction such as artificial intelligence, machine learning, datamining, to undergraduate and postgraduate students, revising and developing courses plans. He published 10 referred journal papers and 24 conference papers with a low acceptance rate (between 22% and 34%).

Ines Gasmi obtained her PhD in computer science from the University of Manouba, Tunisia. She received her Master degree in 2015, from the higher institute of management of Gabes, Tunisia. She works in the LARIA lab. Her main interests are in human computer interface design, mobile user interfaces and software engineering methods. She published two referred journal paper and two conference papers.

Nesrine Mansouri received her Master degree in Computer Science from the higher Institute of computer science and Multimedia of Gabes, Tunisia in 2020. Her research areas include computer vision, machine learning, deep learning, data processing, semantic segmentation, object detection, GANs, and autonomous driving. She has been serving as a reviewer for many highly respected conferences. She published two referred journal paper and one conference papers.

Manel ben rajeb received her Master degree in computer science from the higher institute of management of Gabes, Tunisia in 2020. Her research areas include, machine learning, deep learning, intelligent transportation system, multi-agent system, and object detection. She has been serving as a reviewer for many highly respected conferences.

Marouane Kessentini is a recipient of the prestigious 2018 President of Tunisia distinguished research award, the University distinguished teaching award, the University distinguished digital education award, the College of Engineering and Computer Science distinguished research award, four best paper awards, and his AI-based software refactoring invention, licensed and deployed by industrial partners, is selected as one of the Top 8 inventions at the University of Michigan for 2018 (including the three campuses), among over 500 inventions, by the UM Technology Transfer Office. He is currently a Professor and Chair of Computer Science and Engineering Department at Oakland University. He received his Ph.D. from the University of Montreal in Canada in 2012. He received several grants from both industry and federal agencies and published over 110 papers in top journals and conferences. He has several collaborations with industry on the use of computational search, machine learning and evolutionary algorithms to address software engineering and services computing problems.

Karthik Srinivasan. Assitant Professor in College of Computing and Informatics Saudi Electronic University. He holds many academic certificates such as IBM Predictive Analytics Modeler, IBM Business Intelligence Analyst. She published five referred journal paper and five conference papers. Her research areas include computer vision, machine learning, deep learning, data processing, semantic segmentation, and object detection.

14

A Look Into the Ethics of Autonomous Vehicles Systems (AVS)

Jia Hui (Sherry) Feng[1] and Xiaoxin (Carol) Shen[2]

[1]Engineering, Computer & Mathematical Science, Auckland University of Technology, New Zealand
[2]Heinz College of Information Systems and Public Policy, Carnegie Mellon University, United States of America
E-mail: jiahui.feng@autuni.ac.nz, xiaoxin2@andrew.cmu.edu

Abstract

What choices should autonomous vehicle systems (AVS) make when they are faced in moral dilemma situations? Various research in both academic and industry has been conducted to construct an ethical framework for these autonomous systems. However, there are still ongoing discussions on the ethics and concerns of such systems. This chapter aims to explore and discuss in-depth of the current and advances in ethical issues and framework, both from an internal and external perspective of the AVS.

Keywords: Autonomous vehicle systems, ethics, privacy, policy, self-driving cars, morals, machine learning, artificial intelligence.

14.1 Introduction

Artificial intelligence (AI) has revolutionized the way machines can be applied to society. Compared to traditional machines, AI aims to perform tasks requiring humans by mimicking human intelligence. Nowadays, the impact of AI is evidently seen through many systems in industries such as healthcare, production, and many more (Stead, 2018). In recent years, the advancement of AI such as machine learning and neural networks has made many ideas imagined in history possible. One idea in particular is self-driving cars, also known as **autonomous vehicle systems (AVS)**.

Based on research, it seems that the first recorded idea of AVS dates to around 1478, when Leonardo Da Vinci sketched a self-propelled cart, with a control mechanism to take the vehicle through a predetermined course. From then until now, there have been many attempts and significant achievements in creating such a self-driving vehicle. In 1964 there was the Stanford AI Laboratory cart that could navigate through unseen environments, and then The Navlab1 from Carnegie Mellon University's Robotics Institute at the School of Science (Earnest, 2012; Thorpe et al., 1988). These notable achievements are illustrated in Figure 14.1.

Recently, the biggest players in the automotive industries such as Tesla, Ford Motors, General Motors, Toyota, BMW, Mercedes-Benz, Audi, and Volkswagen have introduced self-driving car models. Although some models are mixed with different levels of autonomy such as the "advanced driver-assist systems" (ADAS) that performs partial self-driving capabilities. Technology nowadays seems to be capable of building such a machine, and some believe that these vehicles will fill the roads *very soon*.

With the recent advancement of AVS, would you be able to ride in a fully self-driving vehicle? The American Automobile Association (AAA) conducted a survey in 2019 asking this question. They found that 71% of US drivers would be afraid to ride in such a vehicle (Edmonds, 2019). Many are still skeptical of such technology, and there are many reasons for this skepticism. Of the major reasons, there are two of the concerns stated in a separate survey by Othman (2021):

1. Ethical dilemma and challenges from the programming of the vehicle.
2. Liability and regulations and their implications for the public.

Based on the above, we have grouped the concerns into **internal and external ethics of AVS**, Internal being various ethical frameworks and moral decision-making of the AVS system itself, and external ethics being the ethics and dilemmas being impacted on society from such a system. This chapter is a survey piece that will explore the research on this topic.

14.2 AVS Internal Ethics

14.2.1 How do AVS make decisions?

Ethics and morals are what is considered "right" and "wrong" conduct. Moral judgments are social inferences that people make about each other. Socially,

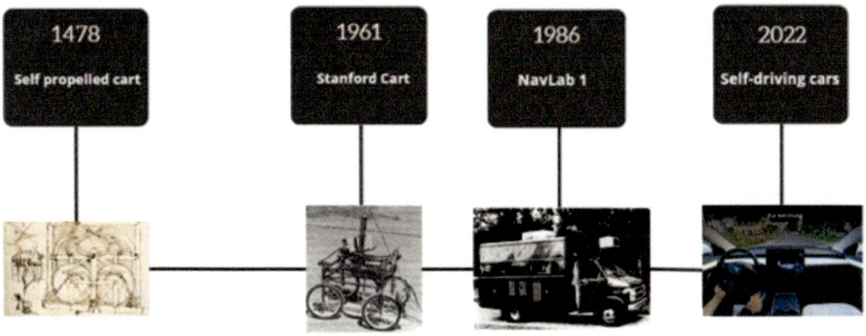

Figure 14.1 Some notable AVS achievements.

those who are deemed immoral are seen as untrustworthy and therefore be avoided (Forsyth, 2019). As we begin to build AI systems that are getting closer to human intelligence, there is a growing discussion of how ethical decision-making and moral judgments should be implemented into these systems, also known as the field of *machine ethics*. In terms of AVS, it has motivated the need to push for more research in machine ethics, since self-driving cars are a pressing and realistic concept. If such a system is already replacing a human driver, how do we trust that they make the right decisions? For the general public, it is more likely that we assume that human drivers should take responsibility and make the right decision when a critical situation occurs, since after all, putting arguable moral dilemmas aside, we as humans should have been taught in general what is ethical and unethical. But how are AVS taught? Most of the general public knows what it is like growing and living as a human, but not as many would understand how an AI system is built.

The method of decision-making of AVS is essentially data and information fed in beforehand, and also gathering and learning while it is on the road. The live data and information are gathered through sensors and cameras of the vehicle. Figure 14.2 is an image of Google's self-driving vehicle, Waymo (2022). Below is additional information of how the example vehicles work.

1. Sensor devices:
    (a) Lidar, a roof-mounted sensor, creates a 3D picture of the vehicle's surroundings
    (b) Cameras that take in a 360° view around the vehicle
    (c) Radar systems to calculate distances to obstacles

2. The “brain” of the Waymo driver, CPUs and GPUs takes in information from the sensors of the car, identifies different objects, and controls actions such as steering and brakes (Waymo, 2022)

The decision-making of the driving actions is quite straightforward, and it is technically feasible with today’s technology. However, the *ethical decision-making* is the tricky part. AVS internal systems are preprogrammed algorithms, therefore the decisions that impact human lives are, in a way, preprogrammed (Goodall, 2014). The concern of how AVS should behave ethically in unavoidable crashes is currently an ongoing discussion (Bonnefon et al., 2016) which will be discussed more in the next section, which explores further the **moral decision-making frameworks of AVS.**

We will begin by discussing the technical aspects of how AVS makes decisions and what is required, then we will explore how ethics are programmed algorithmically. We will then see the complexity of the decision-making of AVS and visit various ideas proposed by researchers and industry.

Although these commercial state-of-the-art AVS algorithms built by these big corporations are currently not available publicly, we explore this through ideas and proposals from published works of how making these ethical decisions are calculated in AVS.

14.2.1.1 Quantifying consequences of decisions

Most AVS are built to make decisions that are preprogrammed, which often involves calculation and probabilities of decisions that it would make (reference). For instance, for an AVS driving in its current path, it calculates a possible trajectory as shown in Figure 14.3. It will perform such calculations in the probability of collision, and at a certain threshold, it will calculate

Figure 14.2 Waymo (2022).

alternative actions that will reduce the probability of collision. Collision probability is a calculation of how likely a collision will occur, and this estimation can come from information obtained from vehicle sensors such as distance, noise, speed, and many more. There are different ways to calculate the collision probability of AV such as probability-based prediction of external trajectories (Hu et al., 2018). It sounds quite straight forward as the AVS could perform actions such as to stop, or to change its driving directions, thereby making the collision probability as low as possible.

In a scenario where there are only choices that are all in the threshold of collision, a simple idea is to choose the probability with the least collision impact. This does not work when you consider the consequences of the collision. If one collision has a higher probability it will not injure anyone, whereas the other collision choice is much lower, but it may harm the passengers, and we need to consider many other factors which will go into our calculations.

Risk in mathematics can be defined as the probability of **potential harm** or **loss**. Geisslinger et al. (2021) utilized this and proposed a formula used for AVS in calculating risk:

$$R = p(u)H(u) \tag{14.1}$$

Risk is represented by R, is the product of *collision probability* represented by p and *harm* represented by H, where p and H are functions of the trajectory (path taken by a moving object) represented by u (Geisslinger et al., 2021).

Naturally, most would agree to make the choice that would be the least harmful on human passengers. Many researchers have proposed different methods on calculating "harm." This is a bit more difficult to represent mathematically from a moral perspective due to its abstractive and subjective

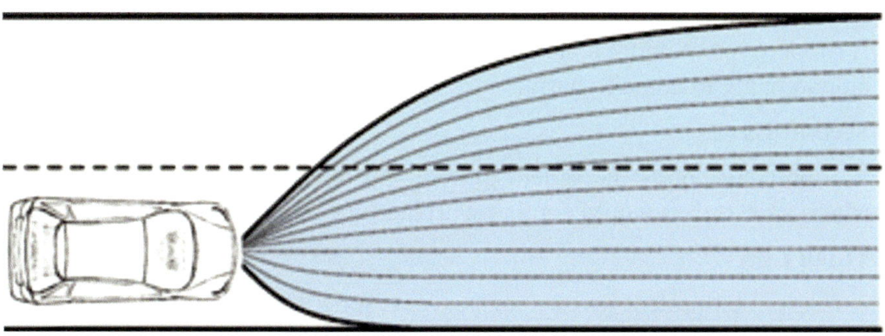

Figure 14.3 Example possible trajectory of an AV (Geisslinger et al., 2021).

conceptions of it. For instance, how do we quantify the harm of mental trauma caused by an accident? Even when we are able to quantify one type of harm, the same harm on different people will differ in consequences and personal impact. Some researchers have avoided estimating harm from this perspective and measure through quantifying the severity of injuries calculated by objective factors such as kinetic energy (Sobhani et al., 2011; Geisslinger et al., 2021).

Overall, there is no set standard in calculating harm, and there is still ongoing research on quantifying harm, not just in the AVS field. There are various proposed methods, but depending on the method chosen to calculate harm, it consequently affects the decisions the vehicle should make. Although in situations where AVS have a choice between an action that would cause "harm," e.g., a collision versus an action that would prevent this collision, it is quite simple to choose. But the crucial research aspect is how do we measure what is considered less "harmful" when there is a situation where an AVS only has choices that would do harm?

14.2.1.2 What decisions should they make?

It seems that in terms of technical feasibility, technology today is more than capable of programming these systems into making ethical decisions. There have been algorithms created for various ethical frameworks (Leben, 2017). Despite having the capability to do so, the part where creators get stuck is what decision should it make?

"Runaround" by Isaac Asimov (2004) was a science fiction short story first published in 1942 that has inspired a large wave of AI enthusiasts to date. This story introduced the **Three Laws of Robotics** or also known as **Asimov's Laws**. The Three Laws, quoted from the "Handbook of Robotics, 56th Edition, 2058 AD":

> **First Law**
>
> A robot may not injure a human being or, through inaction, allow a human being to come to harm.
>
> **Second Law**
>
> A robot must obey the orders given it by human beings except where such orders would conflict with the First Law.
>
> **Third Law**
>
> A robot must protect its own existence as long as such protection does not conflict with the First or Second Law. (Asimov, 1942)

This was one of the earliest publications on the idea of AI moral guidance, and many researchers still refer to this in recent years. Such rule-based ethics is programmable and easily achievable from a practical perspective (Powers, 2006). However, this cannot be applicable to all types of situations. Take the famous trolley dilemma: A popular scenario that represents a moral dilemma is the trolley problem (Foot, 1967; Thomson, 1985)

You are standing at a switch and a trolley is going towards five people who are tied up on the rails. If you do not do anything, these five people will die. You can pull the switch so that the trolley takes the alternate path. However, on the alternate path, there is one person who is tied up on the tracks, if you pull the switch, it will kill that one person. You can do nothing and the five will die or do something and the one person will die.

What will you do?

The trolley problem presents a dilemma that proves difficult for people to unanimously agree, yet how are AVS supposed to be programmed when faced with such an example? The Three Laws can be foundational guidance, but it is clear that it cannot cover all scenarios. For this scenario, no matter what the robot chooses, whether to not do anything or do anything will conflict with the First Law. Thus, there is a need to decide the approach for AVS to take in such situations. Although the chance of an AVS being in the specified situation of the trolley problem is small, and even if AVS is programmed to minimize accidents, there are scenarios where they need to make such decisions, especially as generally acknowledged by experts, "mixed traffic" is a state that will be around for a period of time. "Mixed traffic" is to describe a state where human-driven cars and self-driving cars will be on the road alongside each other (van Loon and Martens, 2015).

14.2.2 Ethical frameworks...for AVS?

Thus, seeing the need to make such decisions, a few questions immediately arise – should AVS be programmed to prioritize the passengers at all costs? Or always minimize the number of deaths? What moral principles should serve these situations? All these questions have created a new concept, the *21st Century Trolley Problem*.

14.2.2.1 Deontology

Deontology ethics, which also is known as *duty ethics* or *rule-based ethics*, where ethical actions are ones that follow the rules, while actions that do not follow the rules are deemed unethical. One example would be the Three Laws of Robotics, and in such rule-based ethics, an AVS that follows the Three

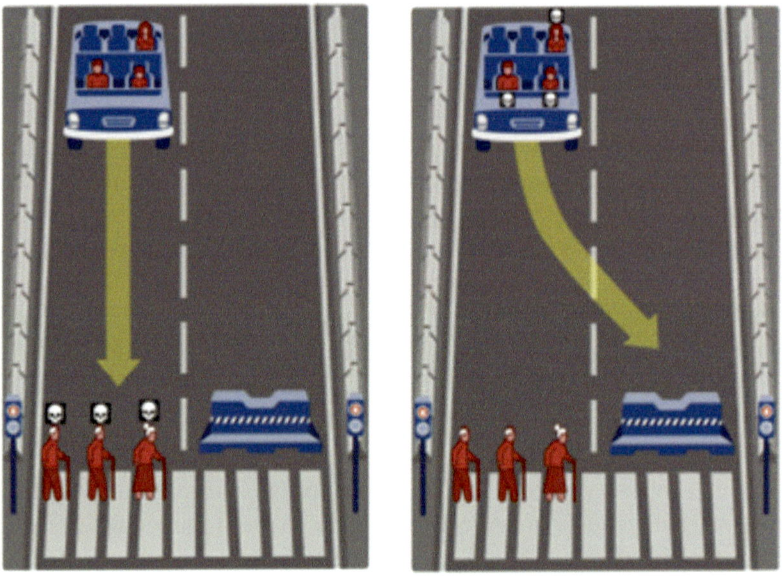

Figure 14.4 Scenario of moral machine example (Awad et al., 2018).

Laws of Robotics will be deemed as doing the right action. But as discussed before, such laws become complex in a situation such as the trolley problem. Some scholars have proposed various other types of rules such as in the form of clusters "forbidden," "permissible," "obligatory actions" (Powers, 2006) and creating a hierarchy of constraints that guide AVs in dilemma situations (Geisslinger et al., 2021). Despite such proposals, there are still some issues

1. **Ignores context-specific situations**Since moral decisions are often not absolute but dependent on the situation. Even if all context-specific situations are imagined and preprogrammed, another issue (below) rises
2. **Representation of reality is limited**The rule-based approach will need to cover *all* types of scenarios, and this becomes difficult due to computing power and representation

14.2.2.2 Consequentialism

Consequentialism is an ethical theory that determines an action is morally correct based on the foreseeable consequences, which is known as *consequence-based* ethics. This theory becomes appealing as it does consider a variety of cases, and the right choices will be different depending on the consequences. The popular statement of consequentialism is "the end will justify the means." There are many different ways to justify what is the *best consequence*. The two main examples of consequentialism are *utilitarianism* and *hedonism*, which both will be explored in the context of AVs in the next section.

14.2.2.3 Utilitarianism

Utilitarianism is a theory of moral philosophy that determines an action is morally correct if it produces a greater quantity of good or happiness (for everyone) than all other actions that are foreseeable. This was introduced by philosopher Jeremy Bentham and aims to maximize human welfare. Bentham believed that human conduct is measured based on "utility" and broadly speaking, such an idea can be expressed in a simplified formula:

$$\text{Amount of good} - \text{Amount of evil} = \text{Utility of act}$$

To apply this in the trolley problem scenario (since this theory considers utility for all), the amount of evil produced could be represented as a cost function algorithm (Lin, 2016). For instance, the cost could be the number of victims, total physical damages, and so forth. Such a theory could advocate for saving five people instead of one. On the other hand, researchers acknowledge that it becomes difficult to judge whether it is right to inhibit the utility of an individual to achieve greater utility for others (Geisslinger et al., 2021). In addition, this again will raise the issue as mentioned before of "quantifying harm," similarly, how do we quantify or evaluate what exactly is the "cost" of five people versus the "cost" of one person when there are many factors to consider?

14.2.2.4 Hedonism

Utilitarianism evaluates actions based on the greater good for all, while hedonism focuses more on the individual. Ethical hedonism claims that only pleasure has worth or value, thereby suggesting that the pursuit of an individual's pleasure should be the ultimate goal.

To the best of our ability, there is currently no current research in attempting to apply hedonism to trolley problems, let alone providing this for AVS. In general, many would deem that hedonists would make a self-choice, not an ethical one. For the trolley problem, the classic hedonist would make the choice of which they believe will derive the most pleasure and least pain regardless of consequences for the overall good. Thus, the idea of applying hedonism on self-driving cars would not be ideal to most, since allowing a hedonist person to make such a decision is not desirable, let alone a hedonistic-based system.

However, it is interesting to note that hedonistic behavior is the foundation of what created reinforcement learning in artificial intelligence systems (Sutton and Barto, 1998). Reinforcement learning systems are programmed to only perform tasks that would give the most pleasure or reward. Nowadays, many advanced self-driving cars adopt the reinforcement learning method (Kiran et al., 2022) due to its effectiveness of it. A simplistic example would be if an AVS hits an obstacle that is considered "painful," so it learns not to perform actions that will give it pain. Whereas if it parks successfully without any collision, that is considered "pleasure."

There is still emerging research in reinforcement learning based AVS to make driving decisions, but there is no current research in applying similar hedonistic nature to make ethical decisions. Since reinforcement learning can be customized to what is considered pain and pleasure, a rudimentary idea is exploring hedonistic nature AVs systems, such as how the reward system of "pain" and "pleasure" be structured? And what choices would they make based on the reward system? There can be further exploration into this.

14.2.2.5 Hedonistic utilitarian

Hedonistic utilitarian is both hedonism and utilitarianism, where one makes choices to maximize their own happiness and those around you. Similar to hedonism, hedonistic utilitarian theory in AVS has not been researched. However, various studies show that most of the general public takes on a hedonistic, utilitarian choice when it comes to AVS and ethical decision-making.

In many different experimental surveys conducted on the general public (from different backgrounds such as countries, age, and gender), it is found that people accept utilitarian self-driving cars (Bonnefon et al., 2015), thereby accepting cars that would sacrifice the passenger if it meant saving more people. However, when asked if they would be in such a utilitarian car and become the passenger, various studies show that the majority of participants would reject using these cars (Bonnefon et al., 2016; Liu and Liu, 2021).

14.2.2.6 Where do corporations stand?

Recently, the big players in the automobile industries have been at the forefront of releasing their achievements in creating self-driving products, and even some of them have already released their AVs, but much less of them have expressed what these systems will choose when faced with such a dilemma. This is perhaps to avoid potentially negative conflicting ideals with their customers. Although staying silent may be a way to avoid publicity conflict for these corporations, the immediate issue is that the general public would not know the decision these systems would make.

In 2016, the United States Department of Transportation issued criteria that must be met for AVS to be on the road, one of the requirements is that these systems can deal with ethical choices, and it needs to be transparent:

Since these decisions potentially impact not only the automated vehicle and its occupants but also surrounding road users, the resolution to these conflicts should be broadly acceptable. Thus, it is important to consider whether HAVs are required to apply particular decision rules in instances of conflicts between safety, mobility, and legality objectives... Algorithms for resolving these conflict situations should be developed transparently using input from Federal and State regulators, drivers, passengers, and vulnerable road users, and taking into account the consequences of an HAV's actions on others (United States Department of Transportation, 2016).

Shortly after, Christoph von Hugo, who is the driver-assistance manager of Mercedes-Benz spoke on this topic in his interview with Car and Driver:

"If you know you can save at least one person, at least save that one. Save the one in the car...If all you know for sure is that one death can be prevented, then that's your first priority." (Taylor, 2016)

Mercedes-Benz was the first corporation publicly announce that their AVS would prioritize the passengers in such a scenario. Others in the industry consider it a "brave" move (Dodgson, 2016).

In the same year when the guidelines were released, a class-action lawsuit was filed in December 2016 against Tesla for their decision not to use their Automatic Emergency braking (AEB) system. The implication is that if a human operator were to drive the car in a way that it would crash, the vehicle would provide warnings but would not override human control and allow the crash to happen. One can infer that Tesla will allow the human operator to make the full decision in any situation at any time (Lin, 2017). To date, there has not been any notable public stance from corporations in regard to what choices their AVs will make.

14.2.3 Growing ideas of other ethical frameworks

Other than applying the major ethical frameworks mentioned above such as consequentialism, deontology, and so forth, many scholars are pushing forward for other approaches. This consists of ethical frameworks that have captured attention in recent years or original proposals.

14.2.3.1 Principle of double-effect

The principle of double-effect believes that actions are morally justified if the harm was not intended. In the case of the trolley problem, the vehicle should continue its original path as it was originally intended. Whereas choosing to change the path will intentionally create harm, which is not morally justifiable. So far, there has not been much research in the algorithm aspect of applying such principles in AVs.

14.2.3.2 Rawlsianism

There is also a growing idea of applying rawlsianism for AVs ethics. Rawl introduced his idea "Theory of Justice" (Rawl, 1971). For a just society, we make decisions behind a "veil of ignorance," meaning we have to pretend we do not know who we are and who we will become in society. Some researchers have attempted to apply rawlsianism to AVs. Leben (2017) created an algorithm where the vehicle estimates the probability of survival for each person in every action. The experiment showed that his algorithm would choose a different path to sacrifice one person for five. Meanwhile, Brändle and Smitch (2021) further argued the use of Rawlsianism in AVs but also raised the idea that more research should be explored in whether it is desirable to authorize AVs.

14.2.4 Do AVS need ethics?

Despite various research and arguments on the adoption of an ethical framework for AVs, there is still no generally agreed-upon standard. After all, one of the ideas is that the trolley problem proves that having one strict type of ethical and moral standard is difficult. There are alternatives that are introduced to solve such problems – Having no ethics at all in AVS and only allowing humans to decide in such situations.

14.2.4.1 No ethics in AVS

Although there is ongoing argument that there is a need for ethical decision-making in AVS especially if they will encounter dilemma situations, Bracanović (2019) argues that having no ethical settings (NES) in AVS is the best solution. The author explains that no matter which ethical framework is chosen, there will always be moral damage. Bravnović uses the example of opposing to weapons at war since it does not "distinguish between legitimate targets from innocent civilians" (Bracanović, 2019) as he states in this paper:

If unintentionally sacrificing innocent lives is a serious reason to reject autonomous weapons in extraordinary situations such as war, it is too unrealistic to expect any serious acceptance of AVs programmed to intentionally choose one innocent human life over the other in ordinary situations like daily traffic (Bracanović, 2019).

However, taking this approach does not clarify what it should do in a dilemma situation. Does it simply allow humans to take control? Does it go into a complete halt when faced with such a situation? Or does it simply have no ethics programmed just like autonomous weapons to simply go and about?

14.2.4.2 Human decision and control

In addition to the NES argument for AVs, some may argue that it is unfair that a machine should make such decisions and that it is the human's responsibility to make moral decisions, thus when a morally conflicting scenario happens, control is transferred to the people in the car, and it is up to them to make the decision. To explore the preference, a survey was done by König and Neumayr (2017) with participants identifying from 33 different countries in the world felt that having a human to be able to make decisions at any time in an AVS is shown to be preferable. Out of all the implications that were identified as the most effective option to increase trust toward self-driving cars, the implication "The driver should have the possibility to take over control whenever wanted" was the top choice. But in reality, as previously mentioned, Tesla was sued for their AVS giving full control to the human passenger.

Some researchers argue against humans taking control in critical situations. Take consideration in the concept of *automation conundrum,* which is when a system has more autonomy and when its reliability and robustness increases, situation awareness of a human operator decreases, which will cause the human operator less likely to be able to take over manual control (Endsley, 2017). In addition, Nyholm and Smids (2016) argue that human reaction times can be slow compared to machines, especially when switching focus from something to another, whereas a car will not react in a panic in ways a human would result in accident scenarios (Endsley, 2017). Therefore, AVs should be preprogrammed to be able to do so (Nyholm and Smids, 2016). If the human operator in the AVS were to make the decision, then the consequences would be evaluated case-by-case to its relative laws with an added dimension of involving the self-driving systems. This part will be further discussed in the section on external AVS ethics.

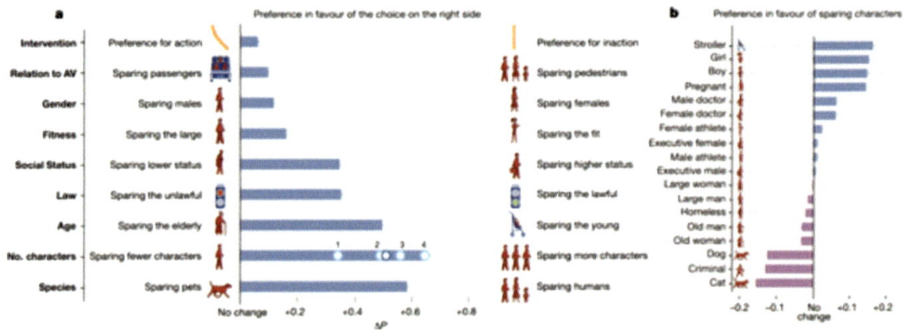

Figure 14.5 Moral machine results (Awad et al., 2018).

14.2.5 What would we choose? The moral machine experiment

A social experiment called "the moral machine experiment" (Awad et al., 2018) was conducted to discover what the general public would choose. The online platform would show different scenarios and collect data based on their choice. Each scenario would have a different number of people with varying demographics such as age, gender, and more. An example of a scenario given to the user is shown in Figure 14.4, and the result of the experiment is shown in Figure 14.5.

The data were processed from 130 countries with *n* range 101 to 448,125 (Awad et al., 2018). Results show that globally participants prefer human lives over animals, and most wanted to spare more lives than less, and young compared to older. The authors acknowledge that there are some limitations despite the large sample size; it does not cover all the complexities such as the hypothetical relationships such as relatives (Awad et al., 2018).

There was also some data to show the preferences based on different countries shown in Figure 14.6. Based on these results, the authors claim that "broad regions of the world displayed relative agreements" (Awad et al., 2018) and further suggest that consensual machine ethics is still potentially possible.

There are still very much conflicting arguments in this aspect for both researchers and public opinion. If the human operator were to take full control, the reality is that human operators could make choices that may bring harm even when the AVS recognizes the consequences, is that still considered ethical? After all, is not AVS meant to make the roads safer?

14.3 AVS External ethics

14.3.1 Overview of AVS impact

With the development of AVS and its growing prevalence in everyday life, internal ethics, defined as ethical decision-making, garnered large amounts of attention in media and academia due to its importance in avoiding collisions or situations where lives would be at risk. External ethics concerns, such as AVS and its effect on the labor market, were also much discussed due to its potential socioeconomic impact. Compared to the previous topics, AVS' data collection ability and its impact on privacy risks received little to no attention. The following sections aim to discuss the socioeconomic impact of AVS, as well as its impact on user privacy and current laws and legislation that may impact data collection in AVS.

14.3.2 Socioeconomic impact of AVS

As of 2022, global investment into AVS technology exceeds $200 billion USD, with governments throughout the world aiding in AVS production and facilitating AVS adoption (Yeruva, 2022). With large corporations and governments behind its development, AVS has the potential to have a massive impact on society and people's lives.

One of the major positive impacts AVS can have on society is improving overall mobility and accessibility for all parties, especially those who ordinarily would be physically limited. Driverless vehicles make mobility more accessible for the elderly, handicapped, children, and other individuals who may find driving physically challenging. Looking at the US population alone, there are around 49 million Americans who are over the age of 65, with around 53 million who have some sort of disability (Klaver, 2020). Those who would ordinarily be restricted in the movement now have an alternative service they can use, allowing for higher degrees of freedom.

Other impacts are more debatable, such as AVS and its effects on the transportation system. AVS optimists hypothesize that autonomous vehicles can alleviate the burden of transportation, making roads less congested. Imagine a future where the owner could send their car home after arriving at their destination, which eliminates the need for inner-city parking, freeing up more space for urban planning and potentially changing designs of future cities (Yeruva, 2022; Duarte and Ratti, 2018). With the transportation system becoming more efficient and AVS eliminating the need for a human driver, others argue that this could have the opposite effect and create more burden.

The ease of operating an AVS could propel more vehicles on the road, with some experts claiming that "everybody having their own [autonomous vehicle] is a really dystopian future" (Wisniewski, 2016).

The potential to reduce the need for parking versus potentially more vehicles occupying already congested roads highlights another AVS impact that is heavily debated in academia, which is AVS and its impact on the environment. Currently, experts occupy contradictory positions on this matter. A recent review of current literature and research on the impacts of autonomous vehicles and its impact on greenhouse gas (GHG) emissions concluded that the potential effects of AVS is uncertain. Positive impact of AVS includes reducing GHG emissions by 35%, attributed to factors such as eco-driving

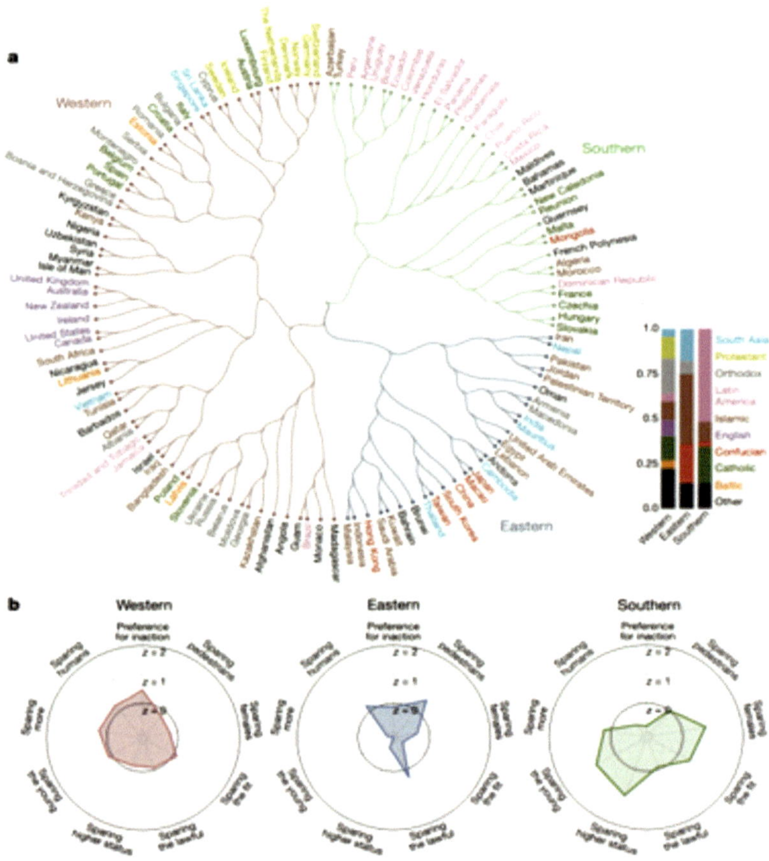

Figure 14.6 Results of geographical location (Awad et al., 2018).

and less need for parking. On the other hand, increased mobility meant for previously "captive passengers," such as the elderly and disabled could increase GHG emissions by 41%. There is also the issue of AVS penetration, where the estimated GHG reductions may not come into existence until AVS adoption and penetration rates reach above 60% (Massar et al., 2021).

Increased safety and reduced accidents are strong positive indicators that support a fully autonomous future. In 2010, the complete cost of motor vehicle accidents to society, including economic damage, reduced workplace productivity, and other cost factors cost the United States around $836 billion dollars annually. Similar situations are seen worldwide, with a 2018 study in the Netherlands estimating that social costs of car accidents cost the country $18 billion dollars (Klaver, 2020). To reduce road incidents by 1% in the United States would mean a cost reduction of $8 billion dollars. Human error is often the leading cause of road accidents, statistics from the Center of Disease Control in the US highlights the top three causes of fatal car crashes being alcohol related, speeding, and distracted driving. These are all behavioral factors that could be eliminated or reduced significantly using AVS technology (Klaver, 2020; Wisniewski, 2016).

As AVS adoption becomes more prevalent in society, its impact on the labor market cannot be overlooked. Looking at the US population, around 2.9% are employed in driving, which amounts to roughly 4 million people. This group, with an average annual income of $45,000 dollars, could lose their jobs to AVS, which adds to a total loss of around $180 billion per year in terms of income (Klaver, 2020). At the same time, AVS would also create new jobs such as AV specialists, remote AV controllers, and other roles and services. In addition, the development and adoption of AV could create new qualifications which would expand the education sector (Yeruva, 2022). The real dilemma is how many of those currently in driving occupations can transition into an AV-related role without needing excessive additional education.

14.3.3 Who is at fault when an AVS crashes?

Fully autonomous vehicles do not have a designated human driver, making it difficult to determine who is at fault when an accident arises. Say a parent order an AVS for their child, and during this trip the AVS crashes with another vehicle or object. In this situation, who should be at fault? Should the company that developed the AVS be responsible, or should the user take responsibility and accept this as a reasonable risk?

The US Department of Transportation's National Highway Traffic Safety Administration (NHTSA) issued a "first-of-its-kind" rule in March 2022 to ensure occupant safety in autonomous vehicles without conventional manual controls, such as steering wheels, brakes, and accelerators (United States Department of Transportation, 2022).

Before today's rule, occupant protection standards were written for common, traditional vehicle features, including steering wheels and other manual controls. The rule updates the standards to clarify what is required of manufacturers when applying the standards to ADS-equipped vehicles without traditional manual controls. (United States Department of Transportation, 2022)

This ruling signifies changes in autonomous vehicle regulations where AVS developers are no longer required to include conventional manual controls inside a vehicle, placing additional pressure on AI ethical decision-making of AVS. Combined with the previous scenario where a legal guardian orders an AVS for their dependent, while this ruling highlights the absolute importance of occupant safety in AVS design and deployment, it does not answer the question of blame attribution, namely, who should be at fault should an AVS crashes?

14.3.4 Should AVS developers take full responsibility for an accident?

Some car markers, the likes of Audi and Volvo, have claimed that they would take full responsibility in the event of an accident if the automated driving feature was in use as intended (Atiyeh, 2015; Maric, 2017). Others, such as Mercedes-Benz and Google, have said in interviews that they are willing to accept liability in the event of a crash (Atiyeh, 2015)

"When the function is operated as intended, if the customer turns the traffic jam pilot on and uses it as intended, and the car was in control at the time of the accident, the driver goes to his insurance company and the insurance company will compensate the victims of the accident and in the aftermath, they come to us and we have to pay them," Dr. Thorsten Leonhardt, Audi (Maric, 2017)

While this sounds like the ideal solution with the car marker offering to take full responsibility, in reality, things are rarely so black and white. On June 3, 2022, a Cruise self-driving vehicle owned and operated by General Motors was involved in an accident with a Toyota Prius operated by a human driver. The incident occurred one day after Californian regulators gave Cruise a permit allowing it to operate without a human driver and collect fares for its rides. This permit was the first-of-its-kind in the state

...the Cruise car turned left in front of the oncoming Prius. It happened at the intersection of Geary Boulevard and Spruce Street. Officials from Cruise claim the autonomous vehicle stopped before proceeding with the left turn, and the self-driving car was actually sitting still when the Prius hit it.

Cruise went on to say that the driver of the Toyota was speeding. Moreover, the company's report claims the Prius was in a lane that required a right turn, though the human driver proceeded straight through the intersection. **A spokesperson for the San Francisco Police Department couldn't confirm or deny Cruise's interpretation of the story since an incident report was either unavailable or not completed in the first place**. (Loveday, 2022)

This incident highlights the difficulty of AVS co-existing with vehicles operated by human drivers. Someone is ultimately responsible for this accident, and with one of the drivers being an AVS, it is hard to discern who is at fault. The interaction between human drivers and the AVS needs to be analyzed before a verdict can be determined, and this would likely be a long and painstaking process due to the lack of precedence and regulations (Loveday, 2022).

These accidents make people question AVS vehicles and its effort toward occupant safety. Just a few weeks after the GM Cruise incident, Nio, a Chinese AV company also suffered from a fatal accident. One of Nio's testing cars drove off the third floor of a parking garage in its Shanghai Headquarters, killing both employees inside the vehicle. It was unclear what caused the crash, but the company remains adamant that the case was an "accident unrelated to the vehicle" .(Shen, 2022)

14.3.5 AVS law and regulations

With the current move in the US toward removing conventional manual controls within a fully automated vehicle, all signs point toward a future where human drivers would regularly share roads and interact with AVS systems. The Law Commission of English and Wales and Scottish Law Commission published a report in January 2022, advocating for an Automated Vehicles Acts. This report recommended that features that assist driving, such as cruise control, be made distinct from fully autonomous features such as self-driving. This would help make prosecution easier in case of accidents and recommends a new system of accountability. When a self-driving feature is engaged, the occupant in the driver's seat is no longer classified as a driver, but the "user-in-charge," who "cannot be prosecuted for offenses which arise directly from the driving task." In the situation where the AVS is authorized to operate without a human driver or a user in the driver seat, the AVS operator is "responsible for the entire journey" (The Law Commissions, 2022)

The development of self-driving vehicles in the UK has the potential to revolutionise travel, making every-day journeys safer, easier, and greener.

This Government has been encouraging development and deployment of these technologies to understand their benefits. However, we must ensure we have the right regulations in place, based upon safety and accountability, in order to build public confidence.

That's why the Department funded this independent report and I look forward to fully considering the recommendations and responding in due course." Trudy Harrison, UK Transport Minister (The Law Commissions, 2022)

This report and the proposed Automated Vehicles Act are aimed at helping regulators discern legal liability and promote AVS adoption. While this report is aimed at building public confidence by minimizing the possibility that an individual would be responsible for decisions made by an AVS, it highlights another concern – what is the current public perception of AVS in the event of a crash?

14.3.6 Public perception of AVS in an accident

While there have been few studies on public perception of fully autonomous vehicles, we can look to perception studies of semi-autonomous vehicles to gauge public sentiment. Liu and Du (2021) conducted experiments measuring participant response to hypothetical car accident scenarios caused by a human driver or a semi-autonomous vehicle. Researchers found that in the situation of a crash, people are more likely to judge the automated system more harshly than they would a human driver. The results highlight what Liu and Du named as "blame attribution asymmetry bias," where participants tend to judge the crashes caused by semi-autonomous vehicles much more harshly than that of a human driver and demand that victims in a crash caused by semi-autonomous vehicles should be compensated more than those harmed in an accident created by human drivers.

Relating Liu and Du's experiment to the Automated Vehicle Act proposed by The Law Commissions in the United Kingdom highlights the

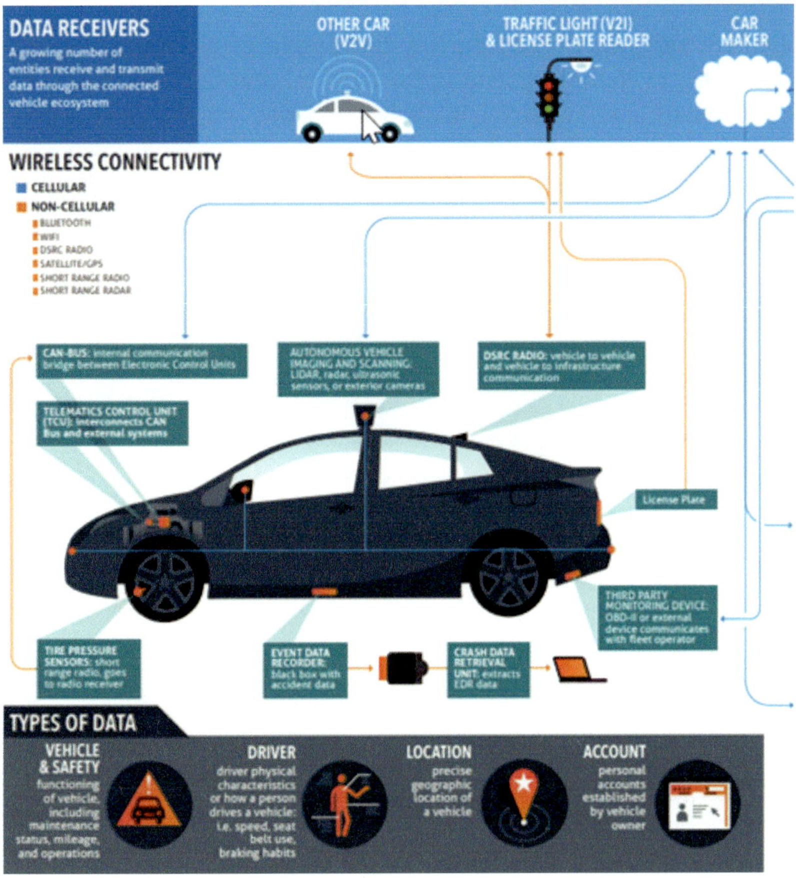

Figure 14.7 Diagram showing data flow of exterior AVS sensors.

delicate nature of the relationship between the public and autonomous vehicles. A before-and-after sentiment analysis of the 2018 Tesla pedestrian crash showed that negative tweets on Twitter related to "self-driving" or "autonomous" increased by 32%, further reinforcing the public's weariness of AVS technology (Penmetsa et al, 2021). Considering the blame asymmetry bias and how AVS accidents can create massive negative impacts on public perception about AVS, policymakers have a fine line to walk. Should AVS operators take full fault in accidents, punitive damages must be significant enough to ensure AVS safety is front of mind for all developers and operators; however, policymakers also need to be wary of negative impacts in perception, which could slow autonomous vehicle adoption in society.

14.3.7 AVS and its impact on user privacy

Physical safety has always been the key focal point of discussion in AVS development due to its high potential for physical and fatal harm to AVS occupants in unexpected situations. In comparison, one area that has received little to no discussion is AVS and its privacy-invasive capabilities. Given AVS' ability to collect large amounts of data on its occupants, geolocation, and movement, it is prudent to investigate associated security risks further.

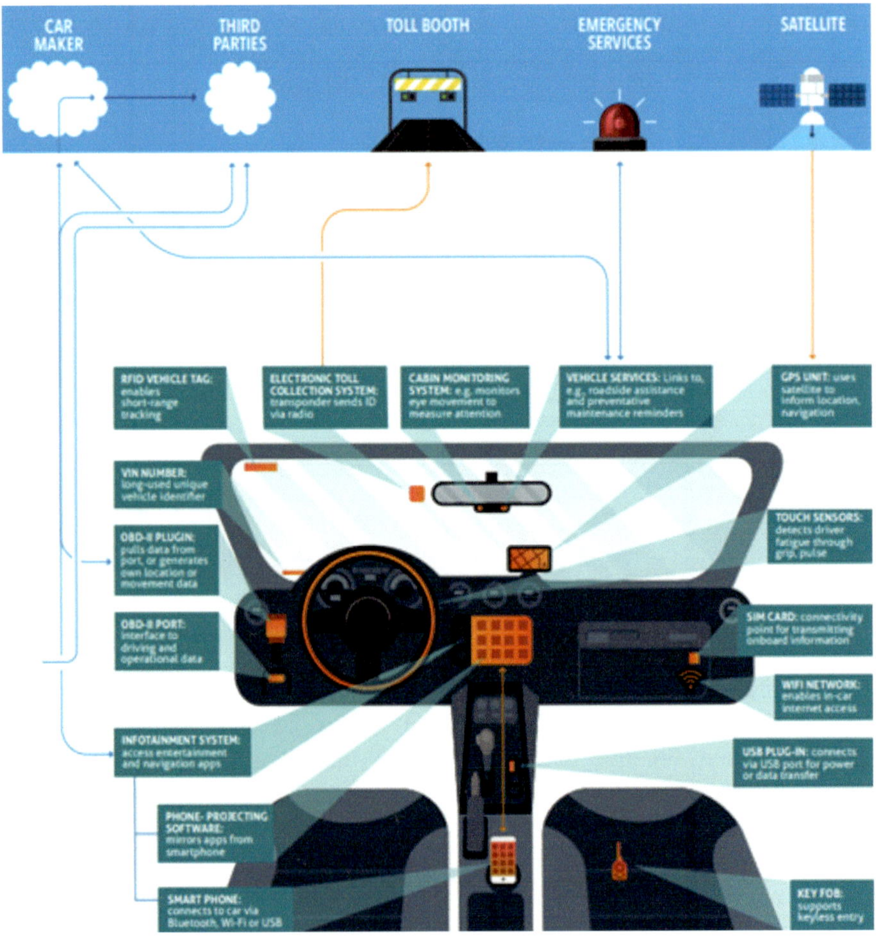

Figure 14.8 Diagram showing data flow of interior AVS sensors.

One of the first pieces of academic literature discussing privacy in autonomous vehicles was written by Glancy and published by Santa Clara Law Review in 2012. In this paper, Glancy (2012) categorized autonomous vehicles into two groups: the first being self-contained autonomous vehicles and the other being interdependent autonomous vehicles. Glancy highlights three technological factors between the two groups, being where the AI is located, how external data is collected and transmitted to the vehicle, and finally, whether if vehicle data is then transmitted beyond the vehicle. While both self-contained AVS and interdependent AVS rely on internal sensors to detect how their parts are operating, the divergent point is how external situational data is collected. Self-contained AVS uses outward-facing sensors such as radar, cameras, and thermal imaging to collect data about road conditions. Interdependent AVS relies on a wireless connection to receive external situational data – it is a vessel that both sends and receives data (Glancy, 2012).

In terms of privacy, both types of AVS would likely have a huge repository of information about its users; however, as self-contained vehicles are not connected to other networks and external information sources, vehicle information is contained within the vehicle until removed. As interdependent vehicles communicate with other vehicles and networks over a wireless connection, it can be controlled over a network and the information it transmits could be intercepted. While self-contained vehicles seem more secure from a privacy standpoint given the fact that it does not transmit data beyond the vehicle, it is important to acknowledge that both types of AVS would likely have detailed and continuous geolocation data, as well as situational information, such as what did the occupant of the AVS saw along their journey, the road conditions, how long the vehicle was parked and its travel time (Glancy, 2012).

Figure 14.7 is an infographic published by Future of Privacy Forum (2017) detailing various sensors and transmitters located on the external perimeter of an interdependent AVS and the type of data the vehicle could collect and transmit. In Figure 14.8, we see systems and transmitters located in the vehicle's interior. As the infographic was published in 2017, it still included conventional manual controls, such as the steering wheel. Given the development of AVS regulations, especially the NHTSA's ruling regarding occupant safety in an AVS without physical controls, we can reasonably anticipate conventional controls slowly disappearing from the vehicle in favor of maximizing space. The infographic also highlights the amount of data and the wide range of devices that may be connected with an AVS, amplifying privacy concerns in regard to information transmission.

Looking at the range of information receivers and transmitters present on an autonomous vehicle, AVS has the capability to collect large amounts of information about its users and occupant, as well as surrounding situational information. Interdependent AVS collects location and movement data using their many sensors, and these data could be analyzed using existing software. The physical surveillance aspect of AVS cannot be overlooked, especially when this feature is owned by a private company. Modern society may have low expectations of privacy due to the prevalence of surveillance technology, such as CCTV. The data set collected is invaluable and could be used to achieve different purposes. While a government may choose to use geo-temporal data to improve its city and infrastructure, a private company with the same data set will likely have a different focus in mind (Bloom et al., 2017)

"If a city like London and a company like Uber have the same data set of geo-temporal points, the former has an obligation to use it to better its constituents and the latter has an obligation to monetize it, bettering its shareholders." (Bloom et al., 2017)

With the combination of different data points, AVS data could be used to create an accurate profile of its occupants. Imagine combining information such as where the vehicle is regularly parked overnight and the average income bracket of the area surrounding the parking. We can potentially estimate the AVS occupant or user's willingness to spend on higher-end goods, which makes the occupant an ideal candidate for certain advertisements (Collingwood, 2017)

"Finally, as a consequence of the pervasive scrutiny of those who travel in autonomous vehicles, surveillance privacy interests are impacted because comprehensive personal information collection could be used to profile, predict, and possibly manipulate the behaviour of autonomous vehicle users" Collingwood (2017)

Given AVS and its surveillance and data collection capabilities, it is not unreasonable to expect a repeat of "The Social Dilemma," where predictive analytics is leveraged to the point where the vehicle could potentially impact its occupants' habits, behavior, and travel patterns. It would be much more difficult for users to "distinguish between acceptable influencing (persuasion and nudging) and unacceptable influencing (manipulation)" becomes much more difficult (Citron and Solove, 2022).

Manipulation could also create societal harm, as individual decisions ripple out and impact society as a whole. Collingwood (2017) hypothesized AVS data and predictive analytics being potentially used to avoid certain parts of the city due to crime, or avoid certain events, which are both likely outcomes. With AVS being interdependent vehicles, this "habit" could be adopted by a fleet of vehicles and dispel efforts to create a more integrated society.

With the recent Supreme Court decision to overturn Roe v. Wade, the White House is recommending women to "be careful" with period apps, with the fear of app data being used against women seeking abortion (Wright and Vazquez, 2022). Now imagine a case where an individual is prosecuted for seeking an abortion, and the court subpoenaed information from app service providers, then cross-referenced this with their ridesharing or AVS geolocation data. The potential for privacy harm is significant, especially with AVS sensors being very comprehensive.

"I wouldn't say we're directing people (to stop using the apps), but I think people should be really careful about that," Jen Klein, co-chair and executive director of the White House Gender Policy Council (Wright and Vazquez, 2022)

Looking at the above situation, exposing AVS data makes impacted individuals susceptible to various privacy harms. With AVS, individuals are already more likely to be impacted by autonomy harm, which is "restricting, undermining, inhibiting, or unduly influencing people's choices" (Citron and Solove, 2022). AVS essentially acts as a delegated agent, with the user asking the AVS to make decisions or choices that the driver would usually take. While some users welcome being alleviated from driving, other users could be forced into making a choice where either they accept the AVS tracking their every move, or having little to no individual mobility (Glancy, 2012; Collingwood, 2017).

Aside from various privacy harms and violations, it is important to assess public perception of reasonable privacy on public roads in hopes of improving the acceptance of connected AVS. In general, people do not reasonably expect privacy on public roads, which is where AVS operates. The main dilemma for AVS is that due to its external sensors to facilitate driving functionalities, AVS are constantly collecting information about nonusers, such as passers-by walking along the road. Nonusers, unlike AVS users and occupants, have not consented to data collection, yet given the conditions, there is no way of notifying nonusers of data collection (Bloom et al, 2017).

With this conundrum, one option for AVS developers is to consider privacy-by-design. It is possible for AVS to be designed in a way that minimizes personal information collection and renders data anonymous to protect its users and avoid violating privacy of nonusers (Collingwood, 2017). Gurumurthy and Kockelman's (2020) study explored Americans' willingness-to-pay (WTP) for privacy enhancing features (such as anonymizing location data) in an AVS. An alternative perspective would be if users should be paying for privacy-enhancing features at all? With the establishment of the California Consumer Privacy Act (CCPA) and Europe's General Data Protection Regulation (GDPR), AVS developers need to recognize the importance of incorporating privacy features into their systems, allowing users to maintain control over their AVS data.

14.4 Conclusion

Overall, there are still ongoing discussions regarding the internal ethics of AVS and the framework they should adopt for making decisions in moral dilemma situations. Although various automobile organizations have introduced new and advanced AVS technologies, only a few corporations have been transparent about their AVS ethical decision-making processes. In the research field, differing opinions prevail; some argue that AVS should not be bound by explicit ethical considerations, while others have demonstrated through research that reaching a mutual agreement on the ethical framework is possible.

Looking at the external ethic issues, the majority of public focus is centered around AVS and its socioeconomic impacts. With AVS making mobility itself more accessible for handicapped individuals and the promise of freeing up more of our finite urban real estate, the positive impacts AVS could have on society are immense. However, as AVS has the capabilities of leveraging AI to influence human behavior potentially, the overall impact remains ambiguous as we are unable to estimate what the final outcome would be.

AVS and its privacy-invasive data collection capabilities are also other subjects that should receive much more attention. While physical security will always be prioritized due to its capacity for fatal physical harm, it is prudent that we begin to look at personal data and privacy as a hidden cost in exchange for AVS services as it has the potential to develop into a powerful network of surveillance owned by private entities. To ensure physical security, it is pivotal that AVS corporations and governments alike begin to assess privacy implications in AVS and develop relevant policies that guard consumer privacy rights.

Acknowledgments

We would like to thank Dr. Lorrie Cranor at Carnegie Mellon University, for providing insight into usable privacy and security. We would also like to thank Ms. Cara Bloom from MITRE for her research and discussions on user privacy expectation for autonomous vehicle interactions.

References

[1] Asimov, I. (2004). *I, robot* (Bantam hardcover ed.). Bantam Books.

[2] Atiyeh, C. (2015, October 8). *Volvo Will Take Responsibility If Its Self-Driving Cars Crash*. Car and Driver. https://www.caranddriver.com/news/a15352720/volvo-will-take-responsibility-if-its-self-driving-cars-crash

[3] Awad, E., Dsouza, S., Kim, R., Schulz, J., Henrich, J., Shariff, A., Bonnefon, J.-F., & Rahwan, I. (2018). The Moral Machine experiment. *Nature*, *563*(7729), 59–64. https://doi.org/10.1038/s41586-018-0637-6

[4] Bloom, C., Tan, J., Ramjohn, J., & Bauer, L. (2017). *Self-Driving Cars and Data Collection: Privacy Perceptions of Networked Autonomous Vehicles*. 19.

[5] Bonnefon, J.-F., Shariff, A., & Rahwan, I. (2015). Autonomous Vehicles Need Experimental Ethics: Are We Ready for Utilitarian Cars? *Undefined*. https://www.semanticscholar.org/paper/Autonomous-Vehicles-Need-Experimental-Ethics%3A-Are-Bonnefon-Shariff/13d456d4c53d7b03b90ba59845a8f61b23b9f6e8

[6] Bonnefon, J.-F., Shariff, A., & Rahwan, I. (2016a). The social dilemma of autonomous vehicles. *Science*, *352*(6293), 1573–1576.

[7] Bonnefon, J.-F., Shariff, A., & Rahwan, I. (2016b). The social dilemma of autonomous vehicles. *Science*, *352*(6293), 1573–1576. https://doi.org/10.1126/science.aaf2654

[8] Bracanović, T. (2019). No Ethics Settings for Autonomous Vehicles. *Hungarian Philosophical Review*, *63*(4), 47–60.

[9] Brändle, C., & Schmidt, M. (2021). Autonomous Driving and Public Reason: A Rawlsian Approach. *Philosophy & Technology*, *34*. https://doi.org/10.1007/s13347-021-00468-1

[10] Citron, D. K., & Solove, D. J. (2021). *Privacy Harms* (SSRN Scholarly Paper ID 3782222). Social Science Research Network. https://doi.org/10.2139/ssrn.3782222

[11] Collingwood, L. (2017). Privacy implications and liability issues of autonomous vehicles. *Information & Communications Technology Law, 26*(1), 32–45. https://doi.org/10.1080/13600834.2017.1269871

[12] Da Vinci, L. (1478). *Codex Atlanticus*. https://codex-atlanticus.it/

[13] Dodgson, L. (2016). *Why Mercedes plans to let its self-driving cars kill pedestrians in dicey situations*. Business Insider. https://www.businessinsider.com/mercedes-benz-self-driving-cars-programmed-save-driver-2016-10

[14] Duarte, F., & Ratti, C. (2018). The Impact of Autonomous Vehicles on Cities: A Review. *Journal of Urban Technology*, *25*(4), 3–18. https://doi.org/10.1080/10630732.2018.1493883

[15] Earnest, L. (2012, December). *Stanford Cart*. https://web.stanford.edu/~learnest/sail/oldcart.html

[16] Edmonds, E. (2019, March 14). *Three in Four Americans Remain Afraid of Fully Self-Driving Vehicles*. AAA Newsroom. https://newsroom.aaa.com/2019/03/americans-fear-self-driving-cars-survey/

[17] Endsley, M. R. (2017). From Here to Autonomy: Lessons Learned From Human–Automation Research. *Human Factors: The Journal of the Human Factors and Ergonomics Society*, *59*(1), 5–27. https://doi.org/10.1177/0018720816681350

[18] Foot, P. (1967). *The Problem of Abortion and the Doctrine of the Double Effect*. 6.

[19] Future of Privacy Forum. (2017, June 29). Infographic: Data and the Connected Car - Version 1.0 - Future of Privacy Forum. *Https://Fpf.Org/*. https://fpf.org/blog/infographic-data-connected-car-version-1-0/

[20] Geisslinger, M., Poszler, F., Betz, J., Lütge, C., & Lienkamp, M. (2021). Autonomous Driving Ethics: From Trolley Problem to Ethics of Risk. *Philosophy & Technology*, *34*, 1–23. https://doi.org/10.1007/s13347-021-00449-4

[21] Glancy, D. J. (2012). Privacy in Autonomous Vehicles. *SANTA CLARA LAW REVIEW*, *52*, 70.

[22] Goodall, N. (2014). *Machine Ethics and Automated Vehicles* (pp. 93–102). https://doi.org/10.1007/978-3-319-05990-7_9

[23] Gurumurthy, K. M., & Kockelman, K. M. (2020). Modeling Americans' autonomous vehicle preferences: A focus on dynamic ride-sharing,

privacy & long-distance mode choices. *Technological Forecasting and Social Change*, *150*, 119792. https://doi.org/10.1016/j.techfore.2019.119792

[24] Hu, Y., Zhan, W., & Tomizuka, M. (2018). *Probabilistic Prediction of Vehicle Semantic Intention and Motion* (arXiv:1804.03629). arXiv. http://arxiv.org/abs/1804.03629

[25] Kiran, B. R., Sobh, I., Talpaert, V., Mannion, P., Sallab, A. A. A., Yogamani, S., & Pérez, P. (2022). Deep Reinforcement Learning for Autonomous Driving: A Survey. *IEEE Transactions on Intelligent Transportation Systems*, *23*(6), 4909–4926. https://doi.org/10.1109/TITS.2021.3054625

[26] Klaver, F. (2020). The economic and social impacts of fully autonomous vehicles. *Compact*. https://www.compact.nl/en/articles/the-economic-and-social-impacts-of-fully-autonomous-vehicles/

[27] König, M., & Neumayr, L. (2017). Users' resistance towards radical innovations: The case of the self-driving car. *Transportation Research Part F: Traffic Psychology and Behaviour*, *44*, 42–52. https://doi.org/10.1016/j.trf.2016.10.013

[28] Leben, D. (2017). A Rawlsian algorithm for autonomous vehicles. *Ethics and Information Technology*, *19*(2), 107–115. https://doi.org/10.1007/s10676-017-9419-3

[29] Lin, P. (2016). Why Ethics Matters for Autonomous Cars. In M. Maurer, J. C. Gerdes, B. Lenz, & H. Winner (Eds.), *Autonomous Driving: Technical, Legal and Social Aspects* (pp. 69–85). Springer. https://doi.org/10.1007/978-3-662-48847-8_4

[30] Lin, P. (2017). *Here's How Tesla Solves A Self-Driving Crash Dilemma*. Forbes. https://www.forbes.com/sites/patricklin/2017/04/05/heres-how-tesla-solves-a-self-driving-crash-dilemma/

[31] Liu, P., & Du, Y. (2021). *Blame Attribution Asymmetry in Human–Automation Cooperation—Liu—- Risk Analysis—Wiley Online Library*. https://onlinelibrary.wiley.com/doi/10.1111/risa.13674

[32] Liu, P., & Liu, J. (2021). Selfish or Utilitarian Automated Vehicles? Deontological Evaluation and Public Acceptance. *International Journal of Human–Computer Interaction*, *37*(13), 1231–1242. https://doi.org/10.1080/10447318.2021.1876357

[33] Loveday, S. (2022, June 7). *GM's Cruise Self-Driving Car Crash Injured Multiple People*. InsideEVs. https://insideevs.com/news/596827/gm-cruise-self-driving-crash-injuries/

[34] Maric, ul. (2017, September 10). *Audi to take full responsibility in event of autonomous vehicle crash*. Drive. https://www.drive.com.au/news/audi-to-take-full-responsibility-in-event-of-autonomous-vehicle-crash/

[35] Massar, M., Reza, I., Rahman, S. M., Abdullah, S. M. H., Arshad, J. & Al-Ismail, F. S. (2021). Impacts of Autonomous Vehicles on Greenhouse Gas Emissions—Positive or Negative? *International Journal of Environmental Research and Public Health*, *18*(11), 5567. https://doi.org/10.3390/ijerph18115567

[36] Nyholm, S., & Smids, J. (2016). The Ethics of Accident-Algorithms for Self-Driving Cars: An Applied Trolley Problem? *Ethical Theory and Moral Practice*, *19*(5), 1275–1289. https://doi.org/10.1007/s10677-016-9745-2

[37] Othman, K. (2021). Public acceptance and perception of autonomous vehicles: A comprehensive review. *AI and Ethics*, *1*(3), 355–387. https://doi.org/10.1007/s43681-021-00041-8

[38] Penmetsa, P., Sheinidashtegol, P., Musaev, A., Adanu, E. K., & Hudnall, M. (2021). Effects of the autonomous vehicle crashes on public perception of the technology. *IATSS Research*, *45*(4), 485–492. https://doi.org/10.1016/j.iatssr.2021.04.003

[39] Powers, T. M. (2006). Prospects for a Kantian Machine. *IEEE Intelligent Systems*, *21*(4), 46–51. https://doi.org/10.1109/MIS.2006.77

[40] Rawls, J. (1971). *A Theory of Justice: Revised Edition*. 562.

[41] Scottish Law Commission. (2022). *Automated Vehicles: Joint report*.

[42] Shen, J. (2022, June 24). *Nio says vehicle not at fault in fatal 'accident' in Shanghai ů TechNode*. TechNode. http://technode.com/2022/06/24/nio-says-vehicle-not-at-fault-in-fatal-accident-at-shanghai-headquarters/

[43] Sobhani, A., Young, W., Logan, D., & Bahrololoom, S. (2011). A kinetic energy model of two-vehicle crash injury severity. *Accident; Analysis and Prevention*, *43*, 741–754. https://doi.org/10.1016/j.aap.2010.10.021

[44] Stead, W. (2018). Clinical Implications and Challenges of Artificial Intelligence and Deep Learning. *JAMA*, *320*. https://doi.org/10.1001/jama.2018.11029

[45] Sutton, R. S., & Barto, A. G. (1998). *Reinforcement Learning: An Introduction*. 352.

[46] Taylor, M. (2016, October 7). *Self-Driving Mercedes-Benzes Will Prioritize Occupant Safety over Pedestrians*. Car and Driver. https://www.caranddriver.com/news/a15344706/self-driving-mercedes-will-prioritize-occupant-safety-over-pedestrians

[47] The Law Commissions. (2022). *Legal reforms to allow safe introduction of automated vehicles announced | Law Commission*. https://www.lawcom.gov.uk/legal-reforms-to-allow-safe-introduction-of-automated-vehicles-announced/
[48] Thomson, J. J. (1985). The Trolley Problem. *The Yale Law Journal, 94*(6), 1395–1415. https://doi.org/10.2307/796133
[49] Thorpe, C., Hebert, M. H., Kanade, T., & Shafer, S. A. (1988a). Vision and navigation for the Carnegie-Mellon Navlab. *IEEE Transactions on Pattern Analysis and Machine Intelligence, 10*(3), 362–373. https://doi.org/10.1109/34.3900
[50] United States Department of Transportation. (2016). *Federal Automated Vehicles Policy: Accelerating the Next Revolution In Roadway Safety*. United States Department of Transportation United States National Highway Traffic Safety Administration. https://www.hsdl.org/?abstract&did=795644
[51] United States Department of Transportation. (2022, March 10). *NHTSA Finalizes First Occupant Protection Safety Standards for Vehicles Without Driving Controls | NHTSA* [Text]. https://www.nhtsa.gov/press-releases/nhtsa-finalizes-first-occupant-protection-safety-standards-vehicles-without-driving
[52] van Loon, R. J., & Martens, M. H. (2015). Automated Driving and its Effect on the Safety Ecosystem: How do Compatibility Issues Affect the Transition Period? *Procedia Manufacturing, 3*, 3280–3285. https://doi.org/10.1016/j.promfg.2015.07.401
[53] Waymo. (2022). *Waymo*. https://waymo.com/
[54] Wisniewski, M. (2016, July 4). *Driverless cars could improve safety, but impact on jobs, transit questioned – Chicago Tribune*. Chicago Tribune. https://www.chicagotribune.com/news/ct-driverless-cars-getting-around-20160703-story.html
[55] Wright, J., & Vazquez, M. (2022, July 8). *White House says Americans should be "really careful" about using period tracker apps—CNNPolitics*. https://www.cnn.com/2022/07/08/politics/white-house-period-tracker-apps/index.html
[56] Yeruva, V. (2022). *Autonomous Vehicles And Their Impact On The Economy*. Forbes. https://www.forbes.com/sites/forbestechcouncil/2022/02/14/autonomous-vehicles-and-their-impact-on-the-economy/

Authors' Biography

Sherry Feng (www.sherryfeng.com) is a Ph.D. candidate in the Engineering, Computer, and Mathematics Department at the Auckland University of Technology. She has an M.Sc. in Learning Analytics from Columbia University (2019) and a B.Sc. in Computer Science from the University of San Francisco (2017). She is currently a lecturer for the Software Engineer Department at Media Design School, New Zealand. Her previous experience includes working as an Artificial Intelligent Consultant for the United Nations. Her research interest includes computer science with an emphasis on artificial intelligence.

Carol Shen is a Graduate Student from Carnegie Mellon University, majoring in M.S. in Information Security Policy and Management. Previously, she graduated with honors from Ecole Hoterliere de Lausanne with a B.Sc. in International Hospitality Management. Her current research interests include user impression in data breach notifications and usable privacy and security.

Index

About the Editor

Ishwar K. Sethi is currently a professor in the Department of Computer Science and Engineering at Oakland University in Rochester, Michigan. He has over 45 years of experience in applying neural networks including deep learning, machine learning, pattern recognition, and image, video, and text analytics to a broad range of projects. He has authored or coauthored over 175 journal and conference articles and has graduated 25 doctoral students. He has served on the editorial boards of several prominent journals including *IEEE Trans. Pattern Analysis and Machine Intelligence*, and *IEEE Multimedia*. He was elected IEEE Fellow in 2001 for his contributions in artificial neural networks and statistical pattern recognition and achieved the status of Life Fellow in 2012. He most recently organized the 20th International Conference on Machine Learning and Applications (ICMLA) in December 2021.